How Maps Work

HOW MAPS WORK
Representation, Visualization, and Design

ALAN M. MACEACHREN
Department of Geography
Pennsylvania State University

THE GUILFORD PRESS
New York   London

To Fran

© 1995 The Guilford Press
A Division of Guilford Publications, Inc.
72 Spring Street, New York, NY 10012

Marketed and distributed outside North America
by Longman Group Limited.

All rights reserved

No part of this book may be reproduced, stored in a retrieval system, or transmitted, in any form or by any means, electronic, mechanical, photocopying, microfilming, recording, or otherwise, without written permission from the publisher.

Printed in the United States of America

This book is printed on acid-free paper.

Last digit is print number: 9 8 7 6 5 4 3 2

Library of Congress Cataloging-in-Publication Data
MacEachren, Alan M., 1952–
   How maps work: representation, visualization, and design / Alan M. MacEachren.
      p.   cm.
   Includes bibliographical references and index.
   ISBN 0-89862-589-0
   1. Cartography.   I. Title.
GA105.3.M32   1995
526—dc20   94-31138
                                                                                 CIP

Preface

HOW MAPS WORK offers a general framework for the study of maps and map use, a plan for finding out why particular maps do or do not work in particular ways, and (perhaps) an approach through which we might achieve "better" maps. Maps are powerful tools, and have been for centuries, because they allow us to *see* a world that is too large and too complex to be seen directly. The representational nature of maps, however, is often ignored—what we see when looking at a map is not the world, but an abstract representation that we find convenient to use in place of the world. When we build these abstract representations (either concrete ones in map form or cognitive ones prompted by maps) we are not *revealing* knowledge as much as we are *creating* it. Understanding how and why maps work (or do not work) as representations in their own right and as prompts to further representations, and what it means for a map to work, are critical issues as we embark on a visual information age. The goal of *How Maps Work* is to provide a foundation for building this understanding.

If I knew then what I know now, *How Maps Work* might not exist. "Then" is when this book began—an indefinite date somewhere between 1985, when I first started thinking about the book, and 1989, when I signed a contract to produce it. Throughout that time, my vision was quite different from, and less ambitious than, what the book became. If presented with the final outline in 1989, I probably would have decided not to attempt the task. As I began developing ideas for this book, "scientific visualization" was a term that had not yet entered the lexicon—thus the three final chapters on "geographic visualization" were not foreseen. Semiotics, while already introduced to cartography, was not yet an approach that I viewed as central to cartographic inquiry. However, as I am undoubtedly not the first to discover, there is a point at which a book seems to take on a life of its own. Initial outlines get tossed aside and the book evolves toward what are sometimes unexpected ends. That point for this book was at the beginning of 1992.

During my sabbatical year in 1991–1992 (the first after 12 years of teaching), I had the good fortune to spend 9 months (together with my wife and dog) in a small beach cottage on the Outer Banks of North Carolina. During that time, I read voraciously and was able to write what ultimately became *Some Truth with Maps* (published in 1994 by the Association of American Geographers), along with the first third of *How Maps Work*. It was during that year that I became convinced of the viability of taking a combined cognitive–semiotic approach to maps and mapping. The essential elements of this approach are detailed in Parts I and II of the present book. The rapid evolution of scientific visualization (as a research field and tool of science) and geographic/cartographic visualization (as a new challenge for cartography) prompted me to select geographic visualization as a "case study" through which to demonstrate how the cognitive–semiotic perspective might be applied.

As should be clear from the above comments, this book is not a how-to book on map design. Neither is it a text in map use, nor a review of existing literature on mapping. I do review a wide variety of literature from many disciplines. My approach, however, has been a highly selective one. I have tried to weave a coherent story about how meaning is derived from maps and how maps are imbued with meaning. This story is an often intricate one that must overcome (or persevere through) the jargon of the diverse disciplines from which I borrow. If successful, the story I have constructed may provide a basis for graduate seminars dealing with human–map interaction, cognitive and semiotic issues in spatial representation, and a range of topics related to scientific visualization. The story is replete with contentions about how maps and other spatial representations work. Most of these contentions, while derived from evidence culled from a diverse literature, have not been directly tested and thus could serve as the basis for experimental research or a prompt to critical analysis.

Beyond the obvious cartographic/geographic/information design audience, I hope that the story I unfold is useful to cognitive psychologists, linguists, human factors engineers, sociologists, semioticians, cognitive scientists, and others who have explored various aspects of spatial representation in general or maps as representation specifically. If the story is sufficiently convincing (or at least thought provoking) that it causes these readers to take a step back from their own disciplines to consider how their work relates to that of others, I would consider my story successful. In the context of scientific visualization, this book is intended to stimulate discussion of a range of issues beyond those of the computer technology used to create ever more realistic displays. Questions concerning what visualization tools are intended to accomplish, how they interact with cognitive structures for spatial thinking, what the implications of realism in representational tools might be, and how to determine what counts as evidence in an increasingly visual world are among those I urge visualization researchers and students to consider.

Writing this book has been an exciting process. The process has changed my perspective on many aspects of maps, mapping, and spatial representation. That perspective continues to evolve. Cartography is a field with a long practical history but a short academic one. We should, therefore, not expect final answers to all aspects of how maps work, and I do not presume to provide them. What is offered is a sampling of what has been learned concerning how maps work, a set of questions that must now be confronted, and a range of linked approaches to seeking answers. The main tenets of this book are that *maps are spatial representations which can in turn stimulate other spatial representations* and that *representation is an act of knowledge construction*. It is by considering the question of representation at various levels from various approaches that we may begin to gain an insight into how maps work and put this insight to use in map symbolization and design.

As in any effort of this scope, credit is due to a wide array of people. I owe much to my early mentors—Hugh Bloemer, who kindled my initial interest in cartography, and the two Georges (McCleary and Jenks), who fanned the flames. Between them, they instilled a suspicion of single solutions to any problem—a suspicion that should be all too obvious in the pages that follow. In addition, their persepctives on maps and mapping remain at the root of many of my own current perspectives.

In relation to the book itself, Greg Knight provided considerable encouragement for the project and introduced me to Seymour Weingarten, Editor in Chief of The Guilford Press, who was willing to take a chance on a cartographic book that was not an introductory text. At Guilford, Peter Wissoker came on board as Geography Editor about the time I began writing in earnest. He has provided encouragement and suggestions throughout the process, and gave me wide latitude in deciding what was important and what was not.

While the book was being written, I was fortunate to find many colleagues who were willing to comment on drafts and discuss aspects of my overall approach. Mark Detweiller, a colleague from the Psychology Department here at Penn State, read and commented on early versions of the entire manuscript (as sections were produced). He provided invaluable advice on interpretation of the psychological literature and kept me well supplied with suggestions on literature to consult. Cindy Brewer,

David DiBiase, Pat Gilmartin, Mark Monmonier, and Tony Williams all provided constructive feedback on complete drafts of the book. Mike Wood provided suggestions on issues in Part I, and Hans Schlichtmann gave me detailed feedback on Chapter 1 plus all of Part II. His comments were particularly crucial in relation to the interpretation of semiotic literature, literature that I was often forced to consult in translation. In addition to these individuals, there were literally dozens of students (both graduate and undergraduate) who read chapter drafts and gave me their reactions (and helped find many errors in the text). I thank all of them for putting up with the process. Among these students, a particular acknowledgment needs to be given to John Ganter, who was as much a colleague as a student during the time when I was making my initial forays into visualization. The initial graphic depiction of the pattern ID model for visualization introduced in Chapter 2 and the initial version of the graphic to image continuum presented in Chapter 6 were his, and our discussions and collaborations were instrumental in defining many of the visualization problems that I have subsequently investigated. Among my colleagues, special thanks are due to David DiBiase for a range of contributions over the years—as a coleader of several seminars, as a coauthor on several papers that were fundamental to the development of ideas presented here, and as the person to whom I could always turn for technical advice on design of illustrations or creative use of software used in producing those illustrations.

The book would never have seen the light of day, of course, without the assistance of many Guilford staff members involved with editing, cover design, production, and promotion. My gratitude to them all.

Finally, I would like to thank my wife, Fran, and her family. Fran suffered patiently through 2 to 3 years of "I don't have time this weekend, because I have a deadline for my book," and was supportive throughout. Her parents, Gertrude and "Rog" Rogallo, offered us use of their beach cottage (mentioned above) where much of the conceptual basis for the book was worked out. Along with Carol and Norman Sparks, they helped make our stay in North Carolina a pleasant one.

Contents

1. **Taking a *Scientific* Approach to Improving Map Representation and Design**   1
   Toward Functional Maps, 2
   Cartography as Graphic Communication, 3
   Objections to Scope and Method, 6
   Art and Science, 8
   Deconstructing the Discipline, 10
   Taking a Fresh Approach to Symbolization and Design Research, 11
   Organization of This Book, 16

   I. HOW MEANING IS DERIVED FROM MAPS   21

2. **An Information-Processing View of Vision and Visual Cognition**   25
   Marr's Approach to Vision, 27
   Visual Cognition, 33
     Processing of Visual Stimuli, 33
     Processing of Imagery, 46
   Conclusion, 49

3. **How Maps Are Seen**   51
   Eye–Brain System, 53
     The Eye, 54
     Eye to Brain, 63
     Brain, 63
   Perceptual Organization and Attention, 68
     Grouping, 71
     What We Attend To, 80

Selective Attention and Separability of Visual Dimensions, 81;
  Divided Attention and Variable Conjunctions, 87; Associativity
  of Graphic Variables, 91; Indispensable Variables, 92
  Where We Attend, 94
    Location, 94; Scale, 96
  Scanning the Visual Scene, 101
  Figure–Ground, 107
    Heterogeneity, 110; Bottom-Up versus Top-Down
    Processing, 117
  Visual Levels, 120
 Perceptual Categorization and Judgment, 123
  Detection, 124
  Discrimination, 127
    Text Discrimination, 127; Point Feature Discrimination, 128;
    Pattern Discrimination, 130; Color Discrimination, 132;
    Motion Discrimination, 133
  Judging Order, 134
  Judging Relative Magnitude, 135
 Perceiving Depth from a Two-Dimensional Scene, 136
  A Taxonomy of Depth Cues, 137
  Applying Depth Cues to Maps, 139
    Physiological Approaches, 139; Perspective Approaches, 139;
    Nonperspective Approaches, 141
 Summary, 147

4. **How Maps Are Understood: Visual Array → Visual**   150
   Description ←→ Knowledge Schemata ←→
   Cognitive Representation
 Mental Categories, 151
  Prototype Effects, 153
    Family Resemblance, 155; Fuzzy Categories, 156; Typicality
    Effects, 158; Maps as a Radial Category, 160
  Basic-Level Categories, 162
  Natural versus Cultural Category Structures, 167
  Multiple Representations, 168
    Dual Representations: Common and Scientific, 168; Fuzzy
    Representations of Well-Defined Concepts, 169
 Knowledge Representation, 170
  Kinds of Knowledge Representation, 171
  Kinds of Knowledge Schemata, 174
    Propositional Schemata, 176; Image Schemata, 185; Event
    Schemata (Scripts and Plans), 190
 Development and Application of Cognitive Schemata, 193
  How Map Schemata Are Developed, 193
    Physiological Bases for Map Schemata, 194; Developmental
    Bases for Map Schemata, 195; General-to-Specific Map
    Schemata, 198

How Map Schemata Are Selected, 202
How Map Schemata Are Used, 205
Conclusion, 209

II. HOW MAPS ARE IMBUED WITH MEANING 213

5. A Primer on Semiotics for Understanding Map Representation 217

The Nature of Signs, 218
   Models of the Sign, 219
   Typology of Signs, 222
   Typology of Discourse, 225
   How Signs Signify: Specificity or Levels of Meaning, 228
   Typology of Comprehension (or Miscomprehension), 232
The Nature of Sign Systems, 234
   Dimensions of Semiosis, 234
   Systemology, 238
      Semiotic Economy, 239; Simultaneity versus Articulation, 240; Combinatorial Relations, 241
   Application of the Semiotic Approach to Map Representation, 242

6. A Functional Approach to Map Representation: The Semantics and Syntactics of Map Signs 244

The Nature of Map Signs–Map Semantics, 245
   Sign-Vehicle as Mediator, 246
   Referent as Mediator, 250
   Interpretant as Mediator, 256
The Nature of Map Sign Systems–Map Syntactics: Logical Interrelationships, 269
   Visual Variables and Syntactic Rules, 270
      Static Visual Maps, 270; Static Tactile Maps, 276; Dynamic Visual Maps, 278; Dynamic Audio Maps, 287
   Sign-Vehicle Sets, 290
   Multiple Linked Sign Systems, 295
   Maps as Signs, 302
   Map Sign Comprehension, 305
Discussion, 307

7. A Lexical Approach to Map Representation: Map Pragmatics 310

Meaning *in* Maps, 312
   Space, Time, and Attribute Denotation, 312
      Denoting Spatial Position, 313; Denoting Temporal

Position, 315; Denoting Attributes of Position in
    Space–Time, 317
  Specificity of Signs, 321
    Singular versus General Signs, 321; Unambiguous versus
    Ambiguous Signs, 323; Monosemic versus Polysemic Sign
    Systems, 325
  Directness of Reference: Literality of Interpretants, 325
  Concreteness of Signs: Concept versus Phenomenon
    Representations, 327
  Etymology and Cultural Specificity of Meaning, 329
Meaning *of* Maps, 330
  Connotative Meaning of Map Signs, 331
    Extrasignificant Codes, 332; A Typology of Map
    Connotation, 336
  The Map Itself as an Implicit Code, 338
    Connotation of Veracity: Truth and Reality, 338; Connotation
    of Integrity: Map Ethics, 340; Valuative Connotations:
    Judgments, 342; Connotations of Power: Territorial Control,
    345; Incitive Connotations: Persuasion to Action, 348; Can
    Connotations Be Measured?, 349
Synopsis and Directions, 351

III. HOW MAPS ARE USED: APPLICATIONS IN GEOGRAPHIC VISUALIZATION   355

8. GVIS: Facilitating Visual Thinking   361

A Model of Feature Matching, 362
Linking Perceptual Organization and Map Syntactics, 367
  Indispensable Variables, 368
    2-D Space, 369; Simulated 3-D, 370; Time, 376
  Scale and Resolution, 380
    Space, 380; Attributes, 384; Time, 385
  Spatial Feature Enhancement through Graphic Variable
    Manipulation, 386
    Using Monochrome Variables, 387; Using Color Variables, 389
The Role of Categories and Schemata, 392
Conclusions, 398

9. GVIS: Relationships in Space and Time   401

Feature Comparison: Looking for Relationships in
  Multidimensional Data, 401
  Space, 402
  Orientation, 409
  Color, 411
  Time, 416
  Focusing, 418
  Sound, 419

Space–Time Processes, 422
  Categorizing Space–Time Phenomena, 423
  Mapping Temporal Entities to Display Variables, 425
  Exploring Space–Time Processes: Kinds of Interaction, 427
    Process Tracking, 428; Postprocessing, 429;
    Process Steering, 432
  Discussion, 433

10. GVIS: Should We Believe What We See?   435
How to Judge "Truth" in GVIS, 435
  Truth of Signs in the Display, 436
  Truth of the Display as Sign: Seeing Wrong versus
    Not Seeing, 444
What "Truth" Is in GVIS, 447
  Visual Thinking and Cognitive Gravity, 448
  Public Presentation and Implicit Connotation, 452
Discussion, 456

Postscript   459

References   463

Author Index   491

Subject Index   497

CHAPTER ONE

Taking a *Scientific* Approach to Improving Map Representation and Design

Cartography is about representation. This statement may seem obvious, but it has been overlooked in our search for organizing principles for the field. Rather than restricting research in cartography to maps that present well-defined messages (and suggesting a single, map-engineering approach to improving the transmission of these messages, as the communication approach did), attention to maps as spatial representation expands the field. Exploring maps as representation forges important links between cartography and a variety of cognate fields concerned with this topic in its various facets (including geographical information systems [GIS] and remote sensing, as well as art, cognitive science, sociology, cognitive and environmental psychology, semiotics, and even the history and philosophy of science). This view of cartography does not discount the importance of communication-oriented research. Some maps do, in fact, serve primarily as vehicles for communicating specific messages. What taking a representational perspective on cartography does do, however, is place this research in a broader context. In doing so, it allows us to better recognize the limits that a communication perspective has as a driving force for the field as a whole.

To promote understanding of the implications that follow from the above contentions, I will begin by taking a brief historical look at the cartographic research of the past four decades. This research had a well-defined direction for much of this time, but during the past decade it became clear that research cartographers, as well as practitioners, were becoming disillusioned with the field's direction. As a result, a number of critiques and suggestions for potential redirection have been put forth. It is within this context of change that the present book, with the perspective identified above, has evolved.

Two developments of the past four decades played crucial roles in establishing a research agenda for the study of map symbolization and design. The first was Arthur H. Robinson's dissertation (published as *The Look of Maps* in 1952), with its call for objective research, and the second was the adoption in the 1970s of a paradigm of cartography as communication science. I begin, therefore, with a brief overview of these events followed by some recent perspectives on the promise versus the limits of a "scientific" approach to cartographic symbolization, and a proposal for an alternative to "communication" as the basis for this scientific approach.

TOWARD FUNCTIONAL MAPS

World War II was crucial in shaping the direction of cartography as a discipline (and a craft).[1] As the result of the war experience of several U.S. geographers, particularly Robinson and his role as a principal player in government cartographic efforts supporting the military, the emphasis of the field shifted from production efficiency and graphic design toward map "functionality."

Beginning with essays from his dissertation, Robinson (1952) pointed out some limits to approaching map symbolization and design from a purely artistic viewpoint, as he suggested was the guiding perspective at the time. Maps, like buildings that are designed primarily for artistic impact, are often not functional (e.g., the 1970s New York subway map won awards for *design* but resulted in a number of lost graduate students when an informal experiment was performed to judge its functionality [Allis, 1979]), and the Renaissance Center in Detroit produced a dramatic change in the city's skyline, but even professional geographers could not keep track of where they were during the 1975 Association of American Geographers annual meeting.

Robinson (1952) argued that treating maps as art can lead to "arbitrary and capricious" decisions. He saw only two alternatives: either standardize everything so that no confusion can result about the meaning of symbols, or study and analyze characteristics of perception as they apply to maps so that symbolization and design decisions can be based on "objective" rules. Although a few suggestions emanated from the international cartographic community concerning standard symbol sets to be used on thematic maps (Ratajski, 1971), that option was not considered seriously by many cartographers, and clearly was not advocated by Robinson (1973). Most academic cartographers took up the second option of formulating "objective" rules.

Robinson's dissertation, then, signaled the beginning of a more objective approach to map symbolization and design based on testing the effectiveness of alternatives, an approach that followed the positivist model of physical science. In his dissertation, Robinson cited several aspects of cartographic method for which he felt more objective guidelines were required (e.g., lettering, color, and map design). He also suggested that this objective look at cartographic methods should begin by considering the limitations of human perception. One goal he proposed was identification of the "least practical differences" in map symbols (e.g., the smallest difference in lettering size that would be noticeable to most readers). This goal was naturally linked by others to psychophysical research in psychology. Psychologists had focused on measuring "just noticeable differences" as a step toward deriving "laws" to "explain" human responses to various stimuli. Fechner's Law, for example, had suggested that just noticeable differences exhibit a logarithmic relationship to actual differences in the magnitude of stimuli, and in 1957 Stevens countered with his "power law." Following from these laws, we would predict that least noticeable type-size differences on a map will increase with increasing lettering size according to a logarithmic or power function, respectively.

CARTOGRAPHY AS GRAPHIC COMMUNICATION

The second dramatic occurrence of the past four decades, in relation to how we address questions of symbolization and design, was the development and elaboration of the concept of cartography as graphic communication. It is this perspective on the cartographic enterprise that is behind much of the empirical research stimulated by Robinson's dissertation four decades ago.

The view of cartography as a communication process has been depicted graphically by many authors. While details of these depictions vary, all models share a basic structure with an information source tapped by a cartographer who determines what (and how) to depict, a map as the midpoint of the process, and a map user who "reads" the map and develops some understanding of it by relating the map information to prior knowledge (Figure 1.1).

Although Robinson did not propose the communication model for cartography, he pointed the direction by arguing for "functional design" and for objective evaluation of map effectiveness as the mechanism to

FIGURE 1.1. A schematic depiction of cartography as a process of information communication.

achieve it. In *The Look of Maps* (1952, p. 15) he made the following comment: "If we then make the obvious assumption that the content of a map is appropriate to its purpose, there yet remains the equally significant evaluation of the visual methods employed to convey that content." In this statement we find the idea that maps have a predefined purpose (rather than the map user having an information need that available maps may or may not meet), and that the goal is to convey (or communicate) the content selected to meet this purpose. We also find the attitude that map content is something that does not have to be questioned, a view that has been disparaged of late. Robinson went on to argue that: "Most scientific cartography is concerned with the dissemination of spatial knowledge" (p. 17). Again, *knowledge* that already exists and that the cartographer has access to is to be *disseminated* through the map, rather than *constructed* by the analyst who uses the map.

Models of cartographic communication did not actually appear until the late 1960s. The initial one was an extremely complicated flow diagram created by Christopher Board (1967). The depiction of cartography as a communication system having the greatest initial impact on cartography, however, was one put forward by a Czechoslovakian cartographer, Koláčný (1969). This model was developed approximately 15 years after Robinson's dissertation. Considering its nature will help us to understand changes in the approach taken to map design and symbolization that succeeded Robinson's initial call to action.

Koláčný's and other graphic depictions of cartography as a communication science reflect contemplation by diverse cartographers over a number of years concerning how to improve maps and how easily and accurately maps can be interpreted. Communication came to be viewed as the primary function of cartography and the map was considered the vehicle for that communication.

Due to the communication paradigm, the purview of cartography expanded to encompass more than mapmaking. It was approached as a *process* of communicating spatial information that had inputs, transmission, and reception of information, and that therefore could be analyzed as a system. From this point of view, authors identified numerous obstacles or filters that information must pass through on its route from reality through the cartographer to the map, and then through the map to the map user. On the cartographer's side of the system, these filters include objectives, knowledge and experience, abilities and attitudes, external considerations such as client demands, as well as the abstraction processes by which information is put into map form (e.g., projection, simplification, generalization, classification, symbolization, etc.). For map use, the following factors were identified as filters: the perceptual and spatial abilities of readers, understanding of the symbol system (e.g., training or ability to understand the legend), goals, attitudes, viewing time, intelligence, prior knowledge, and preconceptions (Figure 1.2). According to communication theory, each of these variables can act to inhibit information transmission, resulting in information loss or communication errors.

Treating cartography as a formal communication system implies that we can improve map communication if we can reduce the filtering or loss of information at various points in the system. An improvement any-

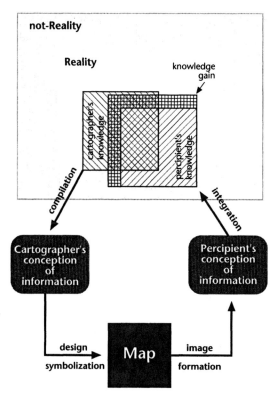

FIGURE 1.2. My own 1979 view of cartography as a process of graphic communication. *After MacEachren (1979, Fig. 1.3, pp. 10–11).*

where in the system should have a positive effect, and an information loss should be impossible to overcome. Most efforts to study cartographic communication have been directed to the middle stages in the system: the cartographer's transformation of selected information into the map and the initial extraction of information from the map by the user. Little attention has been directed to how decisions are made about what should and should not be mapped (a primary concern of the postmodern approach to cartography, discussed below) or to the influence of prior knowledge (map use goals, mental models, schemata, etc.) on what is noticed, how map information is evaluated, or what is retained.

Throughout this book, we will see the consequences of the widespread acceptance of the communication model as a research paradigm. Among them are a concern with the map user, the use of empirical research to investigate the impact of symbolization and design choices, and a belief that objectively derived guidelines can be established for map symbolization and design. Although these and other perspectives inherent in the communication paradigm continue to be accepted by most cartographers, the basic paradigm of cartography as communication science has been attacked on several grounds and few cartographers now accept it in its literal sense.

Among the objections raised are (1) a realization that viewing cartography as communication science omits many ways people use maps and that the particular experimental approach initially selected to evaluate maps as communication vehicles was a barren one; (2) a concern that any (positivist) scientific approach to study and improvement of maps will, by definition, ignore the important contributions of art in the cartographic process; and (3) a philosophical perspective, voiced by an increasing number of scholars, that does not accept the concept of maps as "objective" representations of reality and therefore discounts the idea that objective research is possible.[2] Each of these points will be touched on briefly below to set the stage for the approach offered here to studying how maps work.

OBJECTIONS TO SCOPE AND METHOD

One of the mistakes made by those who adopted the communication paradigm (and I was one of "those") was to place severe restrictions on Robinson's call for research directed toward functional maps. Maps, in relation to the communication paradigm, were judged on a functional basis, but the definition of function was restricted to communicating some predetermined message. Clearly only a small subset of maps are produced to "communicate" a particular message. Topographic maps, maps that depict location of events, and travel maps have a function, but no predetermined message (except perhaps in the rhetorical sense, as Wood and Fels, 1986, have pointed out for state highway maps). Even many thematic maps (e.g., geologic or soils maps, a textbook map of world GNP, etc.) often have no explicit predetermined message (but as Yapa, 1992, points out, the GNP map may have implicit messages). A student, for example, may be asked to use a map of world GNP to compare the wealth of continents, to pick out individual countries with a specified GNP, to study the relationships between latitude and GNP or between GNP and agricultural production, and so on. It is inappropriate to consider the information obtained from this kind of map use as a message that the cartographer was trying to "communicate." In this example, the information extracted by the student is determined by the questions that a teacher poses (and from the meaning and significance assigned to GNP). Such meaning often cannot be anticipated by the cartographer.

The spread of technology for both geographic information and analysis and scientific visualization fosters map use early in the research sequence. Following Tukey's (1977) lead, DiBiase (1990) points out that scientific research progresses through at least four stages: exploration, confirmation, synthesis, and presentation. In the early exploratory stages of a research project, an analyst might create a map to investigate some spatially distributed phenomena. Here, again, there is no predetermined message. The goal of map use is to stimulate a hypothesis rather than to communicate a message. Information is instead "constructed" by the user, from the spatial representation of the world provided by the cartographer.

In addition to ignoring maps that do not have a predetermined message, a second failing of the communication paradigm was its strong link to behavioral psychology. This approach to psychology sought "laws" that relate behavioral responses to stimuli available to our senses. The approach dominated experimental psychology in the United States for several decades but was being supplanted within that discipline just at the time cartography decided to borrow from it. Behaviorists treated humans as black boxes that respond to stimuli rather than as information-processing systems that build knowledge from available input (a view held by many current researchers in cognitive psychology and a basic premise of cognitive science). Along with an assumption that the cartographer's role was to communicate, a behavioral perspective led to an assumption that we could devise rules for manipulating symbols to ensure a desired response. Once these "laws" were worked out, the theory contended, *optimal* maps could be constructed, with "optimal" defined as producing a user response that was as close as possible to the intended response.

During the 1960s and 1970s, when cartographers were embracing the communication model and a behavioral approach to empirical research, psychology was undergoing a revolution in its perspective on what to study and how to study it. Psychologists began to realize that stimulus–response laws do not *explain* human perception or behavior (any more than the gravity models used by geographers can *explain* spatial interaction). While the ability to predict is helpful, both in relation to map symbol perception and intracity trade, by itself it does not provide a basis for dealing with significant changes to the system (e.g., new symbol systems created due to technological advances, or a change in the importance of geographic distance due to similar technological developments). In psychology, a cognitive approach developed in which the focus shifted from predicting behavior to explaining how information is processed.

Although there were calls for cartography to consider cognitive aspects of map reading as early as the mid-1970s (e.g., Petchenik, 1975), it is only recently that cartographers have begun to appreciate what this shift in approach means. Much of the "cognitive" research done by cartographers thus far has retained a neobehaviorist approach of measuring subject reactions without trying to infer either cognitive processes or to draw upon cognitive theory. The approach is similar to that still followed by human factors engineers and might be thought of as *map engineering*. A map engineering approach can solve particular narrow problems (e.g., determining parameters of map interfaces on helicopter display panels), but it is unlikely to result in generalizable theory.[3]

Rather than treating the cartographer and the map as conduits though which information is filtered, it makes more sense to study the perceptual and cognitive processes involved in both map "reading" and spatial information processing to determine constraints and features of the "information-processing device" (i.e., humans) so that symbolization and design can be adapted to it. One of my principal arguments in Part I of this book is that we can facilitate map use by developing models of human–map interaction and human spatial cognition, and through these models identify and more completely understand the most important variables of map symbolization and design.

ART AND SCIENCE

The communication paradigm for cartography (in spite of its dominance in North American cartography during the 1970s and 1980s) is viewed by many as quite sterile. Indeed, when taken to its extreme it is. Some authors have gone so far as to try describing cartography using the formal mathematical/electrical engineering approach of information theory as it was developed to explain the loss of signal quality over lines of electronic communication (Shannon and Weaver, 1949). Attempts have been made to measure map information in terms of information "primitives" that you count at the beginning and end of the communication process. The difference in these totals was considered to be a measure of information loss and the proportion a measure of adequacy of transmission. This approach was doomed to failure, if for no other reason than that the user can combine map information with previous knowledge to produce conclusions that were not part of the initial *map message*.

Information theory and related attempts to treat cartography as a relatively well-behaved physical system have caused some cartographers (e.g., Keates, 1984) to warn us that adopting a scientific approach leads to devaluing the art in cartography.[4] Keates, in discussing the imitative, emotive, expressionistic, and communicative functions of art, makes a rather convincing case that maps do contain artistic qualities that are difficult or impossible to account for through any "scientific" assessment (e.g., Imhof's [1965/1982] terrain shading is surely "imitative," *The War Atlas* clearly is "emotive," and the contrasting styles that we have come to associate with maps from the National Geographic Society, the Central Intelligence Agency, *Time* magazine, and other prominent map producers are clearly "expressionistic"). That maps, like (other?) art, can be "communicative" is probably an acceptable idea to all cartographers. Even in the case of this communicative function of maps, however, the communication model leads us to measure the communication of individual *bits* of information rather than to assess the overall intellectual import of the map and its potential to convey many meanings at multiple levels of analysis.

A new view of the role of art and science in cartography is clearly needed. It is probably a mistake to view maps as objects that contain varied amounts of scientific or artistic content for which we must determine an appropriate balance (as both Keates, 1984, and Robinson, 1952, seem to, with Keates arguing for more art and Robinson for more science). Instead, it makes more sense to consider complementary *artistic and scientific approaches* to studying and improving maps, both of which can be applied to any given cartographic problem. The artistic approach is intuitive and holistic, achieving improvements through experience supplemented by *critical examination* (where critical examination implies expert appraisal of the results of our cartographic decision-making efforts). It draws on science in using perspective, understanding of human vision, color theory, and so on.[5] The scientific approach (emphasized in Part I of this book) is more inductive and often reductionist, breaking the problem into manageable pieces with the assumption that the total picture (in the form of a general theory) will become clear by systematically examining each individual part of the process.[6] A scientific approach draws on art in developing initial hypotheses about light, shading, color, type, and more.

DECONSTRUCTING THE DISCIPLINE

In addition to the concern that a scientific approach to the study of map symbolization and design might lead to the elimination of art from cartography, concern has been raised about other dangers of viewing cartography as an objective activity and evaluating it using the objective, positivist, reductionistic approach of physical science. Borrowing from postmodern thinking, several authors, most notably Wood and Fels (1986), Harley (1988, 1989), and Wood (1992), have pointed to the inherent subjectivity in, and rhetorical content of, maps. Wood and Fels (1986), for example, in their detailed analysis of the seemingly benign state highway map of North Carolina, find subtle propaganda in the presence of the state insect (a busy bee) and blatant bigotry in the choice of photographs to adorn the back of the map.

This perspective suggests that maps are as much a reflection of (or metaphor for) the culture that produces them as they are a representation of a section of the earth or activities upon it. Harley (1989, p. 15), for example, argues that cartographers have created an "epistemological myth" that cartographic method reflects the "cumulative progress of an objective science always producing better delineations of reality."

The contention of these authors seems to be that cartography is neither objective nor a science, and that no amount of research can result in "better" maps because there is no objective way to define "better." Cartography, it is argued, is more akin to literature than to astronomy or geophysics. The appropriate analytical methods, then, should be modeled after literary criticism rather than after experimental methods used in the "hard" sciences. This perspective would direct our attention to philosophy and social theory rather than to psychology, human factors, linguistics, education, or cognitive science for approaches by which map symbolization and design can be assessed. Even artistic appraisal might be viewed as irrelevant or inappropriate to the task of assessing the sociocultural consequences of maps as the product of cartography. The directions pointed by these authors are refreshing, but their apparent insistence on a wholesale replacement of one limiting approach to cartography with another is not.

While the postmodern assessment of cartography has certainly generated a lively debate (see commentaries on Harley's "Deconstructing the Map," in *Cartographica*, 26 (2), 1989), and has reminded cartographers of the social implications of the products they produce, it does not—and by design cannot—provide answers to any fundamental questions about how we should select symbolization or design strategies. What it does provide is a way to assess how these selection decisions impact those individuals, groups, or societies whose environment is represented by the map.

If we accept the premise that maps can "work" (i.e., that they are a useful way of obtaining spatial information), we have an obligation to facilitate their use as information sources. The fact that we cannot eliminate the cultural baggage inherent in any human artifact does not give us a license to ignore the practical consequences of our decisions in designing that artifact. The realization that architecture contains similar cultural baggage, for example, does not reduce the importance of work such as Lynch's (1960) on the image of the city (designed to obtain knowledge by which city planning and building design can be used to make experiencing a city more meaningful or memorable) or ergonomic studies (designed to make working in a building safer, less tedious, or more pleasant). Similarly, research that makes maps used by air traffic controllers or pilots less prone to misinterpretation would probably be valued by anyone who travels by air, perhaps even a "postmodernist." On the other hand, the fact that maps do seem to work does not absolve us of responsibility to consider the kind of work they do, whether explicitly or implicitly, openly or surreptitiously. What is needed, I believe, is a more balanced perspective on cartographic research that attempts to merge the perceptual, cognitive, and semiotic issues of maps as functional devices for portraying space and the sociocultural issues of how these portrayals might facilitate, guide, control, or stifle social interaction. Although this book is largely about how maps work to achieve their explicit goals, attention is also paid to how they work to achieve their implicit goals. Specifically, questions concerning maps as multifaceted representations and as tools of rhetorical discourse along with the social processes by which maps and map symbols acquire their meaning are considered.

TAKING A FRESH APPROACH TO SYMBOLIZATION AND DESIGN RESEARCH

There are three perspectives currently taken toward scientific research on map symbolization and design stimulated by the communication paradigm. One is that a scientific approach to cartography is impractical or irrelevant, either (as noted above) because cartography is an art rather than a science or because the rhetorical content of maps is more important than the *information* they contain (if they are admitted to contain any). At the other extreme is a belief that the communication paradigm is the most promising approach to achieving cartography's ultimate goal of more functional maps, but that a combination of sloppy research, poor selection of initial problems to pursue, misdirected emphasis, wrong methods, and the relative youth of the approach has led to somewhat disappointing results thus far (e.g., Olson, 1983; Dobson, 1985; Medyckyj-

Scott and Board, 1991). The third perspective, and the one adopted here, accepts cartography's function as creating interpretable graphic summaries of spatial information (i.e., representations) and the goal of producing more consistently functional maps, but judges the communication paradigm to be a much too constraining model for the discipline (although it has a function in addressing presentational uses of thematic maps or in evaluating the interpretability of individual symbols or symbol types).

My position is that there is no single correct scientific, or nonscientific, approach to how maps work. As David Marr (1985, pp. 110–111), a noted vision scientist, has asserted, to understand any complex system we must "contemplate different kinds of explanation at different levels of description that are linked, at least in principle, into a cohesive whole, even if linking the levels in complete detail is impractical." The representational approach to maps advocated here is not intended as a call for a single new perspective on cartography to replace another single approach. The intent is to illustrate the importance of understanding representation as a general concept if we are to understand maps. I am contending that the concept of representation is fundamental to all approaches that we might take to cartography.

The map is examined here, then, not as a communication vehicle, but as one of many potential representations of phenomena in space that a user may draw upon as a source of information or an aid to decision making and behavior in space. Emphasis is placed on how the map "represents" in both a lexical and a semiotic sense (see below) and on how vision and cognition represent that representation in forms that allow the map viewer access to meaning. The map user's interaction with the map is viewed as a complex information-processing problem in which a series of neurological then cognitive representations of what is seen are built and these representations are interrogated using schemata (mental representations) that provide a context (or set the limits) within which the conceptual picture derived from the map can be understood.

An integrated view of spatial representation considered at multiple levels provides a major organizing principle for this book. A typology of levels at which representation can be addressed proposed by Howard (1980) has been influential in my parsing of cartographic representation. Howard's typology distinguishes among three perspectives on the study of symbols—the lexical, the functional, and the cognitive—and leads to correspondingly different (but complementary) approaches to the concept of "theory of representation." He deliberately avoids speaking of "*a* theory" or "*the* theory" because of the multiplicity of theories that become possible when we realize the scope of the concept of representation.

The *lexical* approach to representation might be considered the level of meaning in its broadest sense. This approach deals with how symbols achieve their meaning and how we learn to use particular forms of symbolization. Considered are levels of meaning (e.g., specific and general, literal and metaphorical), along with the etymology of symbol meaning and any ethnographic variation that may exist. Among the questions to be considered under a lexical approach to representation, Howard (1980, p. 504) includes "questions of style and interpretation in art or of the nature of artistic, scientific, historical, or religious understanding." Aspects of work by "lexicographers, art critics, cultural anthropologists, epistemologists, and historians" fall within the bounds that Howard delineates for this approach. In the context of cartographic representation, then, the lexical approach corresponds to calls by Harley (1988), Wood (1992), and others to consider both the implicit meaning and the power inherent in maps as well as their explicit meaning—in Harley's (1989) terms, to "deconstruct the map."

Howard labeled the second approach to representation the *functional*. This perspective corresponds "to the broader colloquial view of symbolism as anything that can carry meaning" (Howard, 1980, p. 504). This approach can be embraced under the concept of semiotics: "The logical analysis or plotting of specific differences and kinships among linguistic, logico-mathematical, pictorial, diagrammatic, gestural, musical, and other sorts of symbol systems construed as different ways of using one thing to refer to another." The basic question here is "what are the relations that sustain a particular symbolic function and its contribution to meaning?" (Howard, 1980, p. 504). Drawing on Goodman (1976, p. 143—rather than 144ff. as cited), Howard (1980, p. 504) argues that these relations "form a system consisting of a symbol scheme (the items or inscriptions used to symbolize) correlated with a field of reference." Rather than exploring what symbols mean (as in the lexical approach), the issue becomes "what does it mean to be a symbol and how do they variously provide their meaning?" When we take a functional approach to representation, it allows us to recognize that "different ways of presenting information on a surface can imply different ways of relating to those marks and inscriptions" (Howard, 1980, p. 506). An aspect of this idea that will be considered below is that differences among various forms of visual-information presentations go beyond differences in their surface appearance. More fundamental are the different sets of rules for construction and interpretation (i.e., mental categories and knowledge schemata) that have developed and are understood by users of these depictions.

While lexical and functional approaches to representation look primarily at cultural practices, social processes, scientific practices, and so on, for clues to the meaning of symbols and to the development and application of symbol systems, the *cognitive* approach looks to the individual. At the level of the individual, the issue becomes identification and understanding of "psychological, cultural, or communication processes required or most frequently involved in the acquisition and mature use of many sorts of symbols in thought and action" (Howard, 1980, p. 506). Among the questions that Howard (1980, p. 506) suggests as typical of this perspective are: "How do symbols of different kinds mediate thought and perception?" and "How does information (on some proper analysis of that puzzling notion) admit of multiple symbolic manifestations and which ones are more economical for certain educational purposes or levels of learning?" The cognitive approach, according to Howard, is "concerned with facts and hypotheses about the acquisition and use of symbols in virtually every aspect of life."

To understand how maps work, I believe we should follow Howard's lead and attempt to understand *representation* at many levels (Figure 1.3). How humans represent information mentally determines how groups and societies can develop a consensus about letting symbols (in the broadest sense of the word) stand for objects, relationships, events, and the like, in the "real" world. When the communication paradigm considered representation, it did so in the limited sense of focusing on how cartographers represent the environment with map symbols. In relation to map users, the only questions that seemed relevant following the communication system logic were those related to how users interpret the cartographic representations. If we allow that perception is a representation (e.g., visual perception is the representation of visual scenes before our eyes) and that cognition involves higher levels of representation (of objects, relationships, processes, etc.), then we see that map symbols are not the only representations that should be of concern to cartography. At the other extreme (from individual mental processes) Harley (1989), Wood (1992), and others have made it clear that representation at a social level

FIGURE 1.3. The multiple levels of map representation.

is also an important factor in map understanding. It is not only the map user who mentally represents map information nor only the map author–cartographer who imbues a map with meaning, both explicit and implicit, it is the society and culture within which map author–cartographer and map user coexist that provide that meaning. Society gives meaning (at multiple levels) to the symbols that the cartographer uses to assign meaning (also at multiple levels). How cartographers reach consensus about what should be represented and the meaning of particular symbol types (or even individual symbols), and about the ways in which the cartographer's social context influences these decisions, are issues relevant to a representational approach to cartography. Attention to these issues should complement, not compete with, attention to the visual and cognitive representations derived from the resulting maps.

When we consider Howard's typology of approaches to representation in relation to maps, it becomes apparent that his first two approaches deal with the public realm of how maps are imbued with meaning, from an epistemological–philosophical–sociological–historical perspective in the case of the lexical approach, and from a logical–categorical perspective in the case of his functional approach. In contrast, the third approach deals with the private realm of the individual and how the individual *sees* and *interprets* individual symbols and maps. This public–private distinction is reflected in the first two sections of this book (Figure 1.4). A representational view of cartography, therefore, suggests two primary levels of analysis: the private/perceptual–cognitive (where attention is directed to how human vision and cognition represent concepts about the world and the contents of a visually displayed map, i.e., how meaning is derived from maps) and the public/social (where attention is directed to the ways in which symbols and maps represent, i.e., how maps are imbued with meaning). The private focus is particularly concerned with the processes of vision as a hypothesis about what is seen and the

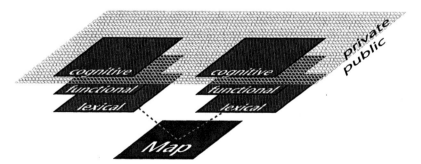

FIGURE 1.4. The public and private issues of maps as representations.

role of conceptual categories and knowledge schemata in assigning meaning to the representations derived by vision. The public focus is concerned with developing logical systems for creating meaningful representations and understanding in a broader context how symbols acquire meaning at multiple levels.

ORGANIZATION OF THIS BOOK

The book is divided into three main sections that address distinct issues of how maps work.[7] The organizing structure for the first two sections of the book is derived from the categorization of approaches to theory of representation described above. The third section applies these approaches to understanding how maps are used in an increasingly critical map application area: geographic visualization.

In presenting the multilevel approach to spatial representation proposed here, the logical place to begin is with the individual. Both social (i.e., public) issues of the mix of explicit and implicit meaning in map symbols and whole maps and personal (i.e., private) issues of how a particular map user interprets a particular map are dependent upon how human perception and cognition represent, and give meaning to, space. I begin by considering how vision and cognition work together as a process that acts upon and transforms representations of sensory input, making use of both unconscious reactions (some innate and some developed with practice) together with mental categories and schemata as the keys to interpreting these representations. From this base (presented in Part I), I go on (in Part II) to consider semiological issues at the functional level of cartographers trying to create logical abstractions of the environment, and the social issues at the lexical level of how symbols acquire their multiple levels of meaning and how map representations are used in social interaction. Understanding of cartographic representation at these three levels, then, allows consideration of how maps are used in geographic visualization, the topic of Part III.

Part I, *How Meaning Is Derived from Maps*, begins with an overview of information-processing approaches to vision and visual cognition and their potential application to the study of cartographic representation. Human–map interaction is considered to have a set of components related to the levels of processing required. These range from largely perceptual processes (e.g., detecting point symbols on the background of a complex highway map, discriminating between colors on a vegetation map, identifying one region as figure on another that is ground) to inferential processes that require substantial contributions from prior knowledge and experience (e.g., visualizing the shape of terrain from a contour depiction, planning a travel route across town, or developing a hypothesis about the impact of temperature on crop damage due to a particular insect). Chapter 2 provides a structure for the two subsequent chapters in Part I, "How Maps Are Seen" and "How Maps Are Understood." The first of these presents a sketch of current understanding of the eye–brain system and the limitations that this hardware may put on what we are able to see. Next, I consider issues of perceptual organization of the visual scene by early (largely preattentive) processes and their implications for "seeing" map symbols and sorting figure from ground. This is followed by some observations about how cartographic research on symbol discrimination, ordering, and estimation can be related to this information-processing approach. The chapter concludes with a section on the methods by which we can trick vision into seeing depth in two-dimensional maps. The final chapter of Part I looks at subsequent cognitive processing of information that vision provides and considers the interaction between preattentive visual processing of maps and the knowledge structures used to mentally organize both general knowledge and knowledge of maps. Recent research on mental category systems and attempts by both psychologists and cartographers to explore the notion of knowledge/problem-solving schemata are considered.

Part II, *How Maps Are Imbued with Meaning*, focuses on the public aspects of representation theory: the functional and the lexical. Both are examined from a semiotic perspective. The section opens with a primer on semiotics for cartography, Chapter 5. This primer is necessary as a way to cut through the daunting morass of semiotic terminology. With the base provided by this synopsis of key semiotic concepts, Chapter 6 is devoted to functional aspects of a semiotic approach to cartographic representation. A framework is presented that underlies a developing rule base of cartographic guidelines for matching both individual symbols and map types to their referents. Potential directions for research dealing with symbolization and design rules are considered. In the section's final chapter, lexical perspectives are drawn on in an effort to explain how map users interpret individual symbols, symbol groups, and entire maps, and how categories inherent in cartographic decisions about symbol–referent relationships dictate a particular view on the reality represented. Both denotative and connotative meaning in maps are addressed.

Throughout the book, I take the approach that maps and, by implication, map symbolization and design, should be evaluated not by how much information they communicate how quickly, but by how well they suit a particular task. With this perspective in mind, Part III, *How Maps Are Used: Applications in Geographic Visualization*, focuses on a key problem domain attracting the increasing attention of cartographers. This part combines the lexical and semiotic perspectives on how maps represent with emphasis on visual–cognitive processes in human–map interaction. Three chapters address prompts to visual thinking, the search for relationships in data, and truth in geographic visualization. The goal of this combined approach is to understand, at various levels, how cartographic decisions about representation influence thinking, problem solving, and decision making that is facilitated by the resulting maps. The topics of noticing unexpected patterns and pattern comparison are discussed in relation to a model for cartographic visualization. The role of both social and cognitive processes in developing knowledge schemata with which map information is processed is considered. Concepts and perspectives from cognitive, environmental, and developmental psychology, as well as from cognitive science, computer-assisted instruction, sociology, and the history and philosophy of science, serve as a framework for understanding how spatial knowledge can be acquired from maps and how maps function in spatial exploration and analysis contexts. Particular attention is given to the role of interactive/dynamic maps in scientific exploration and exploratory spatial data analysis (ESDA).

In the Postscript, I offer a view toward the future that anchors the theory of cartographic representation in the broader context of theory for spatial representation. This approach builds upon the previous chapters to argue for a multilevel, multiperspective approach to maps and map use that recognizes cartography as a field concerned with representation in all its respects.

NOTES

1. An interesting perspective on cartography as a craft is provided by Medyckyj-Scott and Board (1991). They describe the limits to formalizing methods and procedures inherent in a craft-based field and discuss cognitive cartographic research as a complementary approach to deriving operational map design guidelines.

2. Objective is used here to mean "free from personal prejudice, unbiased."

3. Map engineering is defined here as an approach to cartographic design that involves formulating and applying precise rules for decision making. These rules are derived from a combination of the application of scientific principles with iterative refinement through empirical testing. This approach contrasts with a graphic-design-oriented approach to map design. The latter tends to be more holistic, less rule-bound, and focused on aesthetic as well as functional goals.

4. Art is considered here to be more than simply achieving a pleasing appearance. The term is used in its broader sense of grappling with emotions, prompting subjective responses, contemplating "aesthetics" (the branch of philosophy that deals with the principles and effects of beauty), along with concerns for the production of pleasing designs.

5. The artistic approach, as envisaged here, is more systematic in application than the "cartography as craft" approach that Medyckyj-Scott and Board (1991) suggest that cartography has followed.

6. Unless otherwise noted, "science" or "scientific" should be taken to mean following the rubric of logical positivism, thus following methods that involve systematic progress through four stages: observation, theory development, test of theory empirically, and modification of theory in response to results (Harvey, 1969).

7. A brief Postscript summarizes the concepts derived and offers suggestions on where the broader theoretical approach to maps and map use delineated here might lead.

PART ONE

How Meaning Is Derived from Maps

It has been apparent to cartographers for over a century that various characteristics of human vision influence how we "see" and interpret maps (i.e., how maps work). One early example of the application of theories about human vision to map symbolization is the convention of using spectral or part spectral sequences to depict elevation zones on layer tint relief maps. This convention, taken for granted by many atlas and textbook publishers today, was first suggested in 1898 by Karl Peucker (Imhof, 1965/1982). The suggestion was linked to the theory of "advance and retreat," which posits that because different hues will be refracted in inverse relation to their wavelengths, they will appear to be at different distances from us when they are in fact at the same distance. Reds for high elevations to blues for low elevations were therefore expected to take advantage of this feature of our visual-processing system and to give the impression of mountains rising above the plains.

The concept of "advance and retreat" is representative of the perceptual theories that cartographers have drawn upon in attempts to enhance the likelihood that symbols (or entire maps) will be interpreted correctly. In some cases, as in the advance-and-retreat example, we have simply applied psychological and physiological evidence to the map-reading context in intuitively logical ways (without testing the veracity of this logic). Our understanding of figure–ground on maps has been based primarily upon this approach. Perceptual research in psychology and physiology, however, is not designed to make maps work better and seldom addresses cartographic questions. As a result, cartographers over the past four decades have undertaken research designed specifically to evaluate map functionality and the applicability of perceptual (and more recently cognitive) theory to maps. Shortridge (1982), for example, examined the applicability of stimulus-processing models to cartography and concluded that cartography cannot use the models directly because we are primarily interested in the result due to a symbol rather than the process that produces the result.

As Olson (1983) has noted, there is no reason that cartography should direct all its research energy to questions of map functionality. In the long run, a more complete understanding of how maps work is an equally worthwhile goal. Limiting cartographic concern about map reading to measuring reader responses to stimuli is similar to the approach of past regional geographers who viewed their role simply as describing places, and made no attempt to understand the processes leading to what they described. As in that case, description of outcomes is an important step in understanding, but it becomes intellectually barren if we do not follow up with questions of why and how. As long as cartographers are content with making good maps (and doing research leading to better maps) rather than with understanding how maps (and other spatial representations) work, their role will be restricted to the processing of other people's data. Broadening our scope to consider the whys and hows of what makes maps work opens up many new lines of inquiry—and will in the long run be more likely to meet the traditional goal of producing functional maps.

A logical starting place for understanding how maps work might seem to be with detection and discrimination of objects in the visual scene. These processes are essential to further processing. Recognizing the critical role that detection and discrimination play in human–map interaction, Robinson (1952) placed identification of "least practical differences" near the top of the research agenda. Much of the earliest empirical work by cartographers interested in human–map interaction focused on this problem or the related perceptual issues of visual search and magnitude judgments (of lines, circles, gray tones, etc.).

It has become clear that human vision is highly variable and difficult to predict, particularly when the visual task involves a stimulus as complex as a map. At this point, our efforts to explain map reading in relation to constraints of vision and the resulting influence on discrimination and other low-level processes have resulted in an idiosyncratic set of case-specific conclusions that we cannot confidently extend to other applications. One of the weaknesses of most research dealing with aspects of early vision has been a failure to build upon a firm theoretical base, along with a failure to link what we know about low-level perceptual aspects of map reading to the growing knowledge of higher level processing being uncovered by cognitive psychologists and cognitive scientists. As indicated in Chapter 1, perceptual cartographic research, until recently, followed a behaviorist model of discovering "laws" that attempt to predict responses to stimuli without an understanding of why the observed responses occur.

Human vision and visual cognition remain incompletely understood. A variety of competing theories exist to explain how the process works and the extent to which parts of the process operate at preconscious versus conscious levels. The idea, however, that vision is an information-processing system and that information is "constructed" from sensory input (rather than communicated via visual pathways) represents the dominant current viewpoint in psychology and has been adopted by several cartographic researchers (e.g., Eastman, 1985a; Peterson, 1987). This approach seems particularly suited to explaining how humans interpret two-dimensional graphic depictions (the stimuli for much of the psychological research conducted thus far, and the category into which most maps can be placed).

Gibson's (1979) "ecological" approach to visual perception is the dominant alternative to the information-processing perspective. Gibson, however, discounted attempts to study non-natural, two-dimensional stimuli as largely irrelevant to the understanding of a visual system that evolved to deal with the natural environment. It is therefore difficult to relate his theory to perception of the abstract symbolic two-dimensional representations that characterize most maps, and only one cartographer (Castner, 1990) seems to have tried. More importantly, however, Gibson's view of perception as a process in which visual scenes "afford" meaning to the viewer is at odds with commonsense knowledge of map reading. Gibson does consider perception to be an active process (a perspective lacking in some information-processing models of vision). He contends, however, that the meaning derived by vision is embedded in the "optical flows" produced by these active visual processes (rather than being brought to the visual scene by a user who matches her knowledge to input from the scene).

One of the major flaws identified in the communication-model paradigm for cartography was that it failed to account for the active role of the map user in *deriving meaning* from maps. Olson (1979) has, in fact, suggested that readers actively "give meaning to maps." This view contrasts with that of Gibson (1979), but corresponds to suggestions about how knowledge schemata, derived from experience with specific problem contexts, exert control over what can be obtained from a visual display (see the "How Map Schemata Are Used" section of Chapter 4 for details).

Part of the information-processing approach to vision and visual cognition is based on the neurophysiological advances of the past decade. Researchers in these areas are gaining confidence concerning their eventual ability to explain how the brain works because of their success in explaining how individual and groups of cells respond to visual and other input. Some of the most recent computational theories of vision, including the widely cited theoretical work of the late David Marr (1982), has underpinnings in neurophysiology.

A second issue that seems to be regaining prominence in psychological explanations of human vision and visual cognition is the work of Gestalt psychologists. The highly reductionistic approach to explaining visual processes that assumed that all recognition of objects in visual scenes (whether symbols on maps or automobiles on a busy highway) begins with component features simply cannot provide the entire answer. Principles of perceptual organization must also be at work.

A third issue at the core of attempts to understand information processing at higher (i.e., cognitive) levels is the issue of knowledge structures. Two particular aspects of knowledge structures seem relevant to cartographic representation: mental categorization and knowledge schemata. Recent views of mental categorization diverge from the long-held belief that mental categories are relatively well-defined nonoverlapping units to a realization that categories are potentially overlapping, have fuzzy bounds, and are best modeled as groupings of related items around prototypes to which category members have varying degrees of similarity. These "categories" represent elements or nodes in knowledge schemata that define links and relations among these nodes.

Part I sketches the current information-processing approach to human vision and visual cognition, and pays particular attention to what it tells us concerning our potential to process information about the primary visual variables used on maps: location, size, value, hue, and the like. Emphasis is placed on linking map-reading tasks to what is known about how the eye–brain system works, on the role of Gestalt principles and other aspects of perceptual organization in map-reading tasks, and on the importance of mental categories and knowledge schemata to derivation of meaning from maps. Fundamental issues of an information-processing approach to visual cognition are presented in Chapter 2, and details relevant to maps and specific to different levels of processing are elaborated in the remaining chapters of this section.

CHAPTER TWO

An Information-Processing View of Vision and Visual Cognition
CARTOGRAPHIC IMPLICATIONS

To produce functional maps, we need to know something about what our visual–cognitive system is designed to do and what it is not designed to do, about the process by which vision and cognition allow us to derive meaning from visual scenes, and about the representations that are created at various stages of the process. The goal here is not to provide a comprehensive review of what is known about eye–brain functions and visual cognition nor about the cartographic research that has been linked to this knowledge (see Bruce and Green, 1990, and Pinker, 1984, for psychological overviews and Eastman, 1985a, and Medyckyj-Scott and Board, 1991, for cartographic perspectives). Rather, I will sketch the outline of a broad theoretical perspective on human processing of visual scenes and discuss how this theoretical approach can be applied to the process of extracting meaning from maps. This sketch is a selective one in which I draw on a limited set of current conceptions that I see as particularly applicable to the multilevel, multiperspective approach to cartographic representation advocated in this book.

Past efforts by cartographers to consider the implications of research in vision, visual perception, and spatial cognition for map symbolization and design have focused on details of specific low-level task abilities (e.g., discrimination, identification of shape, rank order, extracting figure from ground, etc.) without much concern for what vision and cognition are for, why vision and cognition work in the way they do, or the complex tasks to which maps are actually put. We have not been alone in taking this limiting approach. Much of psychology has pursued this path, and we have followed them.

Experimental psychologists have typically employed a reductionistic approach focused almost entirely on controlled, two-dimensional, laboratory test stimuli that are far removed from real-world experiences. These psychological studies have been quite influential in cartography, perhaps because they dealt with abstract symbols that seemed similar to those on maps. Recently, psychologists and cognitive scientists trying to understand vision and visual cognition have started to extend their focus from human processing of abstract stimuli to issues of how we see (and comprehend) the real world.

Gibson (1979), arguably, led the way in this regard with his *ecological* vision. His view, that the stimuli in the world "afford" meaning to what he considers a visually active but cognitively passive receptor, however, is not accepted by those who consider vision to be an information-processing system (a position that I adopt here). Gibson, in fact, argued strenuously against the idea that vision "processes" information. His conception was that vision "reacts to" information.

Contrary to Gibson's view, neurophysiological, neuropsychological, and psychophysical evidence provides strong support for the contention that vision is modular and involves a series of linked processes (see Chapter 3). In spite of the apparent incompatibility between Gibson's ecological optics and information-processing approaches, information-processing researchers (most notably Marr) have borrowed at least one of Gibson's principles: that to understand vision we need to consider what it is for in the real world. Following from this premise, Marr (1982) contended that several levels of theory development are needed and that computational theories may be most fundamental because they take into account what a system should do, and why, before trying to isolate how or with what mechanism.

There are two reasons why cartography should pay closer attention to research in psychology and cognitive science that is directed to perception and cognition of the real three-dimensional dynamic environment. The practical reason is that cartography is changing. Our "maps" have the potential to become more realistic by simulating 3-D and changing in real time. Eventually they may become embedded in virtual realities that allow the viewer to interact with them, much as the viewer interacts with real objects in the world. Perhaps more importantly, for the near term, attention to what vision and cognition are for in everyday experience may be critical to understanding how vision and cognition work in any context. It is only logical to assume that fundamental structures have gradually evolved for recognition and identification of real-world objects and patterns. The evolutionarily recent development of abstract visual tools such as maps and graphics make it unlikely that special visual processes have evolved that allow us to read them. Understanding representations and processes used to grapple with the real world, then, is likely to take us farther toward understanding how vision and cognition react to stimuli that are as unnatural as maps than has trying to understand these stimuli in isolation.

MARR'S APPROACH TO VISION

David Marr's (1982) approach to vision has had a dramatic impact on understanding both vision and information-processing systems at various levels of analysis. This impact is due to Marr's clear delineation of the levels at which an information-processing task must be addressed if we are to understand it completely. He distinguished between three levels of understanding: the level of computational theory (at which we describe what a process must do and why, along with a logical strategy by which the process might be carried out), the level of representation and algorithms (dealing with how the theory might be implemented), and the level of processing device or hardware implementation (that considers how a particular representation might be implemented in the available device). Marr contends that some observable phenomena may find explanation at only one level, and thus that it is critical to consider the appropriate level of analysis whenever we evaluate evidence about how different processes function. Some failures of map symbols, for example, might be due to limits at the hardware level of neurophysiology, while others might be due to the representation forms or algorithms our brain applies to extracting meaning from a visual scene. Marr claims that the computational theory level is the most fundamental. If we recognize that, logically, our nervous system evolved to meet certain needs, rather than that our perceptual processes evolved to make use of fixed predetermined neurological hardware, it becomes clear that understanding what vision is for is more important than understanding the neurophysiological mechanism by which it works. As a modifier to this de-emphasis of the hardware level, we should remember that the representations and algorithms used to implement the operations posited by computational theory evolved to meet survival needs of behavior in a complex three-dimensional world and, as a result, may not transfer well to the task of interpreting an abstract two-dimensional display. Hardware, in the form of our eye–brain system, can set limits for tasks (such as map reading) that vision has not evolved to accomplish.

Marr (1985, p. 103) defined vision as "the *process* of discovering from images what is present in the world, and where it is." Marr contends that to understand any information-processing system requires us not only to learn about the process, but to consider how information is represented. With vision it is, after all, a representation of the world formed on the retina that must be processed; and if Marr is correct, this retinal representation is transformed in a series of subsequent representations that lead from the two-dimensional retinal representation to a three-dimensional object-centered representation of the structure and organization of the viewed object or scene. From a cartographic perspective, Olson (1979) suggested a similar need to consider not only processes but "that which is processed" when she advocated giving attention to mental categories, mental organization, and mental constructs as things that change as a result of using a map.

Marr draws upon representational theories of the mind which posit that the mind has access to internal representations, that mental states are defined by what the internal representations specify, and that mental processes act on these representations. In the context of information processing, Marr (1985, p. 111) defines a representation as "a formal system for making explicit certain entities or types of information, together with a specification of how the system does this." As such, Marr's approach to representations used by the human visual system has much in common with the formal (i.e., semiotic) perspective on representation that I describe in Part II. As is true when a cartographer or other information designer selects a symbolic representation system to depict a specific set of data, the particular representations used by various stages of vision will "make certain information explicit at the expense of information that is pushed into the background and may be quite hard to recover" (Marr, 1985, p. 112). In discussing mental images of maps as one higher level representation form, for example, Peterson (1987, p. 38) points out that "mental images seem not to be copies of sensory impressions like 'pictures in the head' but rather they are intellectually processed and generalized representations, much like maps."[1]

Following from these philosophical perspectives, Marr (1982) developed a representational framework for vision in which he proposed three linked processing modules that transform an unstructured two-dimensional representation of the visual scene into a highly structured three-dimensional model representation (Figure 2.1). The nature of this framework was stimulated by evidence that mental representations of object shape are stored in a different place than representations of use and purpose, and that humans can grasp the shapes of things independently of

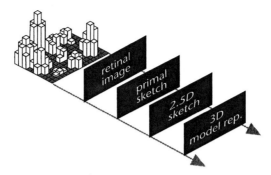

FIGURE 2.1. Marr's stages of vision. *Derived from Marr (1982).*

knowing what the things are named or what they are for (Marr, 1985). This led to a realization of what Marr (1985, p. 123) calls the "quintessential fact of human vision—that it tells about shape and space and spatial arrangement." This led, at the computational-theory level, to delineation of the primary purpose of vision as "building a description of the shapes and positions of things from images" (Marr, 1985, p. 123).

The retinal image provides the input for Marr's three-stage representational framework. This image consists of intensity values at points and has the limited purpose of representing these intensities. The first processing stage transforms the image information into what Marr and Nishihara (1978) termed the *primal sketch*. The primal sketch makes information about the retinal image explicit, particularly intensity changes across the image surface along with the geometry and organization of these changes. Marr contended that computation of *zero-crossings* (changes from light to dark) at various resolutions serve as the main input to the process of extracting a primal sketch from the visual scene (Figure 2.2). Primitive elements of the primal sketch are postulated to include perceptual units such as "blobs" and "edge segments." The primal sketch is envisioned as an array of cells that contain "symbols" indicating the presence of edges, bars, blobs, and so on, and their orientations—primarily features that are invariant over changes in overall illumination, contrast, and focus (Pinker, 1984) (Figure 2.3). As Pinker (1984) points out in a summary of Marr's theory, a crucial assumption (for theories of subsequent processing) is that the features symbolized in the primal sketch are extracted separately for various scales. This allows major features to be distinguished from details and leads to a hierarchical model for storage of shape categories in memory against which information from visual scenes is compared.

The next higher level of processing produces the *2½-D sketch*, a "representation of properties of the visible surfaces in a viewer-centered

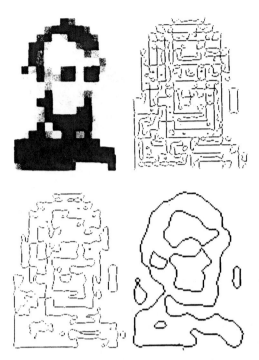

FIGURE 2.2. Zero-crossings from three spatial channels. It is the larger channel depiction (an image similar to one we obtain when squinting the eyes) that allows us to recognize Lincoln in L. D. Hammon's quantized image of Abraham Lincoln. *Reproduced from Marr (1982, Fig. 2.23, p. 74). From* Vision *by Marr. Copyright 1982 by W. H. Freeman and Company. Used with permission.*

coordinate system, such as surface orientation, distance from the viewer, and discontinuities in these qualities; surface reflectance and some coarse description of the prevailing illumination" (Figure 2.4) (Marr, 1985, p. 125). The 2½-D sketch is also conceived of as an array of cells with symbols indicating the various viewer-centered properties. Marr (1982) claims that this depiction has no input from top-down processing nor any global information about shape, only depths and orientations of local pieces of surface. The claim is for a completely precognitive process to this point (although Marr does not discount the possible role of top-down processes for directing attention to particular places in the visual field or for dealing with particularly ambiguous situations).

Finally, processing achieves what Marr (1985) terms the *3-D model representation* (Figure 2.5). This representation is posited to be an object-centered rather than a viewer-centered depiction of the three-dimensional structure and organization of the viewed shape, along with some level

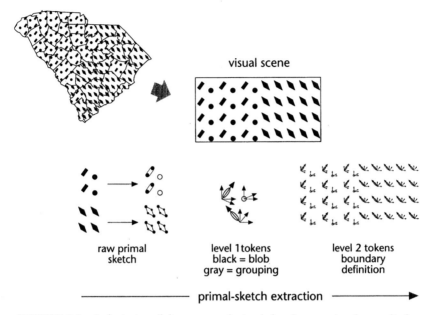

FIGURE 2.3. A depiction of the process of primal-sketch extraction from a display created by two area fills on a map. The raw primal sketch translates map marks to intensity changes in which edges and "blobs" (identified at various scales from zero-crossings) are isolated with their terminations indicated. These features are further processed to produce *place tokens* that are then combined through various perceptual grouping operations. One outcome of this grouping is the identification of boundaries between relatively homogeneous regions. *Derived from Marr (1982, Fig. 2.7, p. 53)*.

of description of its surface properties. The 3-D model representation is considered to be a hierarchical model with each level consisting of a few axes to which volumetric shape primitives are assigned. The hierarchical model of shape descriptions separates information about arrangement of parts of a given size from the internal structure of those parts and information about shape of an individual from shape of general object classes (Pinker, 1984).[2] This modular system allows us to recognize objects (e.g., a bird or a building) as being in a particular class even if we can not recognize the individual (e.g., a duck or a church).

The modularity of processing suggested by Marr has interesting implications for the interpretation of map symbols. For example, when we are visually scanning a National Park Service map looking for a camping site, this modular processing may allow us to sort out the set of winter sports symbols from the set of camping symbols before specific recognition of any symbol takes place. To locate a campsite symbol, we do not

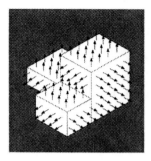

FIGURE 2.4. A depiction of Marr's concept of a 2½-D sketch from a section of the prism map similar to that in Figure 2.1. Emphasis is on orientation of surfaces (shown as dashed lines between object surfaces and solid lines between object and background). *Derived from Marr (1982, Fig. 3.12, p. 129).*

have to understand that a particular symbol means "ski bobbing" to rule it out as a candidate for a closer examination. Ratajski's (1971) system of standardized signs for economic maps seems to fit well with the hypothesis that visual processing is hierarchically organized (although he designed his map symbol system more than a decade before Marr presented the hypothesis) (Figure 2.6).

Marr's approach to shape derivation from visual scenes has had considerable impact on the field of machine vision where it has proved to be

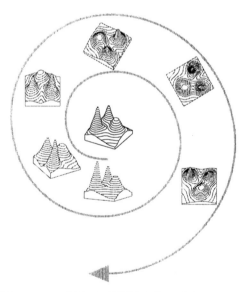

FIGURE 2.5. My concept of Marr's (1982) 3-D model representation. The representation is a complex structural model from which multiple viewpoints can be constructed. The figure symbolizes a mental "flyby" of the 3-D object representation. Not illustrated is Marr's conception of multiple hierachically linked structures that deal with the 3-D representation in parallel at various scales.

more successful than past iterative empirical approaches that tried to engineer a solution with no underlying theory upon which to base that solution. It also has proved superior to highly reductionistic approaches that searched for general principles by severely restricting the scope of problems (e.g., to a world of white toy blocks). Marr achieved a greater measure of success by starting at the level of computational theory. At this level he had to begin by deciding what vision was for before trying to determine how it worked. Marr's impact is due to the interaction of his three-tiered theoretical approach (of computational theory, representation–algorithm, and hardware implementation) to treatment of vision as an information-processing problem, and emphasis on the forms of representation acted upon at various stages of that process.

VISUAL COGNITION

Processing of Visual Stimuli

Visual cognition encompasses issues of how cognitive processes interact with vision to enable us to interpret the world and our apparent ability to mentally manipulate visual information in the form of images. To consider visual cognition and its potential implications for cartographic representation, we look to Steven Pinker. Pinker (1984) suggests that one important implication of Marr's work was that it convincingly illustrated different processes at work and different representations worked on at various stages of vision and visual cognition. As a result, there is a need for theories to distinguish processes of early vision (that are probably dominated by bottom-up processing) from those at higher levels of processing (that use output from early vision in combination with existing knowledge or knowledge structures). That Marr's 3-D model representation is modular and hierarchical is a key feature that allows us to link what is known about early (precognitive) visual processes to cognitive processes for recognition of shape (as Marr attempted) and to higher level visual problems such as graph comprehension (as Pinker went on to do). Marr's overall approach, with its focus on extracting shape and pattern from visual scenes in which there are no predetermined symbol–referent relationships, seems particularly applicable to the image end of the image-graphic continuum that I will describe in Part II. Pinker (1990) has extended Marr's basic structure to the other extreme (i.e., graphics) in developing a theoretical approach to *graph comprehension.*

Pinker's (1990) theory of graph comprehension addresses several fundamental issues that are also likely to underlay map comprehension. The theory will, therefore, be presented here in some detail as a basis

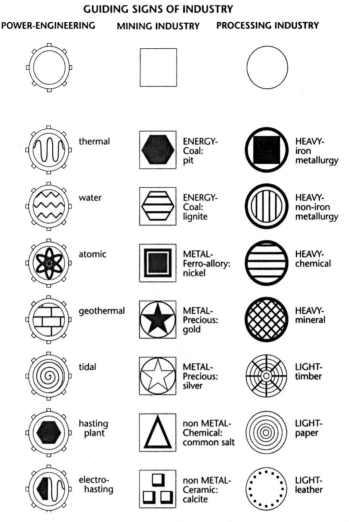

FIGURE 2.6. In preparing a standardized set of symbols for economic maps, Ratajski clearly relied on hierarchical grouping of visual features to assist readers in sorting without having to direct attention to (and waste effort on) the minor details that distinguish symbols in each group. These symbols represent only a sample of the 155 symbols that Ratajski designed for the three economic categories cited here. *Derived from Ratajski (1971, Figs. 14–18 and 21, pp. 153–155 and 157).*

from which to build a similar approach to map understanding. An interesting link to cartography is already included in Pinker's theory. This link is through his reliance on Bertin's (1967/1983) delineation of the tasks that the graph reader must accomplish if successful extraction of information from a graph is to occur. These tasks are (1) to identify the conceptual or real-world referents that the graph is conveying information about, (2) to identify the relevant dimensions of variation in the graphic variables and determine which visual dimensions correspond to which conceptual variable or scale, and (3) to use levels of each visual dimension to draw conclusions about particular levels of each conceptual scale. According to Pinker, these tasks imply that the reader must mentally represent the physical dimensions of marks on the graph. He calls this mental representation a *visual description*. In addition, the reader must draw upon what Pinker labels a *graph schema* to determine how physical dimensions are to be mapped onto mathematical scales.

Pinker (1990) relies on Marr's distinction between precognitive and cognitive aspects of vision to postulate that a *visual array* (equivalent to Marr's primal and/or 2½-D sketch or to iconic memory as described by Phillips, 1974) serves as the input to visual cognition which transforms this array into the *visual description* (analogous to short-term visual store [STVS] as described by Phillips, 1974, 1983, or the visuospatial scratch pad [VSSP] as proposed by Baddeley and Hitch, 1974).[3] This visual description is conceived of as a structural description of the information in the visual array. This structural description involves *variables* that "stand for" perceived entities or objects and *predicates* that specify relationships among, or attributes of, entities. Pinker develops a formal graphical notation system for depicting these visual descriptions, but for our purposes in this chapter the outline of his theory and its implications for map understanding can be appreciated without these details. I will return to the issue of visual descriptions and fill out the details in Chapter 4.

A key issue with Pinker's visual descriptions is that many different ones are possible as representations of a specific visual array. Part of the task of the theory, then, is to predict which visual description is most likely. Pinker (1990, p. 78) sets out to do this "based on knowledge of how the human visual system works." Specifically, Pinker details four factors intended to explain why certain visual descriptions and not others result from particular visual arrays.

The first factor, borrowed from Kubovy (1981), is that certain visual variables are considered *indispensable attributes*, attributes having a dominant perceptual status. These indispensable attributes are *space* (i.e., location) and *time*. Pinker ignores time because his theory is directed to static graphs. For dynamic maps and graphs, however, the fact that time has been demonstrated to be an indispensable attribute is critical. It tells us that change in positions or attributes over time should attract particular attention and serve as a perceptual organizer that is much stronger than hue, value, texture, shape, and so on. Agreeing with Kubovy, Pinker (1990) argues that spatial location will be the dominant factor (over the other visual variables—discounting time) in establishing *perceptual units* (or what the fundamental pieces of a visual scene are). In the example shown in Figure 2.7, there are clearly three basic units. If pattern variables dominated spatial location, however, there could be only two units (the left and middle bar vs. the right and left bar). In addition, Kubovy is cited as arguing that the indispensable attributes have stronger configural properties than other variables (Figure 2.8), are more discriminable, and are less subject to nonlinear functional relations between actual and perceived differences. Nonlinear functions between perceived and actual gray tones, for example, have been the focus of considerable cartographic attention. Kimerling (1985) has demonstrated that the relationship is increasingly curvilinear as the texture used to produce the gray fill in map areas decreases (Figure 2.9). Leonard and Buttenfield (1989) present results that suggest a complex interaction between texture and value when gray tones on maps are created by area fills coarse enough for texture to be noticed.

In addition to the above aspects of indispensable attributes, it is argued that attention is more *selective* to indispensable variables than to other visual variables. This means, for example, that on a map we are more able to attend to a particular location (regardless of symbol shape) than to a particular shape (regardless of location). Selective attention is related to Bertin's ideas about "selectivity" of graphic variables. Bertin (1967/1983, p. 67) seems to accept the primacy of location when he declares that "selective perception is utilized in obtaining an answer to the question: 'Where is a given category?'" Bertin's *associativity* principle, however, suggests that some other graphic variables have an ability to influence the primacy of location. Associativity, in his view, is the ability to visually group all variations on a particular (nonlocational) graphic variable. Shape is associative in this sense, while size is not (Figure 2.10). For the indispensable variable of location to take precedence, it may be necessary for other variation to be associative. Bertin does suggest that the best selection is achieved by relying only on location and "juxtaposing separate images on the plane" (i.e., using something equivalent to Tufte's [1990] small multiples) (Figure 2.11).

Principles associated with indispensability of space and time "place

FIGURE 2.7. Location versus pattern as perceptual organizers. Readers will see three rather than two perceptual units. In fact, even after the two nonspatial units are identified (the pair with lines and the pair with random dots), it is difficult to "see" two units here. *Derived from Pinker (1990, Fig. 4.5a, p. 80).*

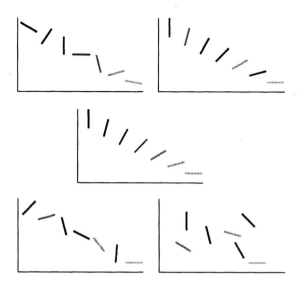

FIGURE 2.8. Configural properties of location, value, and orientation. If both location and value (upper left) or location and orientation (upper right) are ordered, the individual and joint relations are readily apparent. When all three are ordered, the individual and joint relationships are also easily extracted. It is, however, considerably easier to notice ordered location when *both* value and orientation are random (lower left) than ordered value together with ordered orientation when location is random (lower right).

constraints on the parts of an array that variables may stand for, on how numerical variables represent physical continua, and on how predicates are encoded or verified with respect to the visual array" (Pinker, 1990, p. 83). A second factor thought to govern "how atomic perceptual units will be integrated into a coherent percept" is the Gestalt laws of grouping (Pinker, 1990, p. 83). It seems clear that Gestalt laws of grouping play a

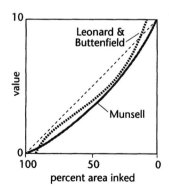

FIGURE 2.9. Kimerling's translation of the Munsell gray scale into percent area inked form (based on 133 lines/inch map output) allows comparison with Leonard and Buttenfield's gray scale for laser printer maps using approximately 50 lines/inch patterns). *Derived from Kimerling (1985, Fig. 4, p. 137) and Leonard and Buttenfield (1989, Table 2, p. 100).*

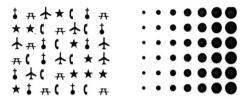

FIGURE 2.10. Shapes can be visually seen as a whole (left), but symbols of different size are seen primarily as different (right). *After Bertin (1967/1983, Figs. 3 and 8, p. 65). Adapted by permission of the University of Wisconsin Press.*

large part in sorting out groups and relationships in maps as well as graphics. McCleary (1981), for example, relies heavily on Gestalt principles in his approach to designing effective graphic and cartographic presentations. In spite of general agreement about the relevance of Gestalt principles to maps and graphics, however, Pinker identifies an issue that has made application of the principles problematic in both a theoretical and an applied context. The limiting factor is that little or no work has been done concerning the relative strengths of these principles (e.g., is it more important for map symbols to be proximate to one another, similar in color, arrayed in linear order, etc. and which of these factors dominates if they conflict?).[4]

A third factor thought to influence which visual description is likely in response to a specific visual array is the mental representation of magnitude. Although magnitudes are probably represented in the visual array as continuous values, there is evidence that higher level representations tend to be encoded as discrete values on an ordinal scale with only about

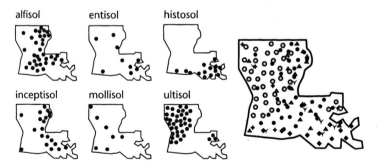

FIGURE 2.11. The selectivity of location is part of what makes Tufte's (1990) idea of small multiples work. With the small multiples shown here, a quick impression can be obtained of the correspondence (or lack of it) between pairs and groups of soils. A single map with various shaped point symbols, however, can be extremely confusing (even at 200% size). *Derived from maps of the entire United States by Gersmehl (1977, Figs. 1, 2, 5, and 7–9, pp. 423 and 425–427).*

seven steps. Pinker (1990) also notes that the likelihood of a value being encoded at all will depend on its context. Extreme values, for example, are more likely to be perceptually encoded.

A final factor related to the extraction of a visual description from the visual array is the coordinate systems used. Pinker (1990) borrows from Marr and Nishihara's (1978) theory to postulate that (1) polar or rectangular coordinates are usually used to represent shape and position, and that (2) different elements of the scene are represented in separate local coordinates that are centered on larger objects that are part of a Gestalt group.

Along with the issue of which particular visual description will be generated to represent a visual array, Pinker distinguishes between *default* and *elaborated* visual descriptions. Default visual descriptions are thought to result exclusively from bottom-up processes. These visual descriptions will be small due to constraints on short-term visual store, which allows between four and nine nodes (i.e., elements) to be kept active at one time. In addition, the contents of default visual descriptions will depend on default encoding likelihoods of predicates (i.e., relations). Although any form of relation that a person can conceive of can be brought to bear on the visual array, the default visual description will contain only those relations that are "just noticed" without conscious thought. The relationships that are just noticed will vary from person to person and are subject to frequency of use (or practice), which partially explains "expert" versus "novice" abilities at recognizing patterns. This accords with cartographic evidence that training can improve performance on what seem to be precognitive tasks (Olson, 1979; Castner, 1983) and with the fact that non-experts are often trained to search for patterns on photographs generated in particle physics laboratories (Judson, 1987). As Schneider and Shiffrin (1977) point out, when someone repeatedly assigns particular visual patterns to specific categories, recognizing these patterns becomes automatic (i.e., preconscious). An elaborated visual description will start from the default description, but will make use of top-down processes to add elements and predicates to that initial description.

Pinker's model of graph comprehension depends on what he calls a *graph schema* to mediate between knowledge representations (or memory) and the visual description derived. These graph schemata are structures for organizing information about variables and relationships in graphs. Graph schemata, according to Pinker, are responsible for (1) specifying how information in visual descriptions is translated into a *conceptual message*, and (2) how queries that involve interrogation of the visual description or that stimulate further interpretation of the visual array are translated into a process that creates or retrieves the required information. Pinker (1990, p. 95) specifically describes a graph schema as "a memory representation embodying knowledge in some domain, consisting of a description which contains 'slots' or parameters for as yet unknown information."

Pinker's graph schemata seem to follow the general idea of cognitive schemata suggested by Neisser (1976, p. 55) as "plans for finding out about objects and events, for obtaining more information to fill in the format." Pinker, however, differs with Neisser in suggesting that precognitive processes involved in creating the visual array determine which schema will be applied, rather than considering schemata as only having a top-down function to direct "exploratory movements of the head and eyes." (Neisser, 1976, p. 55). Pinker's viewpoint seems to agree with that of Antes and Mann (1984) who suggest that the schema applied is often dependent upon how the information is presented to us and at what scale, and Eastman (1985b) who, in relation to maps, demonstrated that map design variables could influence grouping or chunking of information extracted from maps.

Graph schemata can specify what must be true of a particular graph type as well as how it might vary from exemplar to exemplar. Schemata as Pinker describes them can exist at different levels of detail, from those that allow us to distinguish between a graph and other kinds of information display to those that allow us to recognize a particular graph type and interpret the information presented within it (e.g., a line vs. a bar graph and the associated implications of a discrete vs. a continuous function).

Putting the concept of multiple levels of representation (i.e., visual arrays, default and elaborated visual descriptions, and cognitive graph schemata) together with a conception of the kinds of interaction possible between them, Pinker (1990) presents a graphic model of the process of the information-processing stages of "graph comprehension." The model (Figure 2.12) presents graph comprehension as a highly interactive process that makes use of both bottom-up precognitive processes and top-down cognitive processes to gradually build an overall understanding of the graph.

Pinker's model has some parallel's in the cartographic literature, but for the most part cartographic efforts in this direction have been less formalized and only limited attention has been given to information-processing models. This is perhaps because of the surface similarity to information theory models that served as the basis for the communication paradigm (and our disillusionment with that perspective). One of the first information-processing-like concepts suggested in relation to maps was Olson's (1976) hierarchical model of map-reading tasks. This model divided tasks into those involving (1) comparison of individual symbol characteristics (e.g., shape or size), (2) assessment of symbol group characteristics (e.g., pattern), and (3) use of the map as a decision-making or

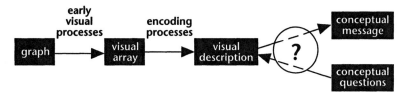

FIGURE 2.12. Pinker's model of visual information processing associated with graph comprehension. The circle with a question mark represents the iterative queries possible between the visual description and prior knowledge. *After Pinker (1990, Fig. 4.14, p. 94). Adapted by permission of Ablex Publishing Corporation.*

knowledge-building device. Olson's approach was largely intuitive and was not linked directly to information-processing theories. Although the model helped suggest some differences to anticipate concerning the relative impact of design research and user training at various task levels, the levels delineated do not match current theories of visual cognition.

At the precognitive level, Dobson (1979b) and Shortridge (1982) have offered information-processing approaches to low-level perception of map symbols. Dobson's goals were applied ones and his work evolved toward a call for a human factors approach to "map engineering." He argued that this approach should use an information-processing perspective as a base from which to design experiments geared toward solving specific map design problems. Shortridge, in contrast, pointed out some cartographically relevant issues that information-processing work in psychology uncovered (e.g., the issue of integral and separable dimensions of multidimensional symbols), but did not share Dobson's optimism for the promise of cartographic research linked directly to information-processing models.

Eastman (1985a), working at a more conceptual level, provides a clear sketch of the information-processing perspective and its potential applicability to user research in cartography. He also deals with the issue of linking a semiotic approach to map symbolization with an information-processing view of map reading. In his presentation of system and process models for map reading, Eastman is careful to point out the fundamental difference between an information theory approach to map communication that measures bits of information transmitted (with the goal being communication of the most bits) and attention to information processing (with the goal being to understand how people actively see and conceptualize about map information). For cartographers to facilitate visual and cognitive information processing, we must understand both the system that does the processing and the processes themselves.

As found in Pinker's model of graph comprehension, Eastman's (1985a, p. 97) discussion of information processing as it might be applied to cartography incorporates the concept of schemata as "a cognitive structure that can be broken down to produce the essential aspects of a typical example of a particular class of objects (called a 'prototype')." He goes on to characterize schemata as "networks of concepts or entities interconnected by a set of relations . . . within a hierarchical framework" (Eastman, 1985a, p. 97). Basing his view primarily on that of the psychologist Palmer (1975, 1977), Eastman contends that entities exist at two levels, the global and the local. Global properties are said to pertain to "chunks as a whole," while local properties are derived from component parts. Chunks and schemata are said to differ only in degree of elaboration, an argument that schemata are hierarchical. This matches to some extent with Pinker's conception of general and specific schemata with the difference being that Eastman allows schemata to exist for parts of a problem context while Pinker's schemata apply to the entire problem (with only the level of detail varying).

As Eastman notes, there is considerable evidence from the study of chess expertise that mental structures stored in memory can facilitate the organization and interpretation of a visual scene. In particular, research by Chase and Simon (1973) has demonstrated that experts organize information into larger chunks, thus enabling them to assess a particular arrangement of the chess board more quickly than novices—but only if that arrangement represents a likely stage in a chess game. Novices, who do not have a well-developed schema for possible arrangements, must process the visual scene at a more local level.

A major difference between Eastman (1985a) and Pinker (1990) is the weight given to top-down processing. Eastman favors the perspective of Navon (1977) who argues that top-down global processes act first to control more local processing of a visual scene. Eastman cites eye movement research in support of this view. Although eye movements when viewing a map have been found to be individualistic, orderly recurrent patterns are common for individuals (Eastman and Castner, 1983). Eastman (1985a) contends that this result can be interpreted as personalized schemata exerting control over how an individual sees a visual scene. He suggests a cyclical process in which a schema directs exploration, which samples the visual scene, and the results of this sampling have the potential to modify the schema (Figure 2.13). While some eye movement evidence does suggest that top-down processes may be dominant in steering attention across the visual field, other results can be interpreted as evidence for automatic precognitive processes that are reacting to the visual array without any conscious control. A recent analysis of differences in viewing strategy for a range of symbolization types seems to support this latter view (Morita, 1991) (Figure 2.14).

Whether or not the eye movement research provides evidence for

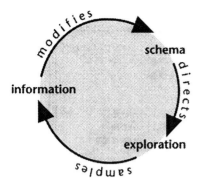

FIGURE 2.13. Eastman's interpretation of Neisser's concept of a schema applied to visual exploration of a map or other graphic display. *After Eastman (1985a, Fig. 3, p. 99). Adapted by permission of* The Cartographic Journal.

top-down processes as a control on attention, this evidence says nothing about the role of top-down processes in translating the visual array into Marr's 2½-D sketch or Pinker's visual description. As with most complex human processes, what is most likely is that both top-down cognitive processes and bottom-up precognitive visual processes complement each other, with each taking precedence some of the time. It may be that behavior in the environment requires more reliance on bottom-up processes as a first sort to ensure that we obtain some information fast enough to do us some good (e.g., when driving a car our eye movements are likely to be attracted to movement with no thought processes involved telling us that we should attend to movement more than to color). In the case of information graphics, however, the problem context for vision is considerably restricted and it is logical that our visual–cognitive processing system can take advantage of this to make better use of expectations in directing where we look or what features we attend to.

Eastman's conjunction of information-processing and semiotic approaches is complemented within the psychological literature by Kosslyn's (1989) efforts toward "understanding graphs and charts." He emphasizes a mix of visual information processing and semiotic principles in the development of acceptability guidelines for graphics. In Kosslyn's information-processing model, a series of factors are identified that influence processing at different stages (and might impair or facilitate graph interpretation). Between the "perceptual image" (Pinker's visual array and Marr's primal and 2½-D sketches) and short-term memory (Pinker's visual description), issues of discriminability (a hardware-level issue), distortion (a hardware and representation issue), organization (a representation issue), and priorities (a representation issue) come into play (Figure 2.15). Short-term memory itself will have capacity limitations and there will be issues of kind of encoding between short-term memory and long-term memory. Knowledge in long-term memory (accessed via Pinker's schemata) acts on the representation in short-term memory, either ac-

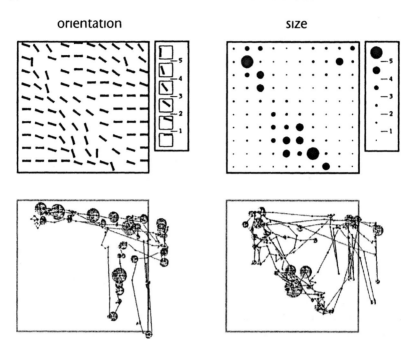

FIGURE 2.14. The influence of symbolization type on visual scanning of mapped patterns. The two schematic maps depict indentical data using two of Bertin's graphic variables: orientation and size. *Reproduced from Morita (1991, Figs. 3 and 4, pp. 4 and 8). Reprinted by permission of the author.*

cepting what is encountered or prompting reorganization or directed search of the visual scene (i.e., reference back to the visual array). This process will obviously be an iterative one.

Eastman's (1985a) view that we should consider maps as facilitators rather than as communicators of information is complementary with Pinker's (1990) view that displays leading to clear perceptual organization will be most effective. In both cases, it is anticipated that characteristics of the display will influence how easily existing schemata are brought to bear on the problem of deriving information through viewing the display. These perspectives fit with my own current ideas about geographic visualization (considered in detail in Part III). Geographic visualization (GVIS), however, with its high level of interaction between viewer and map display and its emphasis on searching for unknown patterns versus interpreting a predetermined message, demands emphasis on somewhat different information-processing tasks. In response to this need, John Ganter and I developed an information-processing model of map-based visualization as a pattern-matching process (MacEachren and Ganter, 1990) (Figure 2.16).

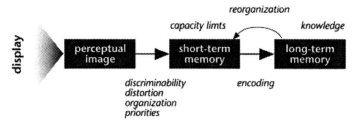

FIGURE 2.15. Kosslyn's information-processing model of graph perception. *After Kosslyn (1989, Fig. 2, p. 188). Adapted by permission of John Wiley & Sons, Ltd., from Applied Cognitive Psychology.*

At the time we developed this pattern-matching approach, we were only vaguely aware of Marr's information-processing model of human vision and we had not encountered Pinker's work on graph comprehension (which was published in the same year). Our model evolved from an overall concern for the impact of scientific visualization on cartography, study of the scientific creativity and scientific visualization literature, and from the perspective on history and philosophy of science presented by Howard Margolis (1987). The fact that two rather similar models evolved from quite different research perspectives indicates a strength for the basic tenet of both models—*that cognitive "schemata" exert some level of control on how we see evidence and that with repeated use, these schemata become ingrained to the point that noticing patterns of particular types becomes automatic, or precognitive.*

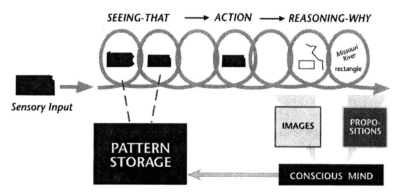

FIGURE 2.16. The pattern identification model of cartographic visualization. *After MacEachren and Ganter (1990, Fig. 2, p. 70). (The initial graphic depiction of this model was developed by John Ganter and appeared in Ganter and MacEachren, 1989.) Adapted from* Cartographica *by permission of University of Toronto Press, Inc. Copyright 1990 by University of Toronto Press, Inc.*

Processing of Imagery

One aspect of the Ganter–MacEachren model that differs from the discussion of graph comprehension by Pinker (1990) or the application of information-processing theory to cartography by Eastman (1985a) is the explicit attention to imagery. Pinker (1984), however, has given considerable attention to imagery in his previous work. His overview of visual cognition presented visual information processing and visual imagery as complementary parts of visual cognition. Imagery, in this context, is seen as a process for interacting with spatial aspects of long-term memory representations. Both Kosslyn (1980) and Pinker (1984) suggest that imagery may use some of the same processes as vision, and that it is analogous to visual processing of stimuli that are present. Finke (1980) suggests that perception uses one set of processes at a neurological level for feature analysis and grouping, another set at a higher level of analysis related to object shape, size, orientation, and the like, and further processes that apply general knowledge and cognitive skills. He goes on to contend that some imagery phenomena reflect the operation of similar (if not identical) midlevel as well as high-level processes. In Pinker's graph-comprehension model (although he does not make this explicit point himself) images may be something equivalent to a visual description and, as such, may be amenable to the same kind of message extraction and interrogation that Pinker postulates for visual descriptions derived from visual arrays.

Within cartography, the concept of mental images is intuitively appealing. All cartographers think visually. It is therefore difficult for a cartographer to conceive of imagery as being an epiphenomenon with no real use, as has been suggested by some anti-image protagonists (see Pinker, 1984, for discussion of the debate). Considerable evidence has been compiled that images of maps can be formed, stored in memory (in some probably nonanalog representation format), extracted from memory, and used in ways similar to a physical map (see, e.g., Peterson, 1985, 1987; Lloyd and Steinke, 1985; Lloyd, 1988; and Kosslyn, 1980). Some of this evidence is reviewed in Chapter 4 where the issue of how information obtained from maps might be mentally represented is considered in detail. For the present, we will accept the existence of images and consider where they fit in an information-processing account of vision and visual cognition.

Peterson (1987) suggests that both maps and images are products of spatial thinking and are dependent upon the process of arranging objects in space. In addition, he contends that mental images, like maps, are not copies of sensory impressions, but intellectually processed and generalized representations. This commonality between cartographic maps and spatial images led him to propose a GIS-like model of how image and propositional information might be drawn upon in spatial problem solving. Peterson suggests that as in a GIS (where data, equations, and mathematical models provide nonvisual information and maps, graphs, and remotely sensed scenes provide visual information) our mental information-processing system has access to both propositional and image information. When we are faced with an interpretation or decision, Peterson suggests that a competitive process (as hypothesized by dual coding theory) is used. The system accesses both kinds of information and the answer to the query will be made using whichever information source can be retrieved most quickly (Figure 2.17). One of Peterson's main points is that the structure or design of maps will influence the extent to which information obtained from them will lead to cognitive representations accessible as images (or, I would add, stimulate generation of images from existing knowledge). Since the kinds of questions and problems for which map information might be useful are spatial, it seems logical that images will be suited to dealing with them. As a result, Peterson argues that good map designs should be evaluated, in part, by the extent to which they prompt memory representations that are accessible as images.

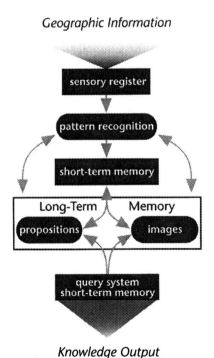

FIGURE 2.17. Peterson's human information system model. *After Peterson (1987, Fig. 5, p. 40). Adapted by permission of* The Cartographic Journal.

Pinker (1984) specifically considers the issue of what images are good for, if we agree that they do exist. He cites four possibilities. Images might serve as a global coordinate system to which information could be related. Similarly, we might use images as a way to compute spatial information (e.g., distances and directions) that we did not encode directly. A third possibility is that images are utilized in some object and pattern recognition tasks, particularly when we are anticipating one of a small number of possibilities. Finally, Pinker suggests that images might provide the best medium for solving certain abstract problems. He contends that images are "representations in a medium with certain fixed properties, and can be subjected to transformations such as rotation and scaling" (Pinker, 1984, p. 57). He suggests that abstract problems could be tackled by "translating their entities into imagined objects, transforming them using available image transformations, detecting the resulting spatial relations and properties, and translating those relations and properties back to the problem domain" (Pinker, 1984, p. 56).

This last potential use for images is exemplified by their purported role in scientific discovery. Visual imagery has been credited with playing a significant part in various scientific discoveries (Hadamard, 1945; Shepard, 1978). Faraday and Maxwell appear to have used imagery in their initial conceptions of electric and magnetic fields; Einstein in a variety of contexts emphasized his own tendency to use visual over verbal thinking; and Kekule's introspective accounts suggest that visual images led to his discovery of the structure of benzene (Shepard, 1978). The image-transformation processes Pinker suggests seem particularly applicable to theorizing in mathematics and physics. Scientific visualization researchers appear to have directed their attention in these fields to facilitating these transformations by making abstract images explicit.[5]

Peterson (1987) has not been alone in suggesting that concrete visual representations can serve to prompt the mental imagery needed for understanding problem situations and creative thinking about those situations (e.g., Beveridge and Parkins, 1987; Larkin and Simon, 1987; MacEachren and Ganter, 1990; MacEachren et al., 1992). Larkin and Simon (1987) emphasize the perceptual inferences that can be made from diagrams and suggest, by calling the information "zero cost," that these inferences are precognitive. Arnheim (1985) offers similar ideas under the heading of "intuition." To Arnheim, intuition is a component of visual thinking that operates like a "gift from nowhere." Both mental images and explicit representations appear to share the advantage of using Kubovy's indispensable variable of space. Both facilitate spatial grouping of information and inferences based upon spatial position. Arnheim's gift from nowhere and Larkin and Simons's zero-cost information, then, may result from the ease with which visual cognition (operating on visual descriptions of an actual scene or on visual images generated with a similar structure) deals with space.

CONCLUSION

The above discussion outlines a view of vision and visual cognition that is achieving growing acceptance in the cognitive psychology and cognitive science communities. Although not all researchers in these fields even agree about the fundamental issue of whether vision and/or cognition can be considered an information-processing system, this perspective is probably the dominant one and makes intuitive sense as one of several avenues to understanding maps as representational devices.

From a cartographic perspective, looking at vision and visual cognition as information-processing activities leads to a focus on a series of representations of what is seen on a map's surface. Representations of different kinds result from each stage of processing. At the lowest level, the representation that results is thought to be a visual array that contains "place tokens" or abstract "symbols" that stand for fundamental components (i.e., edges, objects, discontinuities, etc.) derived from the visual scene (i.e., the map display) and basic information about the components and their relationships (e.g., size, depth in the visual field, brightness, etc.). Higher level processing is judged to result in a visual description of the visual array from the previous stage. Knowledge schemata then mediate between this visual description and knowledge in long-term memory to interrogate the visual description, cause its modification, and eventually derive some level of meaning from the map display. The meaning derived can result in modification to existing knowledge schemata or creation of altogether new schemata, yet another representation of the map display. A key issue that must be emphasized here is that I am *not* hypothesizing an information-"transmission" system (with its concomitant emphasis on ratios of signal to noise) but a modular system in which information is "created" and re-created by a series of interpretive processes.

Applying these concepts to maps, I contend that the structure of visual descriptions derived from viewing maps will be based upon both general and specific map schemata (the latter resulting from expert knowledge or interpretation of legend information). A key factor in map schemata and legend understanding will be the basic human facility for categorizing the world. That this facility probably does not usually result in precisely delineated categories has significant implications for map understanding.

The two main levels of processing and their associated representations form the focus for the remaining two chapters of Part I. At issue will be the processes by which each new level of representation is generated, the implications of these processes for map symbolization and design, and the corresponding implications of symbolization and design decisions for the success at which reasonable cognitive representations are achieved.

NOTES

1. We may even find that the intense efforts of the past decade by cartographers to understand map generalization closely parallel some of the work by information-processing researchers to understand visual cognition. The cartographic principles developed may inform theories of mental image formation (at least in relation to maps and other abstract visual scenes) and work by cartographers and others on visual cognition may suggest some new approaches to those trying to develop a more unified theory of map generalization (rather than the fragmented element-by-element approaches characteristic of most work thus far).

2. A variety of theories have subsequently been proposed to explain how parts are categorized and identified; see, for example, Biederman (1987) and Hoffman and Richards (1984).

3. The VSSP is proposed as a temporary storage location in which (a shelf on which) visual information can be briefly stored until needed by the "central executive." The VSSP complements an "articulatory loop" where phonetic material can be stored (see Baddeley, 1988, for details).

4. For an overview of one cartographic attempt to evaluate the relative impact of specific Gestalt grouping variables, see Chapter 3.

5. See Part III for detailed discussion of computer-assisted visualization in a geographic context.

CHAPTER THREE

How Maps Are Seen

A key aspect of Marr's (1982) approach to vision is his contention that there are three levels of explanation from which to address an information-processing system. The computational level focuses on the what and why. Considering vision at this level, and recognizing that vision involves a series of representations and processes that interpret those representations and build new ones, we begin the task of understanding how maps are seen by asking what the purpose of seeing is. According to Marr, this purpose ultimately focuses on recognizing and identifying shapes in the real world. At intermediate stages, however, there are representation-specific purposes that can be identified. In moving from the initial visual scene as sensed by the retina of the eye to Marr's primal sketch, the purpose can be defined as extracting contrast information (related to differences in intensities and wavelengths) and grouping this information to form edges, regions, and shapes. The purpose of the process leading up to Marr's 2½-D sketch (or Pinker's visual description level) is to make the depth, orientation, and junctions of visible surfaces explicit.

In relation to maps, these two goals imply that the way we establish contrast among map features will be critical at the initial level of vision. At this level, according to Marr, no higher level processes come into play, and therefore the only information available to the map viewer is contrast (from pixel to pixel of the retinal image). Although others have argued that top-down cognitive processes can have an effect even at this early stage of vision, it is clear that applying this top-down control is an effortful process. Sorting out components of a map display will be accomplished most efficiently if the cartographer creates contrasts among those map elements that are most important for the viewer to notice immediately. The second goal, associated with the next stage of perceptual representation, suggests that Gestalt principles of perceptual grouping will play an important role. Again, although top-down processes might be able to facilitate Gestalt grouping (or may interfere with it), the most successful map (at this stage of processing) will be one that elicits grouping that links map elements in logical ways (e.g., areas are seen as homogeneous regions rather than disaggregated individual point features—as might happen with use of a pattern made up of noncompatible elements spaced too far apart).

Cartographically, the goal of research directed to low-level visual–cognitive processes is to understand how the stages of physiological–conceptual representation of a map scene interact with symbolization and design variables. Ultimately, we would like to be able to predict what symbol variables or design choices make differences noticeable (in particular situations or for particular tasks), attract viewer attention, are seen as having equal salience, have an intuitive order, or induce grouping or formation of figures on backgrounds. Vision should, on evolutionary grounds, be good at extracting object shape from the visual scene, assessing depth and relative size, and noticing movement. It must perform these functions from information about contrast on a roughly pixel-by-pixel basis at a retinal level, using neurological hardware to process the retinal image. This hardware appears to rely heavily on spatial filtering and enhancement procedures operating simultaneously at several scales. These filtering procedures take into account sensations received by groups of cells. A key feature of this system is that it emphasizes contrast more than absolute illumination (as it must do if we are to recognize an object as the same at dawn and midday). The system has many more cells devoted to value/brightness difference than to hue or saturation, although those cells devoted to these "color" differences are concentrated in central vision and are agglomerated less as they pass signals to cells in the brain's visual cortex. This concentration means that we have relatively higher acuity for hue (hundreds of differences are discriminable) than for value (tens of differences or less are discriminable). A second key feature of the system is an ability to group the elements that the neurological image processing achieves into "objects," or shapes that higher level processes can match to memory representations.

Within this context, this chapter begins with a brief look at our visual hardware that has evolved to meet the above goals. Once in possession of the basic ideas about how this visual hardware has evolved to meet the needs of vision in general, we can speculate about the limits that it imposes for the abstract task of "seeing" a map. The bulk of the chapter, then, considers selected low-level perceptual processes and the potential implications of cartographic use of the visual variables (i.e., the building blocks of map design) to create contrasts between and relations among map elements. While there is continuing debate about whether the low-level processes discussed in this chapter operate in a completely bottom-up, preattentive fashion, or are controlled (at least in part) by top-down processes, the key commonality of the processes included here is that they are fast (measured in milliseconds) and probably occur in parallel. It is this fast parallel processing that makes visualization such a potentially powerful tool for science in an era of data excess (see Part III).

EYE–BRAIN SYSTEM

The intent of this section is to provide a brief sketch of the eye–brain system's major features and to suggest a few examples of how the eye–brain system puts constraints on the way we see symbols and read maps. Knowing the limits, constraints, and idiosyncracies of vision allows us to avoid presenting map readers with processing tasks that are difficult or impossible to perform. Understanding why such limits exist and what our visual system has evolved to accomplish can give us clues about how we might facilitate processing of map information and also clues about the implications of our decisions concerning symbol form, color, size, texture, and so on, for how the information will be processed. The examples provided may also serve to suggest some possible avenues for cartographic research that draws directly upon the quickly expanding knowledge base concerning how human vision works as an information-processing system.

How human vision works is, of course, incompletely understood. What has become clear, however, is that the system does not transfer little pictures of the world from the eye to the brain. Our "perceptions" are constructed (or reconstructed) from a multitude of fragmentary information, some of which is organized spatially (i.e., a direct mapping from positions in the environment to positions in the brain) and some of which is organized according to other attributes of the stimulus (e.g., color, orientation, texture, movement, etc.). Vision is a complex parallel-processing system in which hundreds of millions of sensing cells react to input of light through the lens of our eyes. Through multiple interconnections, these reactions cause subsequent reactions among the tens of billions of cells in our brain that are devoted to vision. Both psychophysical and neurophysiological research indicates that considerable preconscious processing of the signals occurs between the initial incidence of light on the cornea of the eye and the ultimate perceptual experience.

The Eye

Some common conceptions about how the eye–brain system works evaporate quickly when we take a close look at the structure of the eye. A camera analogy is frequently applied. Like the lens of a camera, the human eye is arranged so that reflected and emitted light passes through a lens and results in an "image" of what is observed on a receiving surface. The extent of the image on the eye's receiving surface is a direct function of the size of the object viewed and its distance from the lens. In comparison to many cameras, the eye contains a rather wide angle lens (focal length of 14–17 millimeters), allowing representation of a scene that extends 60° to either side of the central focus to which vision is directed. Although the camera analogy tells part of the story, it can be very misleading. As the complexity and interconnections of cells in the eye become clear, the camera analogy becomes less useful. The fact that we do not have the sensation of looking at the world through a fish-eye lens is one clue to the complexity of image processing that happens between the eye and our conscious sensation of seeing.

An analogy to image analysis systems used in digital remote sensing might prove useful, at least to cartographers trying to understand implications of the eye–brain system for how map symbolization is "seen." Marr's (1982) computational models of vision will, in fact, sound quite familiar to those conversant with image analysis. His hypothesis is that one of the principle steps in vision is the extraction of "shape contours," and he describes how these contours might be extracted through spatial filtering procedures.

With a camera, a lens focuses an image directly onto a flat piece of film. With the eye, light must pass through a complicated tangle of semitransparent cells on its way to the receptors at the back of the eye, and these receptors lie on a curved surface (Figure 3.1). In addition, unlike a camera, with the eye focusing is achieved by changing the shape of the lens, rather than the distance from the lens to the receiving surface. Receptors in the eye's receiving surface (the retina) vary in density, with substantially more in central vision, and contain distinct kinds of receptor cells that respond to different input.

Two major categories of cells line the retina: rods and cones. The rods are more numerous than cones (about 120 million and 5 million, respectively, in each eye) and will respond to very small changes in intensity of light, but not when light is very bright (Figure 3.2). Rods are insensitive to differences in wavelengths of that light, and therefore to color. Cones need greater illumination in order to react but are sensitive to differences in wavelength. Cones are concentrated in a very small area in the center of the retina (the macula). The fovea, a position that is direct-

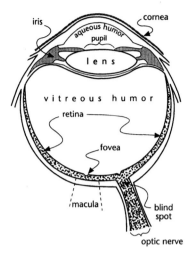

FIGURE 3.1. Structure of the human eye.

ly exposed to light entering the eye, is located at the center of the macula. This is the location of greatest visual acuity and contains no rods, only cones. Cones, by their dominance here, are responsible for our ability to see fine detail.

Cone cells, in persons with normal color vision, can be further distinguished on the basis of the wavelengths of light they respond to. These cone types are generally referred to as L cones (sensitive to long wavelengths), M cones (sensitive to medium wavelengths), and S cones (sensitive to short wavelengths). These different cone cells are unevenly distributed in the eye as well. As a result, the eye's sensitivity to different wavelengths of light varies spatially. Maps of retinal sensitivity to various wavelengths present a complex overlapping picture in which we find, for example, that sensitivity to green is confined to a relatively small hori-

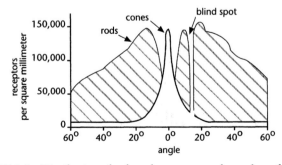

FIGURE 3.2. Distribution of rods and cones across the surface of the retina.

zontally extending band, while that to yellow occurs across a considerably larger, and nearly circular, portion of the eye (Figure 3.3). Sensitivity to blue, although covering a greater area of the retina than red or green, is lowest overall (in magnitude), which makes blue a poor color for small map features.

The retina is the first of three cell layers in the eye (Figure 3.4). The second consists of bipolar, horizontal, and amacrine cells. These in turn connect the receptor cells (i.e., rods and cones) to the ganglia. Many bipolar cells form direct connections. Horizontal cells, however, connect receptors with more than one bipolar cell, and amacrine cells link more than one bipolar to individual ganglion cells. These interconnections mean that a ganglion will not transmit an impulse based on stimulus of a single location on the retina; instead, it summarizes the signals received from a number of inputs, the ganglion's "receptive field." Most of these fields, in humans (as well as other mammals), are roughly circular with sufficient overlap for the foveal areas to overlap slightly (Figure 3.5).

To relate the size of receptive fields to the size of discriminable features in the visual field, the "angle subtended" by the feature is referred to. This is the angle formed from the lens of the eye to the top and bottom of the feature attended to. The angle equals that covered by the image of the feature on the retina (Figure 3.6). If, for example, you were viewing one of the pictorial symbols on a National Park Map (4 millimeters high) from normal reading distance of (approximately 460 millimeters), the image on the retina will cover 30 minutes of arc.

Ganglion cell receptive fields vary in size from the fovea to peripheral areas of the retina. Receptive field centers exhibit particularly systematic enlargement from center to periphery. Near the fovea, where the re-

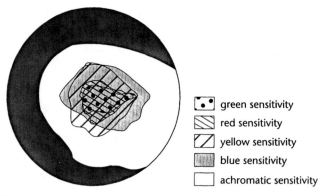

FIGURE 3.3. Diagram of overlapping color sensitivity regions in the eye. *After Wade and Swanston (1991, Fig. 3.20, p. 68). Adapted by permission of Routledge.*

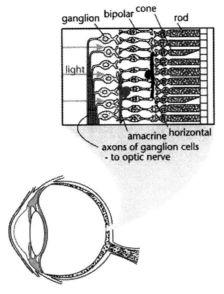

FIGURE 3.4. Schematic depiction of the structure of retinal to ganglion cell interconnections. *After Hubel (1988, p. 37). From Eye, Brain and Vision by Hubel. Copyright 1988 by Scientific American Library. Used with permission of W. H. Freeman and Company.*

ceptive field can be as small as a single receptor cell, the cells are spaced about 0.5 minutes of arc apart (2.5 micrometers). This corresponds to our greatest visual acuity. An example of a single feature that subtends 0.5 minutes of arc is a 0.13-millimeter-wide line on a map at normal reading distance (e.g., representing a road) or a 1-millimeter boundary line on a wall map viewed from about 22 feet away. In contrast to this, receptive field centers near the eye's periphery can be a degree or more. The result is sharply decreasing acuity from the center of vision to the periphery. For maps, this means that a small map symbol, identifiable when we look directly at it, will be less and less clear the further in the periphery it is. For symbols to be recognizable in peripheral vision, then, they need to be larger (Figure 3.7).

If images or parts of images (e.g., skates on the feet of the figure in the National Park Service symbol for skating area) are to be seen and discriminated in peripheral vision, symbols must be considerably larger than required for discrimination with foveal vision. If differences between two symbols are small, therefore, we will require a "fixation" on the symbol to discriminate it from others and identify what it is.[1] In addition, the color sensitivity maps above suggest that the ability to see and recognize a symbol in specific regions of peripheral vision will vary with its hue.

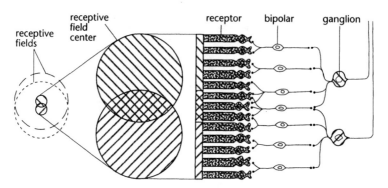

FIGURE 3.5. Diagram of a typical ganglion reptive field. *After Hubel (1988, p. 44). From* Eye, Brain and Vision *by Hubel. Copyright 1988 by Scientific American Library. Used with permission of W. H. Freeman and Company.*

This variation in acuity from central to peripheral vision has express implications for designing general reference and topographic map symbols. On a highway map, for example, map users often try to find particular kinds of features (e.g., points of interest, airports, universities, etc.). Symbols that are clearly distinguishable to the cartographer next to each other in the legend (when both are in the foveal area of vision) may not be different enough to be distinguishable when the map user scans across the map looking for them.

Like most neurons in the brain (discussed below), the ganglia collecting signals from receptive fields of the retina generate impulses of a constant magnitude. What varies is their rate of firing. They exhibit a

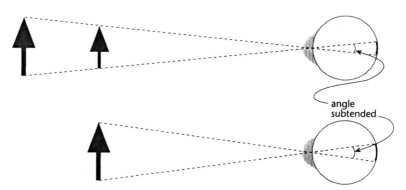

FIGURE 3.6. The angle subtended on the retina by light reflected from an object will depend upon both the size of that object and the distance from the eye.

FIGURE 3.7. Receptive field centers for gangion cells exhibit systematic enlargement from the fovea to the periphery of the eye. The result is that acuity varies across the retina. If you fixate on the central dot (from about four inches away) all map symbols should be equally legible. *Derived from Hoffman (1989, Fig. 2.4, p. 14).*

steady (resting) rate until the combined input from their receptive field reaches a threshold, at which point they either cease firing impulses or increase their firing rate. Whether the firing rate of a ganglion increases or decreases will be a function of the kind of ganglion cell stimulated, together with the spatial characteristics of the stimulus. Most ganglia react differently to stimuli near the center and periphery of the receptive field and, as a result, are termed "center–surround" cells. Both ON-center and OFF-center cells exist. With ON-center cells, a stimulus near the center of the receptive field stimulates an increase in firing rate, while a stimulus from the outer cells of the receptive field inhibits firing. With OFF-center cells, this pattern is reversed.

A constant stimulus that covers an entire ganglion's receptive field will result in competing signals that will partially cancel each other with (usually) a net result of slight inhibition on the ganglion's firing rate. If, on the other hand, the cell's receptive field is exactly centered on a small enough stimulus or it crosses an edge of some type, resulting in a different stimulus for the center and surround, the signals of center and surround can reinforce each other.

An interesting example, relevant to selection of area patterns for maps, of how this center–surround system and the size of receptive fields interact is an illusion called the Hermann grid (Figure 3.8). Most people when viewing this grid "see" dark spots at the intersections of the grid,

FIGURE 3.8. The Hermann grid illusion. Dark gray dots seem to appear at the intersections of the white lines (except at the intersetion you fixate upon). These illusory dots are thought to be the result of center–surround inhibition, with the grid intersections having a greater inhibition, thus the apparent dark spots.

unless they look right at those intersections. If we make use of peripheral vision, the ganglia being used have relatively large receptive fields (about the size of each grid zone). ON-center ganglia with receptive fields centered on the grid intersections will have the same reactions from their central areas as do ON-center cells centered over intermediate points, but an increased inhibition from the surround, resulting in the sensation of a dark spot. If you look directly at the intersection area, the receptive fields for the ganglia now involved are much smaller, and the illusion disappears. While artists sometimes make use of this effect to achieve a feeling of motion or instability in an image, we seldom want such a reaction to maps. We can prevent these distracting effects on maps by avoiding the relatively coarse patterns that match up with peripheral ganglion receptive fields.

In addition to having overlapping receptive fields, ganglia are interconnected and can inhibit each other's firing rate in the same way that a single cell's surround can inhibit the firing rate of its interior. These lateral inhibitory connections are thought to be responsible for the phenomena of "Mach bands," the illusion of shifts in brightness that cause the appearance of two vertical lines in Figure 3.9. Simultaneous contrast is also due to lateral inhibition of ganglion cells. As shown in Figure 3.10, the counties highlighted on the inset map appear to differ in degree of darkness even though they are the same.

Lateral inhibition is important in cartography because it will help accentuate differences between adjacent patterns or between symbols and background. On the other hand, it will make one pattern appear darker when next to a light pattern than when next to a darker pattern. This is one reason that there is an apparently smaller range of gray tones that

(a)   (b)

FIGURE 3.9. (a) The illusion of Mach bands—the dark vertical line toward the left of the illustration and the light vertical line toward the right. These apparent lines do not exist when luminance is measured with a light meter. (b) This illusion causes layer tints on isarithmic maps to appear to gradually change in value in the wrong direction (i.e., between any two isolines, regions that should have a lower data value will end up with an apparently darker color value than regions with a high data value).

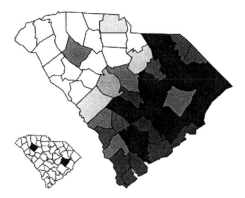

FIGURE 3.10. An illustration of simultaneous contrast for two map zones surrounded by predominantly light versus predominantly dark zones. Both counties (highlighted in the inset map) have the same data value and are filled (on the main map) with the same shade of gray).

people can distinguish on a map versus in gray patch experiments typical of gray scale research (MacEachren, 1982). Evidence of lateral inhibition clearly leads to the prediction that fewer shades of gray will be distinguishable on a map (where context within which any particular gray tone appears will vary) than in side-by-side comparison of pairs of gray patches. Only these out-of-context, side-by-side comparisons, however, have been used in formulating gray-tone selection guidelines. There has been a failure to test gray tones on actual maps because of the expectation that the spatial aspects of gray tone perception might confound results. As a result, cartographers have devised some tightly controlled "clean" experiments resulting in gray scales having unknown applicability for use on maps.

The only attempt that I am aware of to measure gray tone perception on actual maps was an undergraduate term project by one of my students (Terry Idol) several years ago. The experiment used a gridded 20-class choropleth map with gray tones assigned randomly to cells. Subjects (college students) were asked to estimate the actual value (as a percent of black from 0 to 100) of specified cells. The gray scale derived from this experiment was more linear than any of the gray patch-based scales cited in the literature. Because of some printing flaws in the test maps and the small sample used in the study, I would not consider this isolated map-based gray scale experiment conclusive. If replicated, however, the interpretation would be that 0% and 100% anchor the gray scale and simultaneous contrast on actual maps tends to make light grays look lighter and dark grays look darker, thus at least partially compensating for the apparent perceptual underestimation of differences so often cited in the literature. A much more linear gray scale may apply to choropleth maps than we have suspected thus far, one that bottoms out at about 20% reflectance (or 80% black).

In addition to producing simultaneous contrast effects, lateral inhibition between adjacent ganglion cells has a major role in color perception. Ganglion receptive fields for the three types of cone cell include various opponent relationships of center and surround cells. The three general categories of relationships are (1) red–green opponent cells that include ON- and OFF-center arrangements of L and M cells, (2) blue–yellow opponent cells that include ON- and OFF-center arrangements of S with combined L + M input, and (3) dark–light opponent cells that seem to be stimulated by all three cone types. Proponents of *opponent-process theory* (OPT) argue that these opponent relationships are responsible for our full range of hue sensation (Hurvich and Jameson, 1957). The theory predicts that there are four unique hues (blue, yellow, green, and red) and that all other hues result from mixtures of these four basic colors. The theory was developed in the 19th century with neurophysiological support coming in the latter half of the present century. At least one cartographer (Eastman, 1986) has attempted to apply the theory to selection of hue ranges for choropleth maps. This application will be discussed in the next chapter.

The most comprehensive look at one result of lateral inhibition (simultaneous-contrast or surround-induced changes) in relation to color use on maps is Brewer's (1991) dissertation (which dealt specifically with this topic in the context of color maps). Her initial premise was that induction will cause colors on maps to shift in appearance toward the complement of the surrounding hue. She devised an experiment to determine whether this assumption was in fact true and, if so, whether a quantitative model of simultaneous contrast could be used as an aid to selection of easily identified map colors. The opponent-process approach to color was selected as the best starting point for modeling induction. Brewer found the expected shifts in color appearance toward the complement of the surround (e.g., a red surround makes a central color appear more green). An unexpected finding, however, was that center–surround combinations with low contrast exhibited larger shifts than those with high contrast. This result seemed to be related to color saturation. Saturation shifts turned out to be the largest shift identified in the research. Based on her experiments, Brewer devised a model of the buffer around each color that represents its potential appearance with various surrounds. The model, designed to accommodate 90% of potential map viewers, was judged a success. With this model, a cartographer can ensure that colors for map categories are not confused by selecting only colors whose color-space buffers do not overlap.

As we have with visual acuity, we tend to take for granted the similarity of the color-processing system from person to person. Some of the individual variability may, however, be critical to map design (which is why Brewer chose a 90% target rather than designing for the average map reader). As Judy Olson (1989) pointed out, for example, a significant proportion of the population has some level of color deficiency. The deficiency is usually due to the absence of or the failure to function of one or more of the cone types found in the eye. Males are particularly likely to suffer from some level of color deficiency (about 8% for males vs. 0.4% for females). Recent evidence suggests that females, in addition to being less susceptible to color deficiencies of this type, sometimes actually have an extra category of cone in their retina, thus possibly giving them an extra dimension of color vision not shared by any male counterparts. Although about 12% of females may have this extra category of cone, the necessary experiments to determine what impact it has on their color vision have not as yet been conducted.

Eye to Brain

From the ganglia, axons extend that connect these composite signals to the next step in the process: the optic nerves. The optic nerves serve as the connecting link between the eyes and the brain. One of their primary functions seems to be the spatial amalgamation of signals from each eye. After leaving the eye, the optic nerves converge at the optic chiasma where they divide so that information from one side of each eye is directed to the same side of the visual cortex in the brain. Since the image on the retina is reversed, this sends information from each half of our visual field to the opposite side of the brain (Figure 3.11).

There are less than 1 million optic nerve fibers leading from the ganglion cells. In the outer part of the retina, up to 600 rods might be connected to one optic nerve fiber through one or more ganglion cells. In the fovea there is close to a one-to-one match of cones and optic fibers. This is one reason for more acute vision at the center of the visual system.

Brain

In the 1960s neurophysiology predicted the ability to understand thought by understanding our neurophysiological hardware. The slow progress since the breakthroughs that found cells with apparently specific functions (e.g., recognizing edges at specific orientations, and possibly even recognizing faces) has led to a realization that neurophysiological and

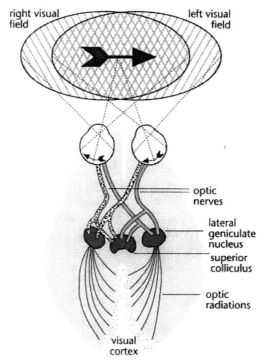

FIGURE 3.11. A depiction of the pathways connecting receptor cells in the eyes to the primary visual cortex in the brain. *Derived from Hubel (1988, p. 60) and Wade and Swanston (1991, Fig. 3.22, p. 70).*

neuropsychological evidence about vision will provide only part of the answer to how vision works. Following Marr's ideas, neurophysiological hardware is best considered as a mechanism that has evolved to meet the needs of vision, rather than as a fixed system that our visual abilities were adapted to. It does, however, exert limits on visual tasks that are not part of everyday behavior in the world (relatively recent visual tasks that human vision has not had time to evolve special procedures for). Reading maps seems to be one such unnatural task, with its typically abstract, two-dimensional static depiction. From a cartographic point of view, then, we are interested in features of how the brain processes visual signals not because this knowledge is likely to tell us how maps work, but because these processes put limits on what symbolization and design variations might work.

As indicated above, the signals sent to the brain by the ganglion cells result from a complex interaction of signals from each cell's receptive field together with the inhibitory interconnections of individual ganglion cells. Cells first reach the lateral geniculate nucleus (LGN) in the brain where neurons behave similarly to ganglia. Receptive fields still correspond to concentric regions in the retina. The LGN is arranged in six layers, each of which contains cells that respond to only one eye. For reasons that are as yet not understood, the six layers are arranged, from the top down, in a left, right, left, right, right, left eye sequence.

Once signals reach the visual cortex at the rear of the brain, linkages back to the retina become much less simplistic. Like the network of ganglia, cells within the visual cortex emphasize the signals coming from the macula. Approximately 50% of each side of the visual cortex is devoted to these signals. In contrast to neurons of the eye, however, those in the visual cortex have been found to be more specialized. Some appear to respond to particular visual elements such as line widths, angles, orientations, and so on, and some to the hue and brightness distinctions found with ganglion cells. The overall consensus of recent work in neurophysiology is that the visual system is composed of a sequence of processes capable of initially detecting edges, lines, and patterns (the processes Marr associated with extraction of the primal sketch) and subsequently analyzing these to result in more complex structures (Marr's 2½-D and 3-D representations).

This research has recently begun to result in maps of the brain in which the spatial arrangements of cells associated with specific functions are depicted (Figure 3.12). It seems particularly apt for a book about how maps work to include maps of the brain as a piece of evidence concerning how the brain might process maps.

Clinical observation beginning in the 19th century was responsible for the first crude mappings of the brain's major sections. It was not until the 1950s, however, when single-cell recording techniques began to yield information about individual and groups of cells that the complexity and intricacy of the neural interactions began to be recognized. Kuffer (cited in Hubel, 1988), in 1952, demonstrated the existence of the center–surround cell receptive fields described above. Much of what we now know

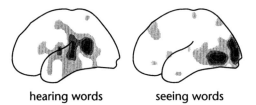

hearing words   seeing words

FIGURE 3.12. Activity maps (derived from positron emission tomography scans) that suggest the varied landscape of functions mapped out across the brain. *After Raichle (1991, color plate 3-1). Reprinted with permission from* Mapping the Brain and Its Research: Enabling Technologies into Neuroscience Research. *Copyright 1991 by the National Academy of Sciences. Courtesy of the National Academy of Sciences Press, Washington, DC.*

about limits to vision imposed by the brain's architecture is due to extensions of Kuffer's methods to examination of other neurons in the eye–brain system (usually of cats and monkeys).

Among the most startling discoveries about cells in the brain was the somewhat accidental finding by Hubel and Wiesel in 1958 that some of the cells in the visual cortex were orientation-specific (Hubel, 1988). These researchers had not been having much luck trying to determine the stimuli that would cause change in the signals of certain cells in the visual cortex of a cat. They had been using mostly opaque glass slides to block light to all but one location in the retina. Suddenly they noticed that the cell they were monitoring was stimulated as they slid a glass slide into place. Eventually they realized that it was the shadow of one edge of the slide that caused the cell to respond.

Experimental work in the intervening years has made it clear that there are at least two kinds of orientation-sensitive cell in the visual cortex. Simple cells respond only to a stimulus that is at a particular orientation *and* at a particular position on the retina (Figure 3.13). Complex cells also respond to single orientations, but are less sensitive to position of the stimulus (figure 3.14). Subsequent research led to discovery of additional cells that respond to "end stopping" (e.g., the end of a line) and to corners. Additional cells seem to respond only to movement, and some apparently only to movement of an edge across the visual scene.

How cells in the visual cortex selectively react to such specific kinds of features is still not completely understood. One likely hypothesis is that orientation-specific cells are linked to a set of ganglion cells that have receptive fields arranged in linear fashion across the retina. Figure 3.15 provides a schematic depiction of how this system might be connected.

Cells in the visual cortex are arranged in layers, and each layer seems to be somewhat distinct. Cells in some layers are binocular (re-

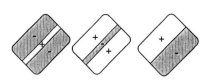

FIGURE 3.13. The response of orientation-sensitive cells in the brain to lines of varying orientation. The stimuli that these cells react to (from left to right) are a slit covering the (+) region, a dark line covering the (–) region, and a light–dark edge on the boundary. *After Hubel (1988, p. 72). From Eye, Brain and Vision by Hubel. Copyright 1988 by Scientific American Library. Used with permission of W. H. Freeman and Company.*

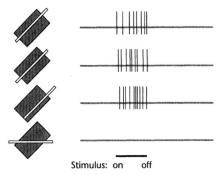

FIGURE 3.14. The response of complex cells to position and orientation. The plots at the right show cells reaction (or lack of it) to slits of light at different orientations and positions. *After Hubel (1988, p. 75). From* Eye, Brain and Vision *by Hubel. Copyright 1988 by Scientific American Library. Used with permission of W. H. Freeman and Company.*

sponding to stimuli from both eyes) and those in other layers are monocular (responding to only one eye). Some layers contain cells that are orientation-selective and other layers have cells that are not. Near the middle of the visual cortex, in the layer know as 4C, cells appear to be arranged in two intersecting slabs, one set in which right and left eye dominance alternates and the other in which orientation selectivity varies systematically (see Figure 3.16).

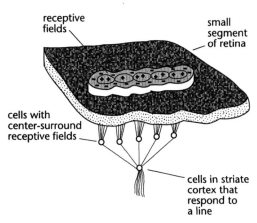

FIGURE 3.15. Hypothesized connection of orientation cells in the visual cortex, through ganglion cells to retinal cells. This particular grouping of cells results in cortex cells that are sensitive to linear stimuli. *After Hoffman (1989, Fig. 5.5, p. 57). Adapted by permission of the author.*

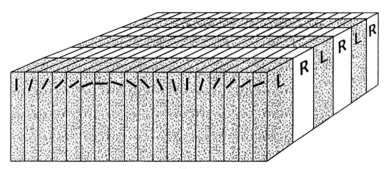

FIGURE 3.16. A schematic view of the arrangement of cells in the visual cortex. This depiction of cell arrangement illustrates the dual divisions of cell function for occular dominance and orientation. As Hubel emphasizes, actual cells in the brain are far less regular in arrangement than depicted with this schematic model. *After Hubel (1988, p. 131). From Eye, Brain and Vision by Hubel. Copyright 1988 by Scientific American Library. Used with permission of W. H. Freeman and Company.*

PERCEPTUAL ORGANIZATION AND ATTENTION

If an information-processing approach to vision and visual cognition is accepted as a useful conceptual structure, then derivation of meaning from maps can be viewed as a linked series of processing modules. A number of authors have made this point and offered their versions of how the overall process should be divided. The first such categorization was probably Olson's (1976) level-of-processing approach in which she delineated three levels: (1) comparing symbol pairs, (2) recognizing characteristics of symbol groups, and (3) using symbols as signals to information about what is represented. Both Phillips (1984), with a "low-level"/"high-level" categorization, and Dobson (1985), with a distinction between "visual-search guidance"/"cognitive-search guidance" emphasize a break between preattentive and attentive or perceptual and cognitive processing. Both posit that map design will have more impact on the lowest level of processing because it is at this level that the system is virtually overwhelmed with input and expert knowledge is least likely to apply. As certain optical illusions demonstrate, early perceptual reactions can often be hard (or impossible) to ignore—an indication that top-down processing (i.e., knowledge) has less control at this level and perhaps in some cases no control at all (Figure 3.17).

Although I do not agree with Dobson that cartographers should direct most of their research energy to the influence of symbolization and design decisions on low-level processes—the higher level processes such as derivation of meaning and decision making are what maps are really

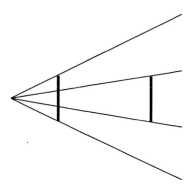

FIGURE 3.17. The length of the two vertical lines is identical, but it is difficult to convince yourself of that.

about—I do agree that failure of a map at this level can make it difficult or impossible to use. A complete understanding of how meaning is derived from maps, then, must begin with an appreciation for the selectiveness of vision in giving us a representation to think about. The *information theory* approach (that treated cartography as a communication system) focused on vision as an information filter with the cartographer's goal being to limit the amount of information that was filtered out in the communication process. This perspective treated perceptual representations as imperfect translations of reality. In contrast, the approach taken here is that perceptual representations are not fuzzy copies of the world, but interpretations of that world. The cartographer's goal is to determine what kind of representation her maps produce and how symbolization and design decisions influence the processes leading to those representations.

The remainder of this chapter, then, considers these initial visual processes from a perspective of how symbolization and design decisions interact with them.[2]

Gestalt psychologists in the early part of this century laid the groundwork for our current understanding of the perceptual organization of visual scenes. The Gestalt approach emphasized the holistic nature of human reactions to sensation. According to Wertheimer (quoted in Ellis, 1955, p. 2), "There are wholes, the behavior of which is not determined by that of their individual elements, but where the part processes are themselves determined by the intrinsic nature of the whole." More specific attention to pattern is seen in Kohler's (1947, p. 103) statement that "the organism responds to the *pattern* of stimuli to which it is exposed; and that this answer is a unitary process, a functional whole, which gives, in experience, a sensory scene rather than a mosaic of local sensations." Wertheimer's initial emphasis was on defining principles of grouping, and he mentions the segregation of figure and ground only briefly at the end of his article. Kofka and Kohler, Wertheimer's contemporaries, however, extended the initial thoughts on formation of figures. Kohler (1947, p. 145), for example, argued that "sometimes it seems more natural to define a principle of grouping not so much in terms of given conditions as in terms of the direction which grouping tends to take." This viewpoint is specifically linked to figure formation in his statement that "a homogeneous field in visual space is practically uniform and, being without 'points,' there are no relations between 'points' within this field. When Gestalten appear we see firm, closed structures, standing out in lively and impressive manner from the remaining field" (quoted in Ellis, 1955, p. 35).

For several decades while the behaviorists held sway in psychology (particularly in the United States), Gestalt psychology and its principles of perceptual organization were ignored by experimentalists. More recently, particularly in response to the needs of computational vision research and the attention to form and structure in vision that it has stimulated, Gestalt principles are being re-examined. Uttal (1988, p. 146), for example, contends that "human visual perception is powerfully driven by the global organization of form." Recent research in psychology that incorporates Marr's basic contentions (that human vision is an information-processing system and that information-processing systems can only be understood if examined at a combination of computational, algorithm–representational, and hardware levels) have drawn heavily upon Gestalt principles as a source of ideas for understanding grouping in early vision and figure–ground separation associated with object and pattern recognition (see Roth and Frisby, 1986, and Bruce and Green, 1990, for overviews of this work).

Pomerantz (1985) points out that the distinction of Gestaltists, between processes of grouping and of figure–ground separation, is significant from an information-processing perspective. Although there is a clear connection between principles of grouping and formation of figures, grouping of as yet unidentified edges, blobs, terminations, and the like, is a requisite step in deriving a primal sketch. Once edge segments are grouped into contours, then it becomes possible for vision to sort out figure from ground. There is considerable evidence that the initial grouping stages are almost entirely preattentive with little or no input from higher level processes. Research results concerning figure–ground segregation are mixed, with some evidence that figures can spontaneously "pop out" of a background, together with demonstrations that input from stored knowledge or expectations does (in some circumstances) effect both the initial appearance of figure and the stability of the figure–ground relationship.

From a cartographic perspective, low-level issues of grouping seem most relevant for exploratory cartographic visualization in which limited attention will be directed to any one map view and the goal is to notice patterns and relationships. Exploratory visualization implies that the outcome is not know, and that knowledge and expectations therefore may often be absent or wrong. Obtaining an immediate impression, before conscious application of knowledge schemata take over, is likely to play a major role in whether the visual displays lead to insight or simply are used to confirm expectations.

Perceptual organization operating at higher levels is important in those situations where particular information is to be emphasized while other information is suppressed. When goals are to create an imagable map (Peterson, 1987) or to ensure that a particular region becomes the focus of attention (Dent, 1972; MacEachren and Mistrick, 1992), issues of selectivity, associativity, and figure–ground become relevant. Most of the references to Gestalt principles by cartographers have been in relation to figure–ground segregation, with only limited attention paid thus far to the underlying processes of grouping.

Grouping

The "pattern of stimuli" mentioned by Kohler occurs due to grouping of elements in the sensory field. In relation to Marr's theory, these elements might be edge parts, blobs, and the like. For maps, viewing these edges and "blobs" will occur in the primal sketch representation in response to map symbols such as points, lines, or elemental parts of textures. Their grouping will determine whether symbols are seen as intended and which kinds and scales of patterns are noticed. Wertheimer (1923; translated in Ellis, 1955) set out the rules for such perceptual grouping in his classic paper, "Laws of Organization in Perceptual Forms." He defined the following factors or rules:

1. *Proximity*: Objects close together form groups. In the abstract, the factor holds that in any array of individual elements, those that are closest together will be seen as part of a group (Figure 3.18). Cartographically, as detailed below, proximity has been postulated to account for the appearance of regions on maps (Figure 3.19). A particularly intriguing part of Wertheimer's argument, in light of current interest in dynamic and animated maps, is his contention that the factor of proximity holds for sound as well as sight. Sounds close together in time will form perceptual groups. This issue (without reference to Gestalt principles) is alluded

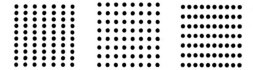

FIGURE 3.18. Grouping by proximity: rows (far right) and columns (far left) of dots versus evenly spaced dots (center).

to by Krygier (1991) in his identification of the audio variable *rhythm* as "the grouping and ordering of sounds."

2. *Similarity*: Like objects form groups. As presented by Wertheimer, similarity relates to nonlocational characteristics of perceptual units. He specifically mentions color, form, and sound. From a cartographic perspective, we might consider the similarity of all graphic variables (color hue, color value, shape, etc.), as well as tactual and audio variables (Figure 3.20). Wertheimer (1923; translated in Ellis, 1955) points out that similarity is not absolute, but can occur in degrees. Thus, judging "more and less dissimilar" becomes an issue in how people experience map symbols.

3. *Common fate*: Objects moving together are seen as a group. For this factor, Wertheimer points out that already grouped units that move together may hardly be noticed, but that units from separate groups moving together can be "confusing and discomforting" and will most certainly be noticed. It is posited that common movement of units from separate (static) perceptual groups will override proximity, similarity, or other factors to achieve a new group held together by their "common fate." Cartographically, of course, this factor applies only to animated or dynamic maps. In this context, however, it may play a particularly strong role in what groups are perceived. A corollary to Wertheimer's common fate in relation to map animation is that objects that change together (even when they do not move) are seen as a group. In our map animation research at Pennsylvania State University, we have used this principle to

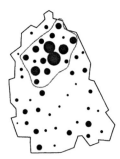

FIGURE 3.19. The importance of proximity in region identification on graduated circle maps. The outlined circles represent the consensus region identified by Slocum's subjects. After Slocum (1983, Fig. 9, no. 14, p. 71). Adapted by permission of the American Congress on Surveying and Mapping.

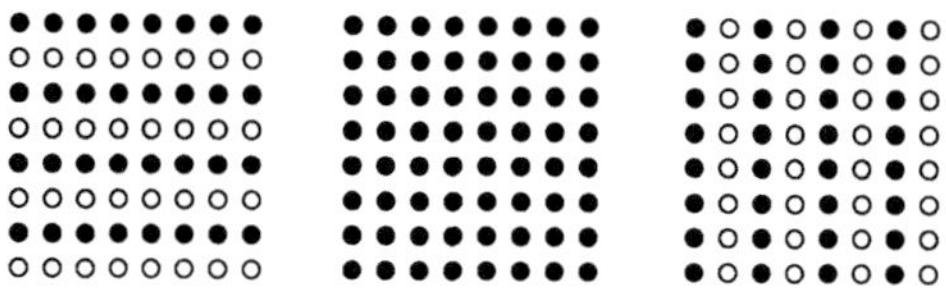

FIGURE 3.20. Grouping by similarity: white versus black circles.

animate static maps that depict existence of a feature with a fixed location (Figure 3.21). Similarly, Monmonier (1992) has employed what he called "blinking" as a method to emphasize the spatial pattern (or lack of it) in the proportion of public officials who are female. Blinking, in this context, involves having a choropleth map class (with values for the United States grouped by quintiles) blink on and off while other classes are turned off. Thus, one at a time, the states in successive quintiles are visually grouped so that the viewer can easily identify regional patterns in exclusion of females from public office.

4. *Pragnanzstufen:* Perceptual groups are characterized by regions of "figural stability." This factor is difficult to translate, but implies that grouping has discrete cases. In relation to proximity, for example, there will be a relative threshold distance at which units will be seen to group or to occupy space in an undifferentiated way. Wertheimer's example sug-

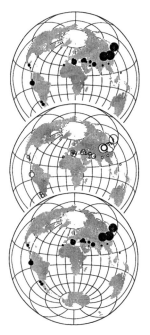

FIGURE 3.21. A sequence of maps simulating earthquake epicenters blinking on and off to highlight their clustering. *Reproduced fom DiBiase et al. (1992, Fig. 3, p. 207). Reprinted by permission of the American Congress on Surveying and Mapping.*

gests that for a row of dots we will see an ab–cd–ef–gh–ij grouping, no grouping at all, or an a–bc–de–fg–hi–j grouping at different possible regular spacings of the dots (Figure 3.22). This concept seems to match with anchor-effect theories of magnitude estimation (discussed below) and ideas about prototype categories (discussed in the next chapter).

5. *Objective set*: With change, there will be a tendency toward stable groups. Following from the above factor, the idea here is that if a set of perceptual units is initially seen as a group and that over time the position of those units changes, perception will try to retain the initial group. In addition, there will be a tendency to see a limited number of states (e.g., grouping A, undifferentiated scene, grouping B). Wertheimer's example is based on a scenario in which seven pairs of dots gradually change relative positions (refer to Figure 3.22). Again, cartographically, this factor applies to animation. The implication is that throughout a movement perception tries to maintain a stable state, resulting in a greater likelihood that we will see a constant grouping on a set of change maps if they are presented dynamically than if they are presented on a page as small multiples. This would be an interesting hypothesis to test. The possibility to be concerned with is that "a certain (objectively) ambiguous arrangement will be perfectly definite and unequivocal when given as part of a sequence" (Wertheimer, 1923; translated in Ellis, 1955, p. 80). The issue here is one of visualization quality and how to determine when a pattern is "real" or illusory (MacEachren and Ganter, 1990).

6. *Good continuation*: Elements that follow a constant direction group. This factor applies not only to straight-line arrangements, but to curves, as illustrated in Figure 3.23. Cartographically, this factor allows contours on a black and white map to be seen as separate curved lines differentiated from roads or rivers that they might cross (Figure 3.24).

7. *Closure*: Closed objects form wholes. There is a tendency to see bounded perceptual units as wholes. Even when bounding edges overlap, there is a likelihood that the factor of good continuation cited above will

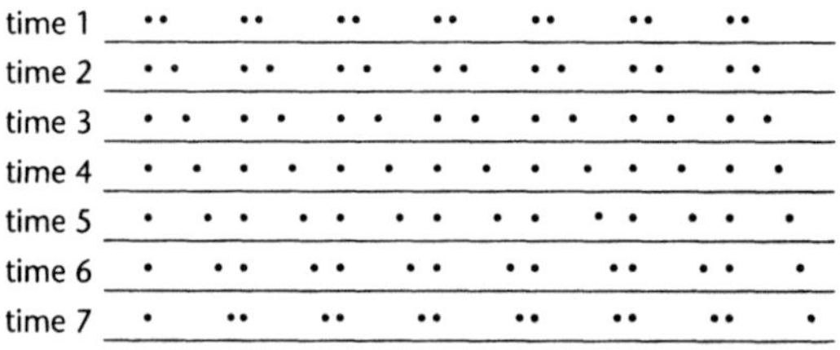

FIGURE 3.22. Grouping due to Wertheimer's Pragnanzstufen factor. If shown as a time series, the original groups (top row) will be seen at time 4, even though all distances are equal at this time.

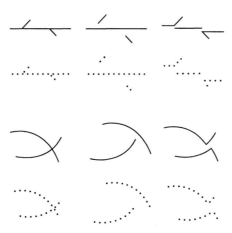

FIGURE 3.23. Grouping by good continuation. On the top, we "see" a long line with two short lines attached, rather than a short line with two angular attachments. On the bottom, we "see" two smooth curves crossing.

allow us to see the separate bounds as units and apply closure to isolate their edges as groups defining wholes. The critical role of good continuation, and its potential dominance over closure was dramatically illustrated by Wertheimer (1923; translated in Ellis, 1955) (Figure 3.25). Cartographically, closure has clear applications to situations such as graduated symbol maps where it has been demonstrated that circle overlap does not prevent readers from seeing the circle segments as whole circles, or from judging circle size (Groop and Cole, 1978). In addition, a variety of graphic methods for emphasizing the closure of a map region have been examined.

8. *Simplicity*: Objects will group in the simplest form. Wertheimer did not specify simplicity as a specific factor, but mentioned it in relation to what he called "good Gestalt." This concept was a basis of

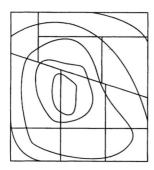

FIGURE 3.24. Good continuation helps map viewers sort out intersection lines on maps. Even in the absence of other contrast, we can visually separate the contours from the county boundaries.

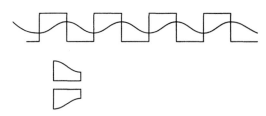

FIGURE 3.25. Dominance of good continuation over grouping by closure. Few observers are likely to interpret the figure above as a set of the irregular shapes shown below. Instead, we see a curved line across an angular one. *After Ellis (1955, Fig. 13, p. 82). Adapted by permission of Routledge & Kegan Paul.*

Wertheimer's "Law of Prägnanz," which Koffka (1935, p. 138) described as follows: "Of several geometrically possible organizations that one will actually occur which possesses the best, simplest and most stable shape." An example relevant to cartography is found with the tendency to interpret ambiguous situations (such as Figure 3.26) as interposition of simple figures rather than more complex adjacent figures.

9. *Experience or habit*: Familiar shapes or arrangements form groups. Many subsequent authors have focused on Wertheimer's contentions that past experience was not essential to perception of groups and that proving a role for past experience would be difficult. As a result, these authors have (mistakenly) characterized Gestalt psychology as disallowing the possibility for knowledge to play a role in both perceptual grouping and figure–ground perception. Wertheimer did, however, contend that experience or habit, in the form of "repeated drill," could play a role and at times cause groups to be seen that are at odds with what the other factors might dictate. Although he placed more emphasis on preconceptual processes, Wertheimer did not rule out the possibility of what we consider in the next chapter as "knowledge schemata" playing a role, even at early low-level stages of visual processing.

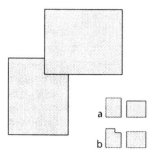

FIGURE 3.26. Grouping by simplicity. It is easier to see two squares overlapping (a) than a square next to an L-shaped region (b)—in spite of the fact that the latter could represent a common geographic feature such as Utah and Wyoming.

A number of cartographers have cited the above Gestalt "laws" (Wood, 1968, 1972; Dent, 1972; McCleary, 1981). For the most part, they have been treated as laws, with attention directed to devising logical guidelines for incorporating the laws into map design. Few cartographers have questioned the principles or considered their relative influence on grouping of map elements. This tendency to take the Gestalt laws for granted is even apparent among psychologists (e.g., Kosslyn, 1989, uses some of the laws as given in developing a procedure for assessing graphic acceptability). Pinker (1990), like Kosslyn, contends that Gestalt principles have a role in the process of translating the initial visual scene to a visual description of a graph (a representation of entities and relationships among those entities). He goes on to suggest, however, that we do not at this point understand how to apply these principles because there has been little empirical research about the situations in which they hold or their relative importance.

Among psychologists, Pomerantz (1985) has provided perhaps the most explicit analysis of Gestalt grouping principles and their interrelations, as separate from the issue of figure–ground segregation. He begins with a convincing argument that grouping is "logically prior to figure–ground segregation" (p. 128). We must group perceptual units into objects and regions before a choice can be made among objects or regions concerning which are the focus of attention.

Although no cartographers, to my knowledge, have explicitly mentioned Gestalt principles in relation to perceptual grouping of map elements, a few have incorporated the principles in their work without crediting them to Gestalt psychology. Olson (1976) alludes to the cartographic importance of grouping with her three-tiered hierarchy of mental processing in map use. Her second level deals with recognizing properties of symbol groups. To recognize these group properties, of course, the visual process must provide groups for which properties can be compared by higher level processes. Olson considers (but does not test) the possible impact of symbol scaling (for graduated circle and dot maps) on regions that might be identified, as well as the role of value contrast among different symbols and between them and the background. In the former examples, the variable of proximity is manipulated and in the latter case similarity is used.

In a somewhat more direct examination of the applicability of Gestalt grouping laws to map reading, Slocum (1983) investigated visual clustering on graduated circle maps. Slocum's stated goal, formulated in behavioral terms, was to develop a method to predict perceived map groups using a combination of the psychological principles he felt were relevant to the problem. He hypothesized that proximity, similarity, and good continuation would play a role in the visual groups seen on graduated circle maps.[3] He was unable to devise a measure of good continuation in the absence of eye movement recordings, so it was not actually tested. In addition to grouping factors, Slocum hypothesized that "figure–ground" would play a role in visual grouping. Figure–ground was limited for purposes of his experiment to a measure of value contrast, with dark areas expected to be seen as figure. The incorporation of this measure was based on prior evidence by Jenks (1975) that value difference had an effect on the groups seen, and its interpretation in terms of "figure–ground" seems to be based on Dent's (1972) emphasis on value contrast as a figure–ground variable.

Slocum's (1983, p. 61) experiment involved having subjects outline sections of graduated circle maps that they "saw" as "visual clusters"— "groups of circles that appeared to belong together and form a visual unit." His analysis indicated that a combination of proximity and figure (defined as relatively dark sections of the maps) provided a reasonable prediction. In fact, 92% of the individual circles on his 10 test maps were correctly classified as in or out of a cluster. Similarity of circle sizes had virtually no effect upon groups seen (Figure 3.27).

Eastman (1985b) has also investigated an aspect of perceptual grouping but, unlike Slocum, did so following a cognitive information-processing approach. Specifically, he examined the effect of several design variations of a typical reference map on the perceptual organization of the map.[4] The goal was to determine whether proximity of locations, similarity of symbolization for those locations, regional inclusion (which can be associated with the closure of country boundaries or road segments), or linear linkage (determined by road connections between cities and at least loosely associated with "good continuation") influenced how map items were grouped in memory. Eastman found that the maps stimulated five different groupings, each of which was primarily associated with one or two of the map designs. All five grouping strategies led to hierarchical memory structures. A comparison of subject groups (that grouped map elements differently) did not support Eastman's hypothesis that dif-

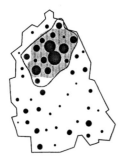

FIGURE 3.27. Grouping as predicted by Slocum (gray region) in comparison to grouping produced by his subjects (outlined region). After Slocum (1983, Fig. 9, no. 13, p. 71). Adapted by permission of the American Congress on Surveying and Mapping.

ferent grouping strategies would lead to differences in learning speed or memory accuracy. Graphic organization of the maps, however, did appear to have an impact, both on how maps were perceptually organized and on whether organization was easy. Lack of graphic organization led to grouping by proximity or horizontal partitions. A map with regions delineated led to regional chunking (Figure 3.28). The map with no graphic organization also proved to be much harder to learn than the rest.

In addition to these relatively direct applications of Gestalt grouping principles to cartography, grouping has been considered less directly in studies of map regionalization. Muller (1979), for example, demonstrated that map viewers arrived at similar regions or groups when asked to delineate regions of high, medium, and low population density on continuous tone choropleth maps. In spite of being presented with many more color values than could be discriminated, subjects were able to group similar values into categories in a consistent way. In contrast to Muller, who asked subjects to delineate high, medium, and low regions, McCleary

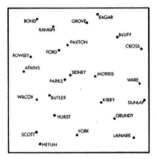

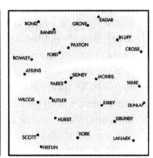

FIGURE 3.28. Eastman's graphically undifferentiated map (top) compared to the map leading to the most consistent grouping strategies (bottom). For subject groups viewing each map (at left), the gray shading represents consensus first- and second-order chunks (middle maps and right maps, respectively). *After Eastman (1985b, Figs. 2, 8, 9, 12, and 13, pp. 5, 15, 16). Adapted by permission of the American Congress on Surveying and Mapping.*

(1975) had subjects delineate any regions they saw on a set of dot maps. An intriguing result of this task was that although grouping by proximity seemed to be at work in all cases, his subjects fell into two quite distinct types that he termed "atomists" and "generalists" (Figure 3.29). The atomists focused on local details. According to McCleary (1975, p. 247), they "seemed obsessed with detail and may have lost sight of the overall pattern of density." For the generalists, on the other hand, "lines are schematic and the 'attitude' expressed by the boundary line drawn suggests a reductionist view of the image." This finding has not been pursued in the cartographic literature but has interesting implications for our current concern with use of cartographic visualization for exploratory data analysis. It is important to determine whether McCleary's atomists and generalists represent general categories of map viewers and whether these tendencies are altered with training or expectations.

What We Attend To

Perceptual grouping is thought to work, at least in part, at a preattentive level. Based on Marr's speculations, some amount of grouping (into edges, blobs, etc., of the primal sketch) is a prerequisite to all seeing. Grouping will interact with visual attention in complex ways. Where our gaze is directed will limit what can be grouped (only global features of a scene in peripheral vision vs. details in central vision). The results of grouping will control what can be attended to and where our gaze might travel next. Where we direct our attention can, of course, also be consciously controlled. As a complement to issues of grouping, then, we must consider

atomist   generalist

FIGURE 3.29. Sample subjects from McCleary's dot map regionalization experiment illustrating the grouping strategies of atomists (left) and generalists (right). *Reproduced from McCleary (1975, Fig. 4, p. 246).*

the combination of processes that fall under the heading *visual attention*. An important issue that Wertheimer considered in relation to grouping is the possibility of more than one factor acting at the same time. Such interaction may enhance visual grouping or may act in opposition to inhibit it (Figure 3.30). In addition to the effect on grouping, the interaction of multiple variables of perceptual units can influence the separability of features of the unit. This has obvious implications for how multivariate symbols are perceived, particularly for which aspects of a multivariate map symbol we can attend to together or separately.

Selective Attention and Separability of Visual Dimensions

Recent research on perceptual organization has emphasized the notion of selective attention as a way to measure the role of different features in the visual scene on perceptual grouping (Pomerantz, 1985). "Selective attention" refers to the ability to attend to one dimension of a display and ignore another. If dimensions or variables can be segregated in this way, they are not grouped. If, on the other hand, it is difficult or impossible to selectively attend to the separate dimensions, they are considered to be perceptually grouped. In a series of experiments, Pomerantz and his colleagues examined selective attention to features of compound stimuli. Their results, in addition to informing us about general perceptual processes relevant to map reading, are likely to be particularly relevant to design of multivariate symbols for maps.

Many of Pomerantz's experiments used sets of simple parenthesis-like symbols that were paired in various ways. These pairings were designed so that some should lead to groups (based upon Gestalt principles

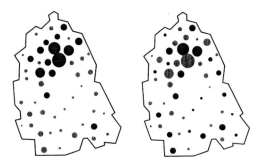

FIGURE 3.30. Similarity and proximity acting together to enhance grouping (left) and in opposition resulting in ambiguous grouping (right). *Derived from Slocum (1983, Fig. 9, no. 13, p. 71)*.

of good continuation, similarity, symmetry, and proximity) and others should resist grouping. One set of stimuli are shown in Figure 3.31.

A typical experiment would match a control case in which subjects had to sort two stimuli (e.g., the top row of each box in Figure 3.31) versus a selective attention case in which subjects had to sort all four stimuli (e.g., the two pairs on the right of each box in Figure 3.31 in one category and the two pairs on the left in the other) (Pomerantz and Garner, 1973). In both cases the task could be completed by focusing on only the left-hand element of the parenthesis pair. If subjects could selectively attend to this element and ignore the other, both groups should accomplish sorting at the same rate. Subjects in Pomerantz's selective attention group, however, took longer to sort their stimuli. This is an indication that the pairs were processed as groups. The control case subjects had the easy task of sorting these perceptual groups into a symmetrical and an asymmetrical category. The selective attention group was forced to treat the four stimuli separately because, as groups, the two columns of parenthesis pairs do not form Gestalt categories (in fact, symmetrical vs. asymmetrical units form categories counter to the ones required by the sorting task). When the same experiment was run with stimuli that should, according to Gestalt principles, not group, there was no difference in response time between the experimental groups. The parentheses did not form perceptual units, and therefore the right-hand parenthesis could be ignored and sorting accomplished by focusing attention on the left parenthesis only. Thus both groups had a two element categorization task and completed it at the same rate.

Cartographically, Bertin (1967/1983) has focused upon issues similar to those that interest Pomerantz, but has not investigated his contentions experimentally. Bertin's hypothesis (which he treats as fact) is that the visual variables can be independently judged on the basis of what he calls *selectivity* and *associativity*, and that these designations are discrete (i.e., a visual variable is either selective or nonselective in all applications). His selectivity is similar to Pomerantz's selective attention. Where Pomerantz focuses on whether conjunctions of two or more objects proximate to one another are seen as a whole (a group), Bertin is interested in whether objects (map symbols) spread across the map can be formed into visual groups. Visually grouping, or attending selectively to, a particular value,

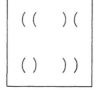

FIGURE 3.31. Test stimuli with good (left) and poor (right) grouping. *Reproduced from Pomerantz (1973, Fig. 6.1, p. 129). Copyright 1973 by the Psychonomic Society. Reprinted by permission of the author.*

for example, seems easier than attending to a particular shape (Figure 3.32). Bertin's concept of selectivity is limited to grouping by similarity (although he does not define it in these terms). The emphasis is on whether visual grouping is "immediate" (a term that can probably be taken to mean preattentive) for all symbols in a category identified by a specific variation of one visual variable (e.g., all blue symbols on a map compared with symbols in various hues). Bertin posits that location, size, color value, texture, color hue, and orientation (of point and line symbols only) are selective variables.

There is empirical evidence for some of Bertin's claims (although not derived from explicit attempts to test those claims). In relation to orientation, for example, Olson and Attneave (1970) demonstrated that a difference in orientation of simple line symbols can cause regions to be discriminated quickly (Figure 3.33). There is even a neurological (hardware level) explanation for why orientation is selective. Research by DeYoe et al. (1986) with monkeys has demonstrated that there are cells in the monkey's cortex (regions V1 and V2) that respond to pattern edges defined by differences in orientations of the texture elements making up the patterns. For this differentiation to occur, orientation differences must be in the center and surround portions of the cell's receptive field.

Nothdurft (1992) found that with limited variation within pattern areas, differences in orientation of as little as 20% were sufficient for a 75% success rate for preattentive pattern segregation. As variability in orientation of individual elements making up the pattern increased, the necessary difference in mean orientation of line segments in the two regions (required to achieve a 75% preattentive selection rate) increased in a roughly linear fashion. Beyond 30% variability in orientation within the individual patterns, pattern discrimination was unsuccessful regardless of magnitude of between-pattern orientation difference.

In contrast to Bertin's sweeping claim, other evidence exists that the

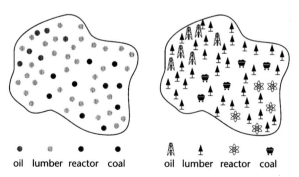

FIGURE 3.32. Value (right) seems to be selective while shape (left) is not.

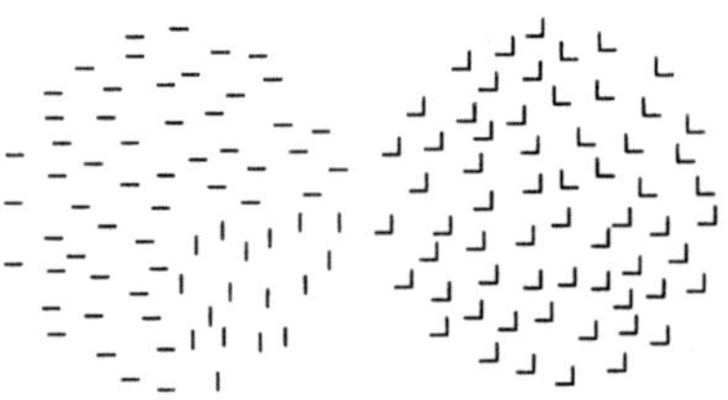

FIGURE 3.33. Differences in orientation result in visual groups, but differences in alignment do not. *Derived from Bruce and Green (1990, Fig. 6.16, p. 117).*

key selective variable is symbol slope rather than orientation. If a symbol is used that has an internal orientation, it becomes clear that we can distinguish (mathematically) between orientations that are 180° apart, but that this orientation difference is not selective in Bertin's terms (Figure 3.34). With Bertin's line segment examples it was not obvious that 180° rotations were not selective because they could not even be detected.

Not all of Bertin's visual variables have been tested for selectivity, and the only empirical tests thus far have been by psychologists (who have not specifically set out to test graphic variables but to study the phenomenon of selective attention). In addition to slope, there is evidence for selectivity of color hue and color value. For both, Julesz (1975) found that regions were easily segregated if distinct value or hue differences exist. An interesting factor for these visual variables was that when pattern elements are small, vision seems to respond to an average signal. A region of mostly black and dark gray squares having a few white and light gray ones mixed in (and as a result mostly dark) is easily segregated from a region of mostly white and light gray squares having a few dark gray and black ones mixed in. Similarly, wavelengths of colors seem to be averaged so that a region of red and yellow squares (and a few green and blue) is clearly discriminated from one of green and blue squares (and a few red and yellow). A red–green region, however, is not easily discriminated from a blue–yellow region.[5] Evidence also exists to support Bertin's contention that shape is not selective, at least in the case of different shapes

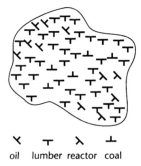

FIGURE 3.34. It appears that slope of parts rather than orientation of an overall shape must differ (in some cases) for orientation to be selective in Bertin's sense.

that have the same number of line segments and terminators (Figure 3.35).

At least one graphic variable that Bertin ignored has also been demonstrated to be selective. Julesz (1965) demonstrated that what he called "granularity" of a pattern (discussed in Part II of this book as pattern arrangement) leads to easy segregation of regions. The general success of Bertin's selectivity claims, along with some discrepancies uncovered by empirical research in psychology, suggests that cartographers need to take a closer look at Bertin's ideas. Studies that empirically test Bertin's hypotheses and investigate the magnitude of differences required along specific dimensions of visual variables (including additional variables that others have added to Bertin's original set) are clearly called for.

Bertin considers visual variables largely in isolation and does not discuss their potential interaction on a map. Experiments along the lines of those conducted by Pomerantz might be used to determine how vision will react to multivariate symbols that are designed to convey redundant information for emphasis and to enhance discrimination or separate information so that interrelationships can be noticed. In the first case, we would want to apply visual variables for which selective attention is difficult; in the latter case we would want the opposite.

No cartographic research (to my knowledge) has been conducted in relation to the issue of selective attention to visual variables in multivariate map symbols. Shortridge (1982) has, however, provided an overview of evidence from psychology and suggested possible applications to map symbolization. In particular, she considered the issue of *integral* versus *separable* dimensions (i.e., visual variables). Separable dimensions are ones for which selective attention is easy; integral dimensions tend to be seen

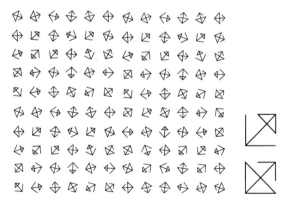

FIGURE 3.35. Shape, with other variables held constant, is not selective. *After Julesz (1981, Fig. 6, p. 95). Copyright 1981 by Macmillan Magazines Limited. Adapted by permission of the author and* Nature.

as wholes, and therefore selective attention is hard. As an example, consider a map that uses line size to indicate temperature at weather recording stations and line orientation from horizontal to vertical to indicate precipitation amount (Figure 3.36). If the two dimensions (e.g., symbol size and line orientation) are separable, selective attention will be possible and a viewer should be able to compare two stations on temperature or precipitation quickly—and not be able to judge temperature–precipitation correspondence easily (a contention that seems to be supported by Figure 3.36).

As Shortridge (1982) points out, psychologists began to distinguish between integral and separable dimensions as a way to explain results of visual search tasks that sometimes indicated processing of multiple stimuli in parallel and sometimes in a serial self-terminating manner (i.e., one symbol at a time until a target is found, at which point processing is halted). For serial searches, if a target is not present the search is exhaustive (relatively long) and will increase in length as the number of stimuli in the scene is increased. When a target is present (and a serial search is used), it will be found (on average) after half of the stimuli have been processed (response times will be 0.5 times that of target-absent cases). If stimuli are processed in parallel (all at once) then processing times will not be effected by the number of stimuli that must be processed. Although predicting whether a serial or a parallel process will be invoked does not seem easy, attention to this question led psychologists to notice the differences between compound stimuli that seemed to differ in the likelihood of serial versus parallel processing. Recognition that some symbol dimensions are integral (i.e., difficult or impossible to attend to separately) led to a *holistic* account of processing as an alternative to the serial–parallel possibilities. This account suggests that integral symbol dimensions create a whole that is processed as a single unit. Evidence for

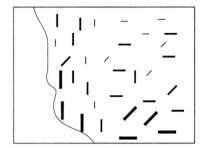
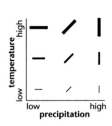

FIGURE 3.36. A map of temperature and precipitation using symbol size and orientation to represent data values on the two variables.

such holistic processing comes from research by Lockhead (1970) and Pomerantz and Schwaitzberg (1975), both of whom found that certain conjunction tasks, in which subjects had to discriminate or categorize symbols on the basis of the conjunction of two dimensions, were performed faster than discrimination or categorization on the basis of either dimension individually.

Divided Attention and Variable Conjunctions

In their research on conjunction tasks Pomerantz and Schwaitzberg (1975) used measures of *divided attention* as a complement to previous selective attention studies of perceptual grouping.[6] They reasoned that if selective attention to parts failed, implying that they were grouped, viewers should find it easier to attend to the groups as a unit. This hypothesis was tested by having subjects try to sort stimuli (of the kind used in their initial experiments—see Figure 3.31) according to groups that were similar while ignoring elements within those groups that would suggest alternative sortings. They found that when Gestalt attributes of element pairs (e.g., combinations of symmetry, closure, similarity) indicated that grouping of the elements into a single whole was likely, sorting by groups was faster than sorting by individual element. When, on the other hand, individual elements did not form "good" Gestalt groups, sorting by individual elements was easy and sorting by group was extremely difficult.

The above evidence indicates that various combinations of map symbol attributes may lead to integral or separable symbol dimensions which in turn may facilitate divided or selective attention. Knowing which will occur in particular cases is clearly crucial to making effective map symbolization choices. Integral combinations should be useful in univariate map applications where the goal is to enhance discrimination while reinforcing appearance of order for quantitative information. One example would be the combination of color value and saturation for area fills on a choropleth map of population density. By combining these variables to produce a wide range of area fills (e.g., from a light, desaturated blue to a dark, fully saturated blue), it may be possible to extend the practical number of categories that can be used. Multivariate symbols with separable dimensions, on the other hand, seem suited to the depiction of multivariate data (either qualitative or quantitative) in which the viewer will want to extract various components of the data separately. Examples include the temperature–precipitation map cited above or a map showing relationships between soils and geology (e.g., Wakarusa quad; Campbell and Davis, 1979). In the latter case, color was used for one variable and pattern for the other. Each was, in fact, a combination of visual variables and neither the combinations nor the conjunction of the color–pattern sets have been examined for selective attention.

In response to our current lack of knowledge concerning how visual variables interact in multivariate symbols, Shortridge (1982) suggests a program of research to evaluate whether specific combinations of visual variables combine in integral versus separable ways. She considers creating a classification scheme based on symbol properties a useful goal. In addition, she presents a hypothesis that integral versus separable conjunctions of visual variables may not be discrete categories, as presented in most psychological literature to date, but may be two ends of a continuum. This proposal allows for some level of integrality to occur between size and color value, a conjunction that Shortridge used with graduated circles to demonstrate the potential advantages of variable redundancy with a quantitative map sequence (but one that psychologists have labeled as separable). Dobson (1983) provides evidence that this particular conjunction of variables (size and value) does improve processing over using size alone. Dobson conducted three experiments in which subjects viewed a graduated circle map of the western United States and responded to tasks requiring location (counting the number of states in a particular category), categorization (identifying the category for a particular state), and comparative judgment (determining which of a pair of states had the higher data value). A control group viewed a map in which black circles scaled by area was presented and the redundant symbol group viewed a map of the same data in which color value as well as circle area was used to represent data values (Figure 3.37). Response times as well as accuracy of responses both indicated significant processing improvements for the size–value conjunction over size alone, an indication that those variables are at least partially integral.

Some psychologists working with integral versus separable conjunctions to study perceptual grouping have recognized a third category of conjunctions that fits between integrality and separability (Pomerantz and Garner, 1973). This intermediate category is termed "configural." Where integral conjunctions refer to two physical dimensions that correspond to a single perceptual code and separable conjunctions refer to two physical dimensions that lead to distinct perceptual codes, configural conjunctions maintain separate perceptual codes, but also code a relational or "emergent" dimension. Both integral and configural dimensions lead to "filtering interference" (interference of the second, nonrelevant attribute in tasks requiring attention to only one attribute) and "condensation efficiency" (improvement on tasks requiring both attributes to be considered as a unit). Integral dimensions differ from configural ones, however, in exhibiting "redundancy gains" (improvements in speed of

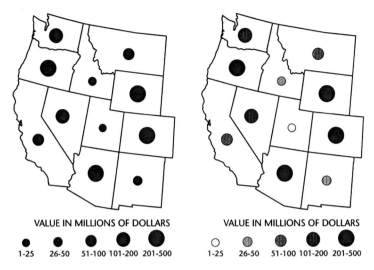

FIGURE 3.37. A pair of Dobson's maps. *Reproduced from Dobson (1983, Fig. 7.1, pp. 156–157). Reprinted by permission of John Wiley & Sons, Ltd., from* Graphic Communication and Design in Contemporary Cartography. *Copyright 1983 by John Wiley & Sons, Ltd.*

performance on tasks in which both attributes provide the same information).

Based on the above definitions, the size–value conjunction that improved performance on Dobson's experiments would be considered integral, but evidence from at least five psychological studies that Shortridge (1982) cites indicates that size and value are at least configural (if not separable). One difference between the psychological studies and Dobson's research is that Dobson's subjects had to assign stimuli to one of five rather than one of two categories. Another difference was that Dobson's subjects had to locate a named state and its circle from the map display containing 11 circles, while the subjects in the psychological studies only saw one stimulus at a time. The apparent redundancy gain in Dobson's experiment may therefore be associated with search time rather than with categorization time. Another possibility, of course, is that Shortridge's continuum hypothesis is correct and that a size–value conjunction is somewhere between the separable and integral extremes.

The concept of an integral–separable continuum of symbol conjunctions has found some support in the psychological literature. Cheng and Pachella (1984) in particular have argued that most phenomenon already categorized as integral or separable actually exhibit "degrees of nonseparability." Further support comes from a recent study by Carswell and

Wickens (1990). They examined 13 stimulus sets involving conjunctions. All were derived from existing graphics. They found that 2 of the 13 commonly used symbol conjunctions contained separable variables, 2 contained configural variables, and the other 9 could not be classified. Rather than interpreting their results as support for a continuum, Carswell and Wickens favor three distinct categories: integral, configural, and separable.

In addition to examining separability of variable conjunctions, Carswell and Wickens (1990) considered whether or not conjunctions were *homogeneous* or *heterogeneous* and whether they used object integration. Homogeneous conjunctions are those in which the same visual variable (e.g., location in space, as on a graph, or orientation as in a wind rose) is used for both (or all) variables. Object integration is the merging of two attributes into a single object (Figure 3.38). Garner (1976) has argued that object integration is more likely to lead to integral or configural conjunctions than will two distinct spatially contiguous objects (e.g., paired bars on a bar chart). Following from these ideas, we might expect that the Carr et al. (1992) bivariate NO_3–SO_2 map (which uses homogeneous conjunctions and object integration) would result in configural conjunctions for which individual attributes and their relationships can be easily extracted from the line slopes, their direction agreement (both up, both down), or the angle between them (Figure 3.39).

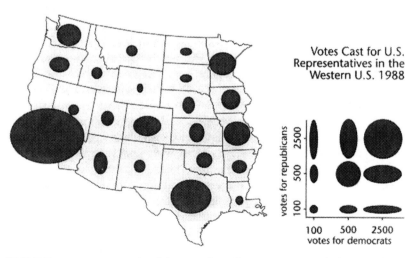

FIGURE 3.38. An example of the use of an ellipse as a map symbol in which the horizontal and vertical axes represent different (but presumably related) variables.

FIGURE 3.39. Bivariate map of NO₃ and SO₄ trends. The original Carr et al. version of this map used a wheel with eight spokes, rather than a simple dot, as the center of each glyph. When large enough, this added feature facilitates judgment of specific values. *After Carr et al. (1992, Fig. 7a, p. 234). Adapted by permission of the American Congress on Surveying and Mapping.*

Associativity of Graphic Variables

As described in Chapter 2, *associativity* exists for a visual variable if variations within that variable (or, in Bertin's terms, the "levels" of the dimension) can be ignored, allowing the units using that visual variable to form a perceptual group. Bertin demonstrates the difference between associative and disassociative variables with a bivariate map composed of point symbols that vary in size (which he considers a disassociative variable) along one axis and shape plus orientation (a pair of associative variables) along the other (Figure 3.40). As is clear here, and for Bertin's original somewhat more complex conjunction of three variables, it is easier to attend to different shapes of the same size than different sizes of the same shape. Bertin's claim is that different levels of particular visual variables retain sufficient similarity that symbols to which these various levels are assigned can be seen as a visual group regardless of proximity. Bertin contends that for his associative variables, this grouping will occur "immediately."

Just as Bertin's (1967/1983) contentions about the selectivity of the visual variables are related to psychological work on selective attention,

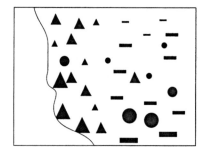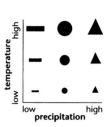

FIGURE 3.40. The bivariate temperature–precipitation map of Figure 3.36, this time using point symbols that vary in shape and size to represent the two quantities.

his arguments concerning associativity are related to research on divided attention. In the case of Pomerantz and Schwaitzberg's (1975) divided attention study, divided attention was easy for pairs of shapes that were in close proximity and formed Gestalt groups. As distance between the elements increased, however, attention to the feature pairs as units became more and more difficult (after 2° of arc separation, response times for divided attention rise markedly). This evidence makes Bertin's contentions about associativity seem unlikely. At the least, associativity will depend upon proximity, decreasing as proximity among symbols increases. At this point, we have no information to suggest the shape of this relationship (whether it might be linear, geometric, or stepped with one or more thresholds), nor do we know whether the associativity–proximity relationship will look the same for all of the visual variables that Bertin claims are associative. Shortridge (1982) suggested that we examine whether the integrality or separability of pairs of visual variables is a discrete phenomenon or is better represented as a continuum. We should perhaps extend this suggestion to all aspects of visual variable combinations and examine whether Bertin's selectivity and associativity concepts also represent continua.

Indispensable Variables

There seem to be differences in dominance among both visual variables and Gestalt grouping principles in various contexts. That position, in both space and time, has a dominant overall role in perceptual organization is the contention of Kubovy's (1981) concept of "indispensable" variables. Both Pinker (1990) in relation to graph understanding and Bertin (1967/1983) in relation to map understanding have chosen to ignore time, the second of Kubovy's indispensable variables. Considering the current attention to map animation and dynamic visualization, however, we can no longer afford to do so.

In a map context, Slocum's (1983) analysis of proximity and similarity as factors in groups seen on graduated circle maps supports the contention that spatial location (in the form of proximity) is a more dominant variable than similarity (of size). In Slocum's study, in fact, circles of different size were more likely to be seen in the same group than those of the same size. His subjects attended to relative location of circles and ignored similarity of size. In the context of multivariate dot maps, however, Rogers and Groop (1981) found that proximity did not overpower color hue. Their subjects were able to identify univariate regions as effectively on trivariate dot maps using different color dots for each of the three variables as they could on individual dot maps. While this result does not necessarily counter the claim that location is an indispensable variable, it does indicate that grouping by a conjunction of color hue plus proximity works as well as grouping by proximity alone.

Humans appear able to segregate the visual scene in terms of both position in X–Y (or the plane of the retinal image) and position in Z (or depth). Research on visual search for objects having conjunctions of two or more variables, for example, has demonstrated that perception can segregate a scene on the basis of depth planes and position in these planes. Nakayama and Silverman (1986) presented subjects with displays in which stereo disparity was used to produce a near and a far visual plane containing colored items. In their experiment, all nontarget items in each plane were a single color hue (e.g., near = red and far = blue). Targets were the opposite color of the depth plane in which they appeared. Subjects were told to locate the colored target and their response times were measured for displays having various densities of nontarget items. The display density did not affect search times, indicating that search was accomplished in parallel (i.e., all potential targets were attended to at once). Since the depth plane that did not contain a target had items of the same color as the target, this result means that subjects were able to direct their attention to one position in Z and to ignore the potentially distracting objects at another position in Z.

Following from these results for position in 3-D space, we might predict that position in space–time will be easily distinguishable (and more noticeable) than position in static space or aspatial time. This makes sense on evolutionary grounds. Our ability to attend to moving objects can be thought of as an ability to focus attention on position in space–time. If the position of an object changes over time, it is very difficult to avoid attending to it. This "fact" is the basis for the Gestalt principle of common fate, which Wertheimer (1923; translated in Ellis, 1955) argued was often dominant over grouping by proximity. Humphreys and

Bruce (1989) cite a number of related studies in which various conjunctions of locational with nonlocational visual variables were tested. It is clear from these studies that visual scenes can be segregated by disparity in both depth and motion (across space over time) and that these aspects of location are dominant over nonlocational variables such as color, form, orientation, or size. Both motion and disparity in depth also seem to dominate position in the plane as a factor in forming perceptual groups. One counterpoint to the argument that disparity in depth is more noticeable than differences in color, texture, and the like, is that natural camouflage of animals and artificial camouflage of military equipment both seem to be effective in concealing, in spite of the presence of depth due to binocular parallax—until movement occurs.

Where We Attend

In relation to visual attention, we began by considering what humans attend to when we look at a particular map (or spatial display) location. In this section, we move on to consider various factors that determine where we look when viewing a map. Two aspects of this question are considered, location within the visual scene and the scale of attention.

Location

Attention to items in the visual scene has been likened to a *spotlight* that highlights a small area making it more visible than its surroundings (Posner, 1980) (Figure 3.41). This spotlight can be directed away from our fixations to objects or events in peripheral vision (without changing the direction of fixation). It is therefore somewhat independent of eye move-

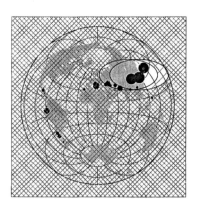

FIGURE 3.41. The attention *spotlight* as a viewer scans a map. Each ellipse represents the region of emphasis by each eye.

ments. In fact, it is probable that the ability to change the spotlight (or location) of attention without eye movement may be how the visual system determines where the next eye movement should fixate (Humphreys and Bruce, 1989).

The view of attention as a spotlight with a focus, a margin, and a fringe can be traced to William James in 1890 (1890/1960), but has regained popularity due to a variety of recent response time studies that show response time decreases for appearance of stimuli at locations anticipated due to a cue and increases for appearance at locations away from location cues (Humphreys and Bruce, 1989). Tsal and Lavie (1988) have also demonstrated that when subjects were prompted to locate targets (letters) of a given color or a given shape, other nearby letters were more likely to be recalled than letters of the same color or shape that were not adjacent to the target. They contend that their findings "strengthen and extend the notion that attention operates as a spotlight" (p. 19).

We seem able to narrow the focus of the spotlight more in the foveal area of vision than in the periphery (Downing and Pinker, 1985). Evidence also suggests that the focus of attention may begin with a wide aperture (but low resolution) and gradually change to a more focused, higher resolution (Eriksen and Murphy, 1987). As a consequence, the analogy of a *zoom lens* has been offered as an improvement on the original spotlight analogy. This multiscale feature of attention with its apparent tendency to begin with a broad view corresponds to evidence for a dominance of global versus local processing of visual scenes (Navon, 1977) and to Marr's contention that 3-D model representations are hierarchical, making recognition of membership in a category possible before recognition of individuals (see Chapter 2 for details).

Cartographically, a key aspect of the way attention works is that initial views, if they take in large segments of a map, will be able to process only gross features. This processing will then guide the narrowing of attention to particular features and objects in order to examine details. Particularly in a visualization context, therefore, graphic design impacts upon the initial wide-scope global view of the map and may dictate what specific details are seen. Also, in the case of reference and travel maps, the ease with which point features, labels, and other small map items can be found by scanning across the map is likely to be controlled to a large extent not by the discriminability of separate features of symbols or text, but by the higher level appearance of symbols and words as a whole and by the overall map structure that may influence attention and thereby guide where attention will be directed (Phillips and Noyes, 1977—see discussion in the "Scanning the Visual Scene" section below).

There is evidence that attention can be directed to objects as well as locations. Duncan (1984), for example, demonstrated that subjects could more easily attend to two attributes of one object (rectangle length plus position of a gap in the rectangle or line type and its orientation) than to elements of two different objects (length of a rectangle plus line orientation). In his experiment, the objects were superimposed, and therefore location was the same. That we can attend to features of objects when location is restricted to foveal vision, however, does not discount the role of location in attention. As mentioned in Chapter 2, when more than one object at more than one location is presented to us, we are more able to attend to a particular location (and to all objects at that location) than to a particular category of objects regardless of position. To attend to an object, we must first attend to its position. It is only then that the features of the object begin to be clear enough to guide our attention to them.

Scale

In the previous section, it was suggested that visual attention may begin with a broad extent but relatively coarse resolution, and then, based on cues obtained from this initial perspective, be redirected to another location or focus in on a particular area or object. Humphreys and Bruce (1989) contend that overall spatial structure is probably available more quickly than is the structure of local details. Humphreys and Quinlan (1987) suggest that both pattern and object recognition might rely on descriptions available from relatively low spatial frequencies—the global features—because patterns at this frequency are more stable over time.

Neurophysiological evidence complements the view of multiple spatial scales of visual processing. Wilson et al. (1990, p. 240) cite research with cats and macaque monkeys indicating that "at each stage of the visual pathway cells with receptive fields in the same part of the visual field can respond to different ranges of spatial frequency." In addition, they contend that "spatial frequency selectivity becomes progressively narrower moving up the system from retinal ganglion cells to LGN cells to simple cortical cells."

The idea of multiple scales of attention is closely associated with research concerning global–local precedence—whether global holistic properties or local components or parts are perceived more readily (Watt, 1988). The divided attention studies of Pomerantz and his colleagues provide one piece of evidence for global structures taking precedence over individual features. As noted above, their results demonstrate that in categorization tasks, certain arrangements of parts are processed more quickly as a unit (a whole) than are either of the individual parts (Pomerantz and Schwaitzberg, 1975). These results seem to support global precedence for "good" Gestalt groups, and local precedence for "poor" groups.

The experiments, however, focus only on the issue of global versus local processing of small perceptual units that are easily attended to because they appear in foveal vision, exactly where they are anticipated. Their concern, then, is not spatial but object-based and their research has little to say about the relative spatial scope of attention and how it is controlled.

In examining a map or other visual scene, one role of visual attention is to determine where to look. Because attention can be directed to various locations at various scales, the issue of whether perception usually begins with a spatially global or a spatially local perspective becomes important. Most studies of global–local precedence seem to implicitly accept the roving zoom-lens analogy for visual attention and have focused on the question of the scale of feature that is most easily attended to initially.

In a now classic study that has stimulated much of the subsequent research, Navon (1977, p. 354) investigated the postulate that "perceptual processes are temporarily organized so that they precede from global structuring towards more and more fine grained analysis [local structuring]." His experimental stimuli (compound letters), were selected so that global and local components could be manipulated independently (Figure 3.42). The stimuli were composed of small letters organized in arrangements to create large letters with the small and composite letters being either the same or different. Subjects were asked to identify either the local stimuli (small letters) or the global stimuli (large letters), and the speed with which they could do so was measured. What Navon found was that identification of global features was faster than identification of local features, and that conflicts between local and global letters interfered with identification of local letters, but not with identification of global letters. Navon's interpretation of his findings was that global processes must necessarily be prior to local ones. What is not clear from this research is whether identification of global stimuli requires a prior grouping (of as yet unidentified local stimuli).

Subsequent research by Paquet and Merikle (1988) has considered situations in which the visual scene is composed of more than one element set. They again used compound letters as stimuli, but presented sub-

FIGURE 3.42. An example of the kind of compound letters used by Navon and other psychologists studying global–local precedence. *Derived from Navon (1977, Fig. 5, p. 365).*

jects with pairs of them rather than a single set (Figure 3.43). Subjects were asked to attend to one of the pair (identified by a surrounding circle or square) and, as in Navon's study, to identify either the local or the global letter. Their results confirmed that the global letters were identified faster and that the global aspect of the attended form was harder to ignore than the local aspect. Beyond this confirmation, however, they found that both global and local aspects of the unattended stimulus could influence identification speed, with local features having an influence if a local identification was requested and global features having an influence if a global identification was requested. Further, Paquet and Merikle (1988, p. 98) found that "it was impossible for observers to ignore the category of the global aspect of the nonattended object." This latter finding seems to add even more support to the idea that space is an indispensable variable—because we anticipate features near one another to be related.

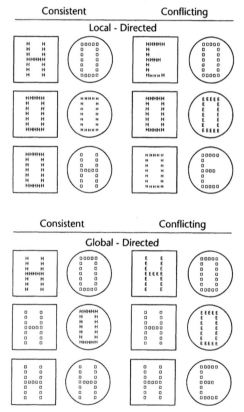

FIGURE 3.43. Sample compound letter stimuli. *Derived from Paquet and Merikle (1988, Fig. 1, p. 91).*

If the zoom-lens analogy for visual attention and the idea that attention varies in acuity from central focal point to its fringes are correct, we can anticipate extensions and modifications to Navon's original ideas about global–local precedence. First, since it is clear that people can attend to local details when directed to, we might anticipate that global precedence will be strongest when we are not already cued to expect some local feature. Second, global precedence can be expected to be stronger on the periphery of attention where the resolution of attention (and of vision) is not sufficient to resolve local details. Third, we might expect to find limits on scale of global elements that will be attended to quickly—if they are too large, the elements will be beyond the bounds of our attentional zoom lens, and if they are too small, they will be local details.[7]

All of the above possibilities have been supported to some extent by empirical research. In an experiment using compound letter stimuli similar to Navon's, Pomerantz (1983) dealt with the first issue. Half of his subjects had to respond to either the global or the local letter when it appeared on the screen at random locations. For the remaining subjects, presentation was always at the center of the screen. For both certain and uncertain presentation locations, global letters were easier to identify than local ones, but the difference was greater for uncertain than for certain locations. In a related experiment, Lamb and Robertson (1988) had subjects fixate on the center of the screen before presentation of a compound letter. Presentation could be central or to either side of center. They found that the global-identification speed advantage was greatest for peripheral presentations.

That size of perceptual units has an impact on global–local precedence is supported in a variety of studies. In research with compound stimuli composed of geometric shapes rather than letters, Kimchi (1988) found that the number of local elements making up the global shape interacted with the strength of global precedence (Figure 3.44). Specifically, when the number of elements was small, thereby making the global figure small, global processing was faster whether or not local and global shapes agreed. With larger global stimuli, composed of more local elements, global processing was faster in situations where there was conflict, but not in situations where the shapes agreed. More direct evidence that the size of a global figure must be within some limit in order to receive attention precedence can be found in research by Kinchla and Wolfe (1979) with compound letter stimuli and research by Antes and Mann (1984) with pictorial stimuli. For the compound letters, Kinchla and Wolfe found that compound figures larger than 8° of visual angle (roughly the size of the United States on a page-sized map of North America at normal reading distance) resulted in a reversal of attention to local prece-

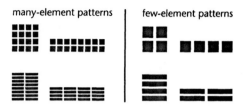

FIGURE 3.44. Nonalphanumeric compound symbols used in testing for global versus local precedence. *Derived from Kimchi (1988, Fig. 1, p. 191).*

dence over global. For pictures, Antes and Mann found that global precedence existed for pictures subtending 4°, local precedence occurred for pictures subtending 16°, and neither global nor local precedence was found for scenes subtending 8°.

An interesting feature of the Antes and Mann study is that their picture stimuli contain global–local dependencies (at a semantic level) not found in the compound letter stimuli. Identification of global scenes in pictures can depend (in part) on identification of local details (e.g., one of their pictures is a farm scene that would be quite difficult to distinguish from many other landscapes unless a local detail, the barn, is recognized). In spite of this interdependency, results concerning the effect of scale on global–local precedence is consistent with that for the compound letter stimuli, for which local and global levels are independent. This comparability of results suggests that global–local precedence effects processing of representations at multiple processing levels (e.g., the low-level primal sketch and subsequent 3-D model representation where recognition can occur).

Most global–local research in psychology, like virtually all cartographic research, has been directed to static displays. There is, however, evidence of global–local processing in the temporal as well as the spatial dimension. The research on temporal context might be considered an extension of the Gestalt grouping principle of "objective set" (that humans tend toward stability over time in perceptual grouping). Objects grouped at one time remain grouped over time, even when changes in proximity would result in no groups or different groups in a static scene. Palmer (1975) examined a similar idea in relation to the effect of temporal context on identification of objects. He hypothesized that global scenes in a temporal sequence would influence the identification of local details presented in subsequent scenes. Palmer found that "appropriate" prior scenes facilitated object recognition and "inappropriate" prior scenes (e.g., a kitchen prior to presentation of a mailbox) impeded recognition. When the subsequent object was similar in appearance to one more logically part of the scene (e.g., a loaf of bread) misidentification was likely.

At this point, we can only speculate upon the implications of global–local research for map understanding. There has been no cartographic research to date that has extended directly from these studies. As Mistrick (1990) points out, however, results of the global–local processing research support the concept that extraction of meaning from visual scenes uses hierarchical structuring of information at multiple spatial scales. This view corresponds quite well to Marr and Nishihara's (1978) ideas concerning how primal sketches are derived from a retinal array and to research directed at higher levels of processing that indicates hierarchical structures for memory encoding of spatial knowledge. As discussed in Chapter 8, issues of global–local precedence may have particular relevance to exploratory visualization with maps—a situation in which an analyst is not entirely certain what patterns to expect and a situation for which dynamic manipulation of display scale will be a significant part of the analysis.

Scanning the Visual Scene

Both what is attended to and the scale of that attention interact with the process of visually exploring a map or other graphic display. It seems clear that both global views and peripheral attention act to steer eye movements toward important information in a visual scene and away from unimportant information. As early as the 1970s, cartographers investigated the use of eye movement recordings as a tool for understanding the visual–cognitive process of map reading and the impact of both changes in map design and training on that process (see Steinke, 1987, for a comprehensive review). In addition, cartographers have studied a variety of visual search problems on maps in an effort to determine how to facilitate search for specific kinds of features (e.g., placenames, point symbols, etc.). Much of the early work in both visual search and eye movement analysis by cartographers suffered from a lack of theoretical perspective. As Steinke (1987, p. 57) noted in relation to eye movement research, early work seemed driven by a "let's see what happens when we put a map in front of somebody and photograph their eyes" approach. Only recently has cartographic research dealing with visual scanning of maps begun to build on a firm theoretical base grounded in perceptual–cognitive theory. Looking back from an information-processing perspective, however, we can identify some links between early cartographic eye movement research and the perspectives, presented above, on vision as a modular system for processing increasingly interpreted representations.

Both Marr's model of vision and Gestalt principles suggest that edges are important elements of visual scenes that are processed early in vision.

Steinke (1987) cites eye movement research by both Thomas and Lansdown (1963), with medical images, and Gratzer and McDowell (1971), with landscape photos, in which foveal attention to edges in the scene was demonstrated to be a common tendency, in spite of highly individualistic scan paths. In the Gratzer and McDowell study, edges defined by the skyline, ridgelines, shorelines, and vegetation boundaries all received particular attention. Not all edges, however, seem to be equal in attracting foveal attention. Mackworth and Morandi (1967), for example, provide evidence that simple contours are noticed using peripheral vision and largely ignored by foveal fixations. It is unpredictable edges, or edges with unusual details, that seem to attract foveal attention.

Although cartographers employing eye movement techniques are clearly aware of psychological research showing attention to edges, no cartographic research based on this knowledge seems to have been done. On maps, eye movement techniques might be used to determine the relative "goodness" of the contour established by different methods of creating contrast between map regions. Alternatively, analysis of eye movements might be used to assess techniques for enhancing the identification of map regions. Although Jenks (1973) found little similarity in map viewers' scan paths when viewing a dot map, he did identify commonalities in relative attention paid to different parts of the map. He did not, however, examine fixation times in relation to regional edges delineated by his subjects. The one example that he does provide of a subject's fixation times for map cells suggests that more attention was given to the edges of dot clusters than to the core of those clusters (Figure 3.45). Dobson (1979a) also found correspondence among subjects about relative attention to different parts of a test map. In examining his data, he found a high correlation between the "informativeness" of map sections and visual attention to those locations. Both Jenks's and Dobson's results suggest that eye movement analysis might be applicable to assessment of the distinctiveness of regions on thematic maps. Measuring attention to transition zones around a region or between regions could be used to assess the strength of regional edges.

One common feature of early cartographic research using eye movement analysis was that subject attention to maps was observed in the absence of defined tasks. Castner and Eastman (1985) point out that we should expect quite different perceptual and cognitive processes to be at work in this situation, which they called "spontaneous looking," and in "task-specific viewing." With spontaneous looking, location of attention will be influenced primarily by the properties of individual map symbols, Gestalt properties of symbol groups, and reader attitudes or expectations about the stimulus and the experimental situation. For this kind of viewing (perhaps typical of early exploratory visualization), then, map design

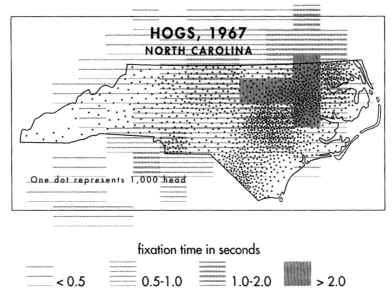

FIGURE 3.45. A map of attention to locations on a dot map. *Reproduced from Jenks (1973, Fig. 9, p. 34). Reprinted by permission of Universitätsverlag Ulm GmbH.*

changes are likely to be particularly important. On the other hand, Castner and Eastman contend that for task-specific viewing, cognitive processes will exert a much stronger control over eye movements and attention. In these cases, eye movement analysis might be used to distinguish different problem-solving strategies (or application of different schemata). Recent evidence by Morita (1991), however, demonstrates that while task-specific viewing can lead to greater similarity in eye movement parameters than spontaneous looking, design alterations can still play a major role. He found very distinct differences in patterns of visual search on schematic maps in which numerical information was symbolized by seven different graphic variables (see Figure 2.14).

Most cartographic research using eye movement analysis has focused on questions of where foveal vision is directed. The technique can also provide information relevant to the issue of global–local precedence. A graduate research project by Guyot (1971; cited in Steinke, 1987) provides one example. Guyot was interested in how map patterns are compared. In an experiment in which subjects were required to select one of two maps that was most like a third, eye movement recordings were used to measure which of the two maps was attended to first, as well as which received the most and the longest fixations. The finding that the first map looked at was generally selected as the most similar to the referent map suggests that peripheral vision played a key role in pattern comparison for these maps. This in turn suggests that global processing of map patterns is quite sophisticated and that it directs eye movements, at least in a task-specific viewing situation.

Recent uses of eye movement analysis in cartography have focused on task-specific viewing. This focus is driven, in part, by the lack of consistent results of past "spontaneous looking" experiments and by the desire to achieve particular application goals. The map-use task to which eye movement evidence seems most applicable is visual search. Phillips and his colleagues in the Psychology Department of University College in London (with input from cartographers Bickmore and DeLucia) were among the first to use eye movement techniques to examine strategies of visual search for information on maps. They focused on the influence of map design on search and gave particular attention to the problem of searching for names on maps (see Phillips and Noyes, 1977). They proposed that reducing either the number of fixations or the duration of individual fixations would speed searches for place labels, and that reducing the number of fixations should be more effective. Three map-design procedures were suggested to reduce the number of fixations: generalization that reduced the total number of names on the map, categorization and visual coding of names so that only a subset had to be attended to, and use of a map grid that could direct attention to a limited section of the map. Adjustments to type placement were used to control individual fixation times. Both procedures reduced search times, but the three techniques that reduced the number of names considered were (as predicted) more effective. Of these, using small map grids had the most dramatic effect.

Eye movement analysis has not proved to be as powerful a tool for cartographic research as originally anticipated by Jenks and others (Steinke, 1987). This limited success is due to both practical and conceptual issues. From a practical point of view, eye movement analysis has been difficult and expensive. Conceptually, it suffers from the problem that there is no simple way to determine whether a fixated location is also attended to. Because of these problems, the question of visual scanning (particularly visual search) has been investigated in a number of other ways. The two most commonly used measures in visual search experiments (other than eye movement analysis) are the accuracy and the speed with which targets are identified. Accuracy is assessed by the number of correct identifications compared to the number not found and/or misidentified.

In the only cartographic study to consider the role of figure–ground in visual search, Lloyd (1988) compared perceptual and imagery processes involved in determining the presence or absence of pictorial point symbols on a simple map like display (Figure 3.46). The display had a central green area surrounded by a blue area and symbols in yellow were dispersed relatively evenly across the map's surface. The central green area (presumably due to centrality, surroundedness, and slightly smaller size) was expected to be a figure on the blue background. Subjects in the perception condition had the task of determining whether or not two different symbols both appeared on a display. Symbols were either both absent, both on the figure, both on the ground, or one on the figure and one on the ground. Lloyd predicted that when both symbols were on the figure they would be found more quickly because the figure was expected to draw the subjects' initial attention. When both symbols were on the ground he expected opposite results, and with one symbol on the ground and one on the figure he expected intermediate results. Although mean response times exhibited this ordering, differences were not significant. The explanation provided, and one that could be predicted from Gestalt principles, is that the yellow point symbols (due to small relative size and greater contrast with either area than the areas had with each other) were the dominant figures seen and that for most subjects both the green and blue areas became ground.

Treisman et al. (1990) (building on work by Cavenagh, 1987, 1988) proposed a model in which five independent visual pathways—luminance, motion, binocular disparity, color, and texture (Figure 3.47)—process different attributes of the visual scene. This model is supported in

FIGURE 3.46. Simulation of the test map from Lloyd's experiment. *After Lloyd (1988, Fig. 4, p. 366). Adapted by permission of the American Congress on Surveying and Mapping.*

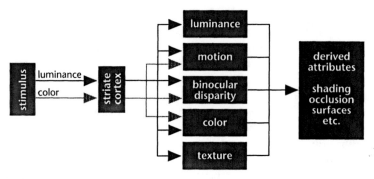

FIGURE 3.47. Treisman's five-pathway model for processing of the visual scene. *Derived from Treisman et al. (1990, Fig. 16, p. 294).*

part by neurophysiological findings that provide evidence for separate color–form and motion pathways and neuropsychological evidence from brain-damaged patients that shows luminance operating separately from either color or motion. Experiments using visual search tasks, examination of after-effects, and identification of shape from contours have confirmed the separate pathways for color, luminance, and motion, and have provided evidence that early vision must also process texture and binocular disparity separately. Each of these visual pathways seems able to code size and orientation information which, in turn, may function as shape primitives.

The model led to Treisman's (1988) feature integration theory (FIT) that posits a series of "feature maps" with one set for each pathway plus one for size and one for orientation. Each set of feature maps consists of individual layers (analogous to the structure of a GIS) to code possible variations along the specific dimension (e.g., red, yellow, and blue color maps or orientation maps for various angles). If the Treisman model is correct, it can explain why some visual search tasks can be conducted in parallel while others require serial processes. If a target differs from other elements of the scene along one dimension or feature, a parallel holistic process can be used. When the search is for a conjunction target that can share features with nontargets, however, location of potential targets must be relied upon to link separate feature maps before determining whether the conjunction occurs. This location-based linking can only proceed in serial.

Recently, Cave and Wolfe (1990) suggested a modification of Treisman's FIT that seems relevant to visual search for map symbols. As originally conceived, FIT predicts that if a target (such as a map symbol) differs from all distractors by a single feature, parallel processes should be able to detect it. If, on the other hand, the target is defined by a conjunction of features, FIT predicts that parallel processes play no role and that the serial processing of each individual element proceeds until the target is found. Cave and Wolfe's "guided search" theory posits that both parallel and serial processing are used in conjunction for all visual search tasks. They suggest that a fast parallel (i.e., holistic) processing stage identifies the basic visual features that are present (just as Guyot, 1971, found that global processing of maps in peripheral vision could determine map similarity). This first stage is followed by a slower serial stage that combines the features identified to produce object representations. With conjunction symbols on maps, then, we would predict that a parallel search procedure would act to limit the serial search stage to elements sharing one of the conjunction variables. Guided search, then, is thought to use parallel processing to steer search by identifying the likelihood that a feature is a target and by allowing vision to bypass those features with low target likelihood.

Visual search remains incompletely understood. Cheal and Lyon (1992), for example, demonstrate that neither FIT, nor guided search, nor similarity theory (Duncan and Humphreys, 1987) can handle all aspects of how parallel and serial processes might interact in the search for even relatively simple shapes. Another alternative for combination of parallel and serial search was actually suggested in earlier work by Treisman (1985). If the visual scene can be subdivided into regions (as it often is on maps) within which a conjunction target differs from nontargets along only one feature, search can be parallel within each region, while remaining serial from region to region. The advantage of regionalizing search is supported for map reading by Phillips's placename search research described above (Phillips and Noyes, 1977; Phillips et al., 1978).

Figure–Ground

The combined information concerning perceptual grouping, visual attention, and global–local processing of visual scenes provides a firm base from which to understand the concept of *figure–ground*. Segregation of figure from ground requires that perception organize the visual input sufficiently for elements of that input to group and attract attention to themselves. In the environment, as well as on maps, distinctions between significant and insignificant elements of a scene need to be made at a coarse holistic level so that they can guide further attention to specific details. figures that attract our attention are distinctive from the background and often appear to be in front of that ground. Hochberg (1980, p. 90) seems to sum up the general concept of figure–ground, and illustrate the fuzziness of figure–ground as a concept, when he states that "figure is thinglike and shaped, ground is more like empty space, amorphous and unshaped."

Figures are defined by their contour or the boundary between object and nonobject. The contour operates on only one field or the other, but not both (Arnheim, 1974). When it is unclear which region of a scene the contour belongs to, an ambiguous figure results in which first one, then the other, part of the field becomes figure (Figure 3.48). Such ambiguity is what cartographers strive to avoid.

Most cartographers who have considered the issue of figure–ground segregation on maps have turned to Gestalt psychology for guidance. This is a good place to start because Gestalt psychology has made the greatest contribution to our current understanding of figure–ground. Gestalt psychology alone, however, does not answer all of the cartographically relevant questions. The discussion that follows will begin with a brief overview of some of the more important Gestalt ideas relating to figure–ground and will then explore more recent psychological and cartographic research that has attempted to extend from this base.

Gestalt psychologists' initial approach to figure–ground began with principles of perceptual grouping because, to see a figure, a perceptual unit must exist and grouping produces perceptual units. All grouping factors have a potential role in defining regions of a visual scene, and the scene must be differentiated in order for figures to appear. Extending from the fundamental grouping principles, a set of related principles has been devised to deal with the visual strength of perceptual groups as figures segregated from a background. Some were suggested directly by Gestalt psychologists and others were derived more recently by researchers following similar logic. The factors below seem to be the most relevant to establishing symbols and regions as figures on maps.

1. *Heterogeneity*: A visual field must be differentiated to form groups before one part of the field can stand out as figure. While not one of Wertheimer's (1923; translated in Ellis, 1955) grouping principles, this idea was offered (with no label assigned) at the end of his paper. The main guiding principles offered for establishment of figure were that an

FIGURE 3.48. An ambiguous map in which subjects were asked to determine which area was the figure. Half of the subjects picked light and half picked dark. *Derived from Mistrick (1990, Fig. 3.3, p. 60)*.

enclosed shape for which there was a color difference between the shape and the background will stand out as a distinct figure. This basic idea was elaborated by a number of other Gestalt psychologists. Among them was Koffka (1935) who introduced the concept of articulation as a figure–ground principle (see below for more details).

2. *Contour*: Objects are more easily seen as figure the more definite the edge between object and nonobject. The establishment of contour follows directly from establishment of heterogeneity. A noticeable difference between areas creates an edge, boundary, or contour between them. If differences are due to relatively coarse features, the edge will be rather fuzzy and the contour (and the experience of figure) may be weak (Figure 3.49). If the differences consist of fine-textured fills or solid colors, however, (or a distinct line separates the regions) the contour will be stronger, as will the experience of Figure (Figure 3.50).

3. *Surroundedness*: Completely surrounded objects tend to be seen as a unit and thus as figure (Bruce and Green, 1990). That is why, even with weak contour, the white area of Figure 3.49 is seen as figure. This principle is probably the single most useful in creating figure–ground distinctions on maps. As a number of conflicting experiments recounted below make clear, it is hard to avoid ambiguity about figure versus ground in displays that do not have a surrounded figure. Centrally located surrounded shapes will enhance figure formation.

4. *Orientation*: Objects with a horizontal or vertical orientation are seen as figure (Bruce and Green, 1990). A rotation of areas on a hypothetical map illustrates this point (Figure 3.51). It is likely that this tendency has to do with relationships between visual displays and real-world visual scenes in which most figures are upright (as are humans, trees, etc.) or aligned with the horizon.

5. *Relative size*: The smaller of two areas is more likely to be seen as figure. This factor is essentially a corollary to the factor of surroundedness cited above. It is particularly relevant when the larger area completely surrounds the smaller. Assuming the object has sufficient size to be easily detected, the smaller the object relative to its surround, the more it is seen as a figure (Figure 3.52).

FIGURE 3.49. A central figure that fades into the background due to a weak contour.

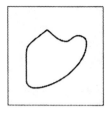

FIGURE 3.50. The previous figure–ground difference enhanced by a stronger contour.

6. *Convexity*: Convexity will be seen as figure. A schematic map of England, for example, has greater convexity than a map using a detailed but smoothed boundary. As a result, it should be more likely to stand out as figure (Figure 3.53).

Heterogeneity

Of the factors thought to lead to "good" figures, it is the first, heterogeneity, that has been given the greatest attention. Issues of contour, surroundedness, orientation, symmetry, and convexity have typically been considered as complements to a primary focus on heterogeneity. Both psychologists and cartographers have investigated how segregation of figure and ground is influenced by various methods of creating heterogeneity between areas. Psychologists have primarily been interested in what figure–ground reactions to differences in texture, value, temporal frequency, and so on, tell us about the visual and cognitive processes underlying perceptual organization of visual scenes. Cartographers, on the other hand, have been most interested in developing guidelines for use of area fills on maps that will lead to consistent identification of specified portions of the map as figure. Psychological research has used a rather limited range of possible area fills, limiting the generality of their findings, and cartographers, while they have tested more kinds of area fills, have not linked their research to psychological theory beyond that of the Gestalt principles developed in the 1920s and 1930s. The discussion below provides an overview of some of the work from both disciplines (with

FIGURE 3.51. When value contrast is relatively equal, map regions that are horizontal or vertical are more easily seen as figure than regions that are diagonal to the map border.

FIGURE 3.52. Small, surrounded map areas are more easily seen as figure.

an emphasis on the role of brightness differences) and uses an experiment by one of my graduate students as an example of how we might integrate these two research streams more effectively.

Both cartographers and psychologists have given considerable attention to the role of brightness (i.e., color value) differences in figure–ground. On maps, it is a particularly important tool because all other means of creating heterogeneity between areas (with the exception of color hue) have a tendency to interfere with other map information (e.g., texture differences require one area to have a coarse enough texture to be noticeable as texture—and coarse-textured area fills make text difficult to read). In psychology, attention to brightness as a figure–ground factor began in the 1920s, with Wever (1927) among the first to address the issue.

In relation to brightness, Wever (1927, p. 222) contended that "a minimum brightness difference is necessary for the experience of figure and ground. As brightness-difference increases, the 'goodness' of the experience increases, though at a constantly diminishing rate." This contention was based on research in which subjects viewed irregular forms presented tachistoscopically. In his experiment subjects viewed 1,060 different black forms on white backgrounds with illumination varied, result-

FIGURE 3.53. A detailed depiction of the coastline of England, Scotland, and Wales compared to a schematic depiction. The latter exhibits greater convexity.

ing in differences in contrast between brightness of the white and black areas.

Much of the psychological research subsequent to Wever has used variations on a simple pie-wedge stimulus. This stimulus allows researchers to experimentally manipulate the actual value of parts of each stimulus, the number of components to the stimulus, their relative size, and the background upon which the stimulus appears. One of the first uses of this stimulus was made by Goldhamer (1934) who examined the influence of relative size on the appearance of white versus black wedges on a gray background (Figure 3.54). He showed that for equal-sized regions, black tended to be seen as figure, but when size varied, it was the smaller shapes that were regarded as figure, regardless of brightness. Oyama (1960) used similar stimuli but came up with somewhat conflicting conclusions. The difference in his experiment was that the surrounds for the stimuli were either white or black rather than gray. Half of the wedges were gray (of varying shades for different stimuli) and half were the opposite of the surround (Figure 3.55). He found that the sectors opposite in brightness to the surround tended to be seen as figure, and that this tendency was strongest when the alternate wedges were closest in brightness to the surround. The situation in which surround and one set of wedges are similar resulted in the other set of wedges appearing as small shapes on a relatively homogeneous background. That the wedges of opposite brightness to the background appeared as figure in this case agrees with Goldhamer and with general Gestalt principles about small surrounded areas being seen as figure. When the alternating wedges were closer (in color value) to each other than either was to the background, however, the white wedges with black surround (and grey alternate wedges) were seen as figure more often than the black wedges with white surround (and gray alternate wedges). This disagrees with Goldhamer's finding that black shapes stand out as figure when size does not differ.

Overall, studies of brightness difference using the pie-wedge stimuli have produced equivocal results. There has been a consistent finding that the smaller of the areas are seen as figure regardless of brightness and that horizontal–vertical wedges are more likely to be seen as figure than diagonal wedges (Bruce and Green, 1990).

FIGURE 3.54. A sample of the kind of stimuli used by Goldhamer (1934).

FIGURE 3.55. Examples of the variations on the standard pie-wedge stimuli used by Oyama (1960).

Most of the remaining psychological research on brightness as a figure–ground variable has made use of stimuli similar to the Rubin's vase–face ambiguous figure (Figure 3.56). This stimuli is somewhat more similar to the situation on a map in which land–water areas are adjacent and the cartographer wants one to be seen as figure. Harrower (1936) created an experimental setting similar to Goldhamer's, but with the vase–face figure on a surround. He used the following combinations: black surround, black face, white vase; black surround, white face, black vase; white surround, black face, white vase; white surround, black face, white vase. Whether there was a light or a dark vase or face did not seem to matter. The figure proved to be whichever differed from the surround, although there was a slight tendency toward face as figure. Harrower also examined a range of brightness differences assigned to surround, vase, and face. In this case, the face part of the stimulus was mounted on a track that allowed the halves to be pulled apart. The subject's task was to attend to the face as figure as long as possible. Results indicated that the face was held as figure longer as brightness difference was increased. Again, it did not matter whether the vase or the face was darker.

More recently, Lindauer and Lindauer (1970) used the Rubin's vase–face figure in a similar experiment involving brightness contrast. They compared a control (an unshaded outline drawing of the vase–face figure) with versions in which one area was white and the other was filled with a 20%, 40%, 60%, 80%, or 100% black pattern. In the unshaded control, the face was seen as figure, a finding that the authors attribute to

FIGURE 3.56. A typical vase–face ambiguous figure.

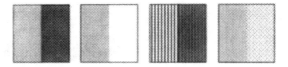

FIGURE 3.57. A typical stimulus from Dent's study. *Derived from Dent (1972, pp. 208–221).*

familiarity. For the test stimuli, responses to the shaded area as figure increased as the contrast between areas increased.

Building upon the rather mixed psychological findings, there have been a small number of published cartographic studies of the influence of brightness on figure–ground segregation. The first was by Dent (1972) as one component of his dissertation. He used bipartite squares, stimuli that were simpler than any of those used in previous psychological studies (Figure 3.57). As part of a larger study, his stimuli used various combinations of area fills on the two sides of the square. These were created with both dot and line patterns, in many cases texture was apparent because Dent was testing for it as well as for brightness. Dent's experiment, unlike the psychological research, was not based on response times (to identify figure or to hold particular areas as figure). Instead, his subjects were asked to examine each stimulus and to mark the side of the square that they saw as figure (or that visually stood out). In general, Dent found the coarser areas to be seen as figure (as the Gestalt concept of articulation would predict). When both areas were shaded with similar-sized dots, however, the finer textured, darker pattern was seen as figure—an indication that brightness might be more important than texture.

On a follow-up to the figure–ground test, Dent assessed subject preferences for maps in which a central focus area (e.g., North America) was shaded and the surround was white. For four different maps he found preference for the shaded maps over unshaded maps ranging from 71% of subjects to 96%. Wood (1976) also examined the "most desirable" brightness differences for maps. In this case all maps used some shading with relative brightness of figure, ground, and surround varied across the range of 16 test maps. Regardless of surround, maps with figures lighter than the ground were preferred.

The two preference studies by Dent and Wood suggest that heterogeneity of value for areas is preferred to homogeneity, but that preferences for lighter or darker areas as figures are inconsistent. Neither study directly addressed the issue of how value differences on maps influence the likelihood of map areas appearing as either figure or ground. Mistrick (1990) designed a study to do just that. In essence, she replicated a portion of Dent's (1970) initial figure–ground choice study using stimuli that were more maplike than his bipartite squares.

Mistrick's test maps, as we reported in our paper on brightness contrast as a figure–ground variable (MacEachren and Mistrick, 1992), depicted the land–water border along the coast of Korea (see Figure 3.48). The test maps were cropped so that the map border was square and the land and water areas of the map both occupied 50% of the area included. A pretest had shown that without labels the coastline was recognized by few students at Pennsylvania State University (the source of subjects for the experiment). The test stimulus, then, was a real map but due to the lack of familiarity with it, prior experience was not a factor in determining figure–ground segregation. In addition, the lack of familiarity allowed the map to be presented at four orientations with any change in response to it attributable to relative position of features and their concavities rather than to variations in recognizability of the map.

Mistrick's (1990) test map was created at four orientations with six combinations of area fill (white–dark gray; dark gray–white; white–light gray; light gray–white; light gray–dark gray; dark gray–light gray) applied to the land and water areas respectively (Figure 3.58). Each subject viewed only one map and indicated the region seen as figure. Two hundred forty subjects participated, half of whom saw unlabeled maps and half of whom saw maps with the labels "land" and "water" outside the map border next to these respective areas. The land–water labels had no effect on identification of one area as figure in relation to the other.

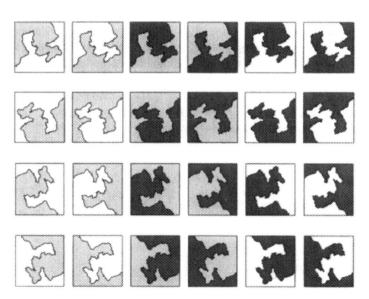

FIGURE 3.58. The complete set of maps used in Mistrick's study (at greatly reduced size). *Reproduced from MacEachren and Mistrick (1992, Fig. 6, p. 96). Reprinted by permission of* The Cartographic Journal.

Somewhat surprisingly, in relation to Dent's earlier findings with bipartite squares, the darkness of area fills had no effect on figure–ground segregation either. Exactly half of the subjects identified the relatively darker area as figure and half identified the relatively lighter area. This "negative" finding does, however, seem to agree with previous cartographic interpretations of Gestalt "rules" by Wood (1968), Spiess (1978), and McCleary (1981) that advocated relative value contrast (heterogeneity) between figure and ground but did not suggest that absolute value of area fills is relevant to what is seen as figure.

That contrast influences figure–ground segregation independently of the direction or sign or the difference is logical on "hardware" grounds. Shapley et al. (1990, p. 438) cite neurophysiological evidence that "fundamental neural mechanisms of pattern recognition are essentially nonlinear because these mechanisms ignore the sign of the contrast." They go on to contend that "form depends on the magnitude of contrast at a border while brightness depends on its sign." They demonstrate the implications of this fact with an illusory contour demonstration which reveals that contour-sensing processes do not rely upon the sign of contrast (Figure 3.59). In a similar vein, Barlow (1990) cites evidence that different sets of brain cells are specialized for light, for dark, and for value contrast. These contentions of independent brightness and form processing at a neurological level support the argument that Mistrick and I make (MacEachren and Mistrick, 1992) that contrast should be much more important in establishing figure on maps than specific brightness relationships.

In addition to research on heterogeneity due to brightness differences in relation to figure–ground, a number of psychological studies have considered heterogeneity created by contrast in spatial and temporal frequency of areal fills. This research has been closely linked to information-processing and computational-vision assumptions about how early vision represents visual scenes.

The influence of spatial frequency (or focus) on figure–ground segregation has been addressed in several psychological studies. Wong and

FIGURE 3.59. Two Kanizsa squares illustrating the dominance of contrast over brightness in establishing figure–ground.

Weisstein (1983) found sharp (high spatial frequency) targets (i.e., symbols) to be detected better against figural regions, while blurred (low to middle spatial frequency) targets were best detected in ground regions. Klymenko and Weisstein (1986), for regions that were otherwise ambiguous, examined the dominance of fills having differing spatial frequency (defined by horizontal sine wave patterns at frequencies of 0.5, 1, 2, 4, and 8 cycles/degree). They used the percentage of viewing time that each region was judged to be figure as a measure, and found high frequency patterns to be consistently seen as figures. In addition, the greater the frequency difference, the stronger the effect was. Significantly, in light of cartographic research based upon bipartite, or divided, squares, they found that a greater frequency difference was required for figure dominance in this situation than with more complex regions (i.e., a vase–face figure or an eight-element pie-wedge figure). This may be due to the importance of contour, or edges, in establishing figure, and because contour is more apparent when it has defined form. Kylmenko and Weisstein (1986, p. 324) link their results to Gestalt principles of articulation by suggesting that "greater detail is more or less correlated with the presence of higher spatial frequencies in the stimulus." They also cite a geometrical demonstration by Pentland (1985) that shows "gradient of blur" to be an ecologically valid indication of relative distance in depth. As a result, it can be postulated that blurred patterns will be seen as ground because they appear to be physically in the three-dimensional background. The idea that sharply defined patterns are seen as figure and fuzzy patterns as ground is relevant to the issue of visualizing uncertainty on maps through manipulation of *focus* (renamed "clarity" below) (MacEachren, 1992b).[8]

Bottom-Up versus Top-Down Processing

Of the Gestalt grouping factors empirically considered in relation to figure–ground, the role of "experience" (i.e., top-down processing), or lack of it, has generated the most lively commentary. Initial Gestalt views on figure–ground segregation treated vision as an immediate process acting on "wholes." As such, Gestalt psychologists discounted any role for top-down processing. Gregory (1990) is in general agreement with this perspective. He contends (based on his own research) that illusory contours occur at a processing stage between the retina and the parietal cortex, thus before a level at which object knowledge could be accessed. Kienker and Sejnowski (1986) cite a study by Ullman (1984) who reported that subjects presented with a closed outline and a small spot that could be either inside or outside that outline were able to determine that in–out location within a few hundreds of milliseconds. They agree with Ullman that subjects must have separated figure from ground, then made a decision about in or out. This contention derives from the Gestalt view that only figures have form, and thus an inside or an outside. Kienker and Scjnowski (1986, p. 198) go on to contend that the speed of processing Ullman found compared to time scales for neural processing and "suggests that figure–ground separation is computed in parallel over the visual field." Based on this evidence, figure–ground segregation is (or at least can be) a preattentive, bottom-up process.

There seems little doubt that figure *can* be found without input from higher level processes. Marr's information–processing approach suggests that the extraction of figure is one of the functions of early visual processing which, in the form of the 2½-D sketch, provides higher level processes with the input about object surfaces and orientations needed in order to recognize the objects. He demonstrated that, computationally, bounding edges and surfaces could be extracted from a scene with no input of prior knowledge (Marr, 1982).

The fact (if it is one) that figure–ground segregation can occur preattentively with no input from higher level processes, of course, does not prove that attention and top-down processing cannot sometimes play a role in figure–ground segregation. The ability of most people to consciously control figure–ground reversal when viewing Rubin's vase–face or other similar figures attests to this. In relation to ambiguous figures, Tsal and Kolbert (1985) have demonstrated empirically that figure–ground segregation is affected by attention.

Another related piece of evidence concerning the possible influence of top-down processes on figure–ground segregation comes from research by Peterson et al. (1991). She and her colleagues began by addressing the potential role of object recognition in figure–ground reversals of ambiguous figures. They had subjects view inverted and upright versions of Rubin's-like ambiguous figures. Their test figures were designed so that one orientation (upright) had highly denotative surrounds (viewers agreed on a specific shape represented) while an inverted version was not denotative (they were not recognized as a particular object) (Figure 3.60). Subjects continually reported which of the two regions appeared as figure during 30-second trials. They found that surrounds were more easily held as figure when they were upright (when the surround orientation was seen as denoting a particular identifiable object). Convergent evidence from four experiments led to the contention that "figure–ground reversal computations weigh inputs reflecting the goodness of fit between stimulus regions and orientation-specific structural memory representations" (Peterson et al., 1991, p. 1086). This finding agrees with Marr's model of shape recognition that posits that perceptual descriptions of shape structure (at the 2½-D sketch level) are matched to the best fitting structural

FIGURE 3.60. A sample pair of upright and inverted Rubin-like test stimuli. The surround on the left denotes a woman while that on the right has no clear denotation. *Reproduced from Peterson et al. (1991, Fig. 2a, p. 1077). Copyright 1991 by the American Psychological Association. Reprinted by permission of the author.*

memory representation (defined as a memory representation that specifies parts of a shape and their relative locations with respect to a canonical orientation for the object). Peterson et al. also interpret their findings to demonstrate that orientation-independent shape representations had no influence on figure–ground reversals.

Although Peterson et al. (1991) support Marr's views on matches with structural memory representations, there is disagreement about whether this matching can occur before figures are isolated. Peterson et al. (1991) contend that their results suggest a mechanism by which object recognition may facilitate initial figure segregation, as well as figure reversals. As they point out, this view seems to create a paradox: How can experience with shapes influence figure–ground organization given that no shape description should exist until figure–ground organization is determined? Their hypothesis is that contours may be evaluated from both sides simultaneously before figure is determined. A parallel set of experiments by Peterson and Gibson (1991) support this contention. In this set of experiments, full and half-versions of figures with denotative surrounds were used and presented in both upright and inverted orientations. The figures created were designed to meet Gestalt principles for establishing the central area as figure in the full versions but to be ambiguous (according to Gestalt principles) for the half-versions. Their expectations were that (1) for the ambiguous half-figures, there would be an exposure duration at which the denotative region of upright versions would be chosen as figure more often (indicating that shape recognition input can facilitate figure–ground segregation when Gestalt variables are missing or ambiguous), and (2) for upright stimuli a dominance of Gestalt variables would lead to an initial identification of the center as figure; a lack of Gestalt variables would lead to initial identification of the surround as figure—for the upright cases; and if both Gestalt variables and shape recognition work together there will be equally many identifications of center and surround as initial figure. Interactions of display time, uprightness, and Gestalt goodness of the central shape were found. Evidence indicated that shape-recognition routines required about 150 milliseconds and that if Gestalt variables were not strong enough to have isolated figure from ground in this time that shape recognition played a role.

Both Marr (1982) and Gregory (1990) concede that cognition (top-down processing) can be employed to deal with ambiguous situations. Gregory contends that there would be an evolutionary advantage to a system that worked in parallel with preattentive perception coming up with a quick interpretation that is usually (but not always) correct and conceptual processes (sometimes) modifying that initial impression. Such a parallel system seems to be well supported by Peterson et al. (1991) and is the process behind the pattern identification model of cartographic visualization cited above (MacEachren and Ganter, 1990). Based on this model and ideas about grouping and figure–ground segregation detailed here, we can predict that manipulating any design variables that influence strength of contour or heterogeneity of regions will have a dramatic effect on the patterns noticed. This issue will be considered further in Chapter 8.

Visual Levels

Closely related to the issue of figure–ground segregation is the concept of *visual levels* in graphic illustrations. The theory behind visual levels is that a viewer of a graphic depiction can group sets of objects into common wholes that are seen as occupying different visual (or conceptual) planes. The concept is related to Bertin's (1967/1983) selective and associative principles in that viewers are believed to see different objects as sufficiently similar that they become a unit and different units as sufficiently different that they are visually segregated. Robinson (1960) may have been the first to discus the principles involved in creating visual levels on maps (although he did not use this term). To introduce novice map designers to how a map might be structured so that items will appear at "differing position on the scale of visual significance," Robinson (1960, p. 223) used an analogy to hierarchical outlines for organizing written essays. He then went on to discuss how contrasts of lines, shapes, colors, and value can be manipulated to achieve this structuring.

Michael Wood (1968) took one of the first systematic looks at the idea of visual levels applied to cartography. Wood's (1968, p. 61) objective was to derive principles that would allow a cartographer to place information "on an imaginary scale of distance planes." The "distance" between these planes (i.e., levels), according to Wood, should be based on the similarity of the data occupying them. The separation into planes is required to "provide for focused attention and a good 'gestalt.'" Wood interpreted the concept of visual planes directly in terms of depth perception (which allows humans to segregate a three-dimensional visual field into many depth planes). Wood's goal was to draw on the psychological literature to derive principles for simulating depth planes on two-dimensional maps. Building on research discussed by Vernon (1962), Gibson (1950), and others, Wood proposed ways in which graphic variables such as texture, hue, and value could be used to create depth cues on two-dimensional maps. Although he was able to develop some suggestions for creating visual planes on maps, Wood saw these suggestions as an interim solution to a question that required empirical research to answer more fully. In subsequent work, Wood (1972) specifically considered the application of several Gestalt principles of figure–ground segregation as they relate to separation of visual levels on maps. He cautions, however, that the map viewer's knowledge and assumptions "can easily reverse" the levels intended by the cartographer.

A number of cartographers followed Wood's lead in looking to psychological literature for ideas about how visual levels might be created on maps (e.g., Dent, 1970; Spiess, 1978; McCleary, 1981). Dent (1970) also looked to the graphic design literature, particularly Bowman's *Graphic Communication* (1968). Dent cites Bowman's concept that graphic depictions can have one of three categories of visual depth organization: planar, multiplane, or continuous. Dent then proposes that most maps should use a multiplane strategy in which information is organized into a small set of discrete visual planes. Following Bowman's lead, Dent suggests that techniques employing contrast, aerial perspective (Bowman's dissimilar focus), and overlay can be used to segregate two-dimensional map information into multiple visual levels. He even goes as far as to propose a formula for predicting how contrast and contour sharpness (an aerial perspective cue) interact to produce position in visual depth:

$$PVH_j = f(I_j, ES_j)$$

where PVH_j is the position in the visual hierarchy of object j, I_j is the intensity of object j, and ES_j is the edge "sharpness" of object j. This hypothesized formula was never empirically tested. Based on more recent evidence concerning brightness contrast cited above, at least one major flaw in the hypothesis seems apparent without testing. An understanding of visual organization suggests that perception of individual features on a map will not happen in isolation from other map elements. Perceptual representations are inferences based on processing relative intensities in the visual field. The intensity (i.e., brightness) of an object, therefore, is probably not related directly to prominence in the visual field. Contrast

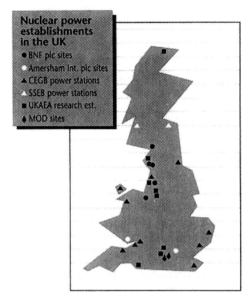

FIGURE 3.61. An example of a map with four (or more) visual levels: a base, areas *on* the base, point symbols *on* the areas, and a legend (which has its own levels) in "front" of all map elements.

in intensity of an object from its surroundings should, instead, be used. In spite of this flaw, the formula Dent offered is intuitively appealing and could be tested quite easily, but no one has yet done so.

Another untested hypothesis concerning visual levels on maps was proposed by Spiess (1978) and included in the ICA-sponsored text on *Basic Cartography* (Spiess, 1988). Spiess (1988) contends (as if it were a fact) that no more than three visual levels should be attempted on maps. This view challenges earlier cartographic proposals by both Wood (1968) and Dent (1970) who suggest that at least four visual levels are possible and Bowman's (1968) contention in relation to information graphics in general that continuous as well as multiplane organization is both possible and useful in some cases (Figure 3.61). Particularly in light of recent technological developments that allow binocular depth cues to be added to the cartographic tool kit, three visual levels for maps seems unduly restrictive. Whether we claim three, four, or more visual levels as a practical maximum, however, visual hierarchies can clearly exist within each level (e.g., a road crossing a stream). We could easily make a case (based on this kind of evidence) for ignoring visual levels altogether and treating visual hierarchy as a continuum. As considered in Part II, however, grouping categories of features into a small number of levels facilitates a semiotic approach to development of a map syntactics (see Chapter 6).

PERCEPTUAL CATEGORIZATION AND JUDGMENT

Underlying perceptual organization are processes of categorization. Hoffman (1989, p. 84) contends that "whatever else it does, the brain must be able to simplify and categorize the structures in the patterns it processes." At the most fundamental level, a bipartite categorization into same–different (i.e., discrimination) is required for early vision to isolate perceptual objects and organize them into groups. All Gestalt principles for grouping or figure–ground segregation are dependent upon heterogeneity of perceptual objects. If there are no differences between objects and background, we will see no objects. If differences are absent among objects, there will be no groups (Figure 3.62).

When vision discriminates between elements of the visual scene, it also generally orders those elements in some way (e.g., one is lighter than another, of coarser texture, longer, closer, etc.). This tendency to order is related to perceptual organization research devoted to visual attention. Items that appear higher on some ordered scale are likely to be more noticeable and thus are probably more often attended to. Eye movement research with maps, for example, has documented the intuitive notion that large map symbols attract more attention than small ones. In addition to ordering perceptual units, psychophysical tests suggest that the judgment of magnitude is also a preattentive process, at least for size and brightness.

The output of perceptual organization can also be the input for further categorization processes. Detection and discrimination among perceptual objects and groups as well as identification of order or relative magnitude are required if features of a visual scene (i.e., map symbols) are to be assigned to more specific categories. From the perspective of an in-

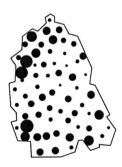

FIGURE 3.62. A map from research by Slocum for which subjects produced no consistent regionalization (i.e., groupings of circles). In this case, differences exist, but they are small. *Reproduced from Slocum (1983, Fig. 9, no. 13, p. 71). Adapted by permission of the American Congress on Surveying and Mapping.*

formation-processing approach to vision and visual cognition, then, detection, discrimination, and ordering on maps are important in relation to perceptual organization of map marks making up symbols, patterns, and regions and in relation to categorization of map symbols (an issue addressed in more detail in Chapter 4).

Cartographers have directed attention to the empirical examination of map symbol discriminability and have devoted considerable energy to a search for "laws" by which judgment of order and estimation of magnitude are performed. The driving force behind these efforts has been the map engineering goals of determining "least practical differences," making the order of ranked information intuitively obvious, and scaling symbols to match perceptions. These goals, first identified by Robinson (1952), have been approached from a communication perspective—for example, if a cartographer wishes to communicate that there is a categorical difference between two map elements, what physical difference is required to ensure that most people will notice it (and interpret it correctly). In spite of this limited perspective, past results can, to some extent, be put in a broader information-processing context where they might inform work in perceptual organization of maps as well as work dealing with symbol categorization, identification, and interpretation. No attempt is made here to provide a comprehensive review of cartographic research on discriminability, apparent order, or magnitude judgments. What is provided are a few key examples to illustrate how this research might be fit into a broader (thus potentially more useful) context. A sampling of recent ideas from the psychological literature is also described to help link the psychophysical approach taken in much of the cartographic research with the overall cartographic-representation perspective presented in this book.

Detection

Discrimination is the ability of vision to recognize a difference. Detection is, in essence, a discrimination problem in which a viewer must discriminate between some signal and the background on (or in) which that signal appears. For some purposes, however, it is useful to distinguish between detection (the ability to notice the presence of an object or feature) and discrimination.

For areas, there is at least one detection issue that has been of interest: detecting texture of area fill patterns. This is an issue because most area fills on maps to be printed (even if the final appearance is of a solid color) are made up of a textured pattern.[9]

One of the things that has become clear from both neurophysiological and psychophysical research is that the visual system is relatively insensitive to high frequency (fine) patterns. This allows us to create the impression of flat gray tones from patterns made up of fine dots. Castner and Robinson (1969) were among the first cartographers to investigate the perceptual thresholds involved. If patterns are coarser than about 40 lines (or dots) per inch, we see them as predominantly a pattern (at normal reading distance) (Figure 3.63). Patterns between 40 and 85 dots per inch are ambiguous (as with ambiguous figures in figure–ground research, these fills can be seen as either gray or textured, but not both at once). We can recognize the pattern easily, however, it does not dominate our impression, and we can see a value difference as well. Above 85 dots per inch we no longer notice the texture (unless we are consciously trying to see it—as you probably are now). We see these patterns generally as a color value or a gray tone. If a viewer is consciously trying to detect a texture, however, we have to go to almost 300 dots per inch before our visual system becomes incapable of detecting individual dots in foveal vision.

In contrast to Castner and Robinson's thorough examination of texture detection, cartographers have given little attention to questions of point or line symbol detection. Keates (1982) discusses the issues and suggests that detection will be a combined function of symbol size and contrast with the background it appears on (generally a contrast of color value or hue). Viewing distance is an additional factor that is often ignored because it is assumed to be normal reading distance (an incorrect assumption for route maps posted in public places, slide presentations, etc.). At normal reading distance with high-contrast symbols (e.g., black on white), size is not an issue because lines or symbols too small to be seen cannot be consistently printed (at least on normal porous paper). It is when color hue or value differences are used that detection of point and line symbols can be compromised. Although Keates (1982), Spiess (1988), and most authors of cartographic texts caution cartographers to use symbols with sufficient contrast to the backgrounds they appear on, no guidelines exist because little empirical testing has been done.

Noting the importance of low-level vision to any higher level pro-

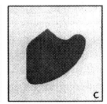

FIGURE 3.63. Map regions with 35 lines/inch dot fills (a), 65 lines/inch dot fills (b), and 133 lines/inch dot fills (c).

cessing of map information, Dobson (1985) advocates a concerted effort by cartographers to investigate questions of symbol "conspicuousness." Vision research offers a few hints about potential detection problems on maps, as well as about particular issues that deserve empirical research. The structure of our eye results in a rapid decrease in visual acuity from the fovea to the periphery. Detection will obviously be best for features (e.g., map symbols) that we are looking directly at. With map search tasks, however, in which the map user wants to find an occurrence of a particular feature, symbols must be detectable with peripheral vision if the search is not to be painfully slow. Engel (1977) has demonstrated that increased contrast can increase detectability in peripheral vision.

For maps, black symbols on white backgrounds are usually detectable. It is when color is added that detection problems become likely. Travis (1990), for example, points out that about 8% of men are congenitally red–green color deficient. For these map viewers, detection of symbols on backgrounds and discrimination between symbols can at times be impossible if this problem is not taken into account. Thus far, Olson (1989) seems to be the only cartographer to explore the issue of color deficiency empirically. Based on her research, she devised some guidelines for color choice that should limit hue detection and discrimination problems for the color deficient.[10]

Beyond issues of color deficiency, all humans have differences in acuity for different hues. Because our eyes have no blue cones in the fovea, our ability to detect blue map symbols is reduced over other hues. As Robinson (1967) noted, the traditional choice of blue for coastlines and depth contours is a poor one in situations for which quick discrimination of these features from their backgrounds is critical (e.g., navigation charts). For logical reasons, and some not so logical, cartography is a tradition-bound discipline. We are therefore unlikely to see a sudden change in color of depth contours or coastlines, in spite of empirical (and neurophysiological) evidence favoring such a change. In producing a detailed map of Georges Bank, for example, cartographers at Woods Hole Oceanographic Institution conducted extensive tests of detectability and discriminability of various point and line symbol colors on a range of background colors, but did not even bother to test anything other than blue for depth contours (Woods Hole Oceanographic Institute, 1982).

Humans are particularly sensitive to motion. We can detect motion of a few seconds of arc (Mowafy et al., 1990). This ability can be put to use in animated displays—if time is available. Similarly, humans are quite sensitive to aspatial change. As Travis (1990, p. 431) notes, "From neon signs in Las Vegas to the blue light atop an ambulance, flickering lights are used in our society to gain attention." On otherwise static maps, blinking symbols can be used in symbolic ways to highlight important features (MacEachren, 1994b). Examples of blinking symbols have been used by a number of animation authors (DiBiase et al., 1992; Monmonier, 1992).

Discrimination

Discrimination (in its usual sense of noticing a difference between two perceptual units rather than between one unit and its background) has been investigated by both psychologists and cartographers using two experimental paradigms, one based on same–different tasks for stimulus pairs, the other on visual search tasks for a target on a background of similar features. Visual search tasks are a less direct measure of discrimination in situations for which the target is both conceptually and visually different from the nontargets. When differences are strictly visual, however, search tasks provide a direct measure of discriminability (Uttal, 1988).

Using visual search methods, psychologists have uncovered a rather unexpected aspect of how vision discriminates. There appear to be natural categories (e.g., circles) from which vision will "notice" differences more readily. Discrimination seems to be asymmetrical. Within a dimension or feature class, some values appear more likely than others. Deviations from these standard values are more noticeable in relation to the standard than the reverse. For example, curved lines among straight lines are discriminated more quickly than straight among curved lines as are tilted lines among verticals, circles with gaps among whole circles, and ellipses among circles (Treisman et al., 1990).[11]

Text Discrimination

For maps, the same–different experimental method has been particularly useful in studies of place labels. Shortridge (1979) used this method to develop guidelines for the minimum point size difference necessary for text labels to be noticeably different (Figure 3.64).[12] In collaboration with Welch (Shortridge and Welch, 1980, 1982), she went on to investigate how both the experimental methodology and the features of town labels other than point size influenced discriminability. In the first follow-up study, they focused on discriminability of dot symbols of different size used to indicate town location. They found that larger size differences were required for discrimination if a simple same–different task was posed than if the task was for subjects to indicate the larger dot. In addition, if subjects were led to expect some dot pairs to be the same (as they would when viewing a typical map), the difference required for discrimination

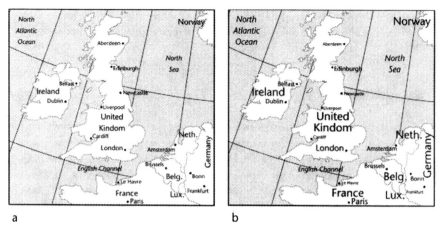

FIGURE 3.64. Based on Shortridge's results, the label sizes on map *b* will be distinguishable for about 75% of readers, while those on map *a* will not. How many sizes do you see?

was even larger. These results indicate that order can be perceived, even in cases for which discrimination is marginal or uncertain, and that expectations have an effect on discrimination—an indication that discrimination is not an entirely bottom-up preattentive process, as is often contended.

In their second follow-up experiment, Shortridge and Welch (1982) examined the issue of whether multiple feature differences between stimuli increase their discriminability. They measured discriminability of place labels distinguished on the basis of point size, type boldness, type case, and location-dot size, individually and in all possible combinations. They found that feature combinations increased discriminability up to three features, but a fourth had no impact. Interestingly, the discriminability of feature combinations was not predictable on the basis of their independent discriminability. Type size and boldness, for example, was a more discriminable combination than type size and case, in spite of the fact that case was considerably more discriminable by itself than boldness (which was the least discriminable feature). This lack of additivity suggests a holistic level of processing for these feature conjunctions.

Point Feature Discrimination

Discriminability of text could be considered a special category of discrimination of point features on maps. A number of psychologists have looked at the interaction between point feature discriminability and visual search for those point features. Quinlan and Humphreys (1987), for example, compared search tasks in which subjects searched for a single-feature target, two different single-feature targets, and a conjunction target combining the two features. Their evidence demonstrated that the rate of conjunction searches is influenced by discriminability of features, but the kind of search process used is not. Regardless of how discriminable the features, they found conjunction searches to proceed in serial fashion, while single-feature searches were executed in parallel when symbols were sufficiently different. Treisman (1988), however, offers some evidence that when two highly discriminable sets of distractors are present, attention can be directed to subgroups of items as a whole and, using inhibition of feature categories that cannot be the conjunction sought, can limit the search in such a way that a parallel process can be used.

There have been several studies that have dealt more directly with discriminability of positional symbols for maps. Most of these studies have measured "confusability" of symbols—a measure that does not separate discrimination from identification (assigning a label to symbols). An example of this kind of study is Johnson's (1983) empirical evaluation of the National Park Service point "symbol" set. He had subjects match symbols with labels (both with and without a legend present). A confusability index of sorts was devised based on the number of misidentifications for each symbol (Figure 3.65). Subjects were then presented with a visual search task in which the number of correct identifications in a limited time was determined. Those symbols that were highly confusable and were judged to be so because they look alike (rather than because they refer to similar things) also rated low on the visual search task (e.g., the

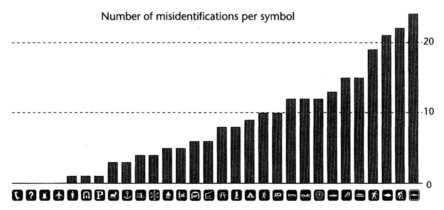

FIGURE 3.65. The part of the National Park Service symbol set that was tested, with number of misidentifications per symbol indicated. *Derived from Johnson (1983, Fig. 23, p. 112).*

lighthouse and the service station symbols). In a similar study with four alternative sets of symbols for tourist maps, Forrest and Castner (1985) found that iconic symbols took longer to locate than abstract symbols, but that fewer identification errors resulted. In addition, they found that the advantages of iconicity (for identification) and of simple abstract shapes (for visual search) could be combined by bounding iconic symbols with geometric frames (triangles, circles, and squares).

Pattern Discrimination

Discriminability of point features is probably most critical in situations where visual search is demanded. In contrast, discriminability of map patterns is probably most important when map readers are faced with general tasks related to pattern analysis (e.g., identification of homogeneous regions).

Based on Gestalt grouping principles, we would expect similar elements in close proximity to group and be seen as a whole. In order for patterns to be discriminable, then, grouping must occur for subsections of the visual scene. Patterns in which one feature (i.e., visual variable) is altered in an obvious way are usually quite discriminable, but patterns differing on a conjunction of features or rearrangement of subcomponents of elements composing them are not (Beck, 1966; Treisman, 1985). For patterns made up of coarse textures of the type used to depict qualitative data on maps, there seems to be a good understanding. Because of their probable importance in early vision (leading to Marr's primal sketch) considerable attention has been directed to how texture boundaries are identified and to algorithmic approaches to solving the same problem in computer vision. Malik and Perona (1990) compared the results of visual search measures of pattern element discriminability with a computational approach to texture discrimination and found nearly perfect correspondence (Figure 3.66).

Julesz (1975) has developed perhaps the most complete theory of texture discrimination. He posits three levels of discrimination (Figure 3.67). First is discrimination on the basis of darkness (or percent area inked). Second is discrimination of the characteristics of pattern arrangement. Both of these processes operate at the global level of the whole pattern. At a third level is discrimination of local geometry. Area patterns for maps, then, could be expected to be most discriminable if differences exist at all three levels. Global-level differences should allow quick parallel processes to be used in discrimination. Patterns that differ only at the local geometry level should be rather difficult to discriminate and will

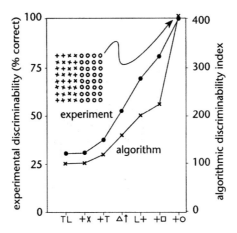

FIGURE 3.66. A comparison of texture discriminability determined by experiment and by computations. Texture pairs are those evaluated by both Kröse (1987) and Malik and Perona (1990), using experimental and computational procedures, respectively. The most discernible texture pair is shown as an inset on the graph and elements making up all texture pairs appear as labels on the X-axis. *Derived from Cleveland (1993, Table 1, p. 335).*

probably require serial processes in which patterns are closely examined one at a time.

Julesz (1981) has formalized the mathematical description of his three levels of texture differentiation and demonstrated that the preattentive visual system is unable to process statistical information beyond the second order. It is possible, however, to create patterns that are discriminable even though they have identical first- and second-order statistics. Discrimination in these cases involves local conspicuous features that Julesz calls "textons" (elongated "blobs" of specific widths, orientations, and aspect ratios). This research, which Julesz has linked to Marr's

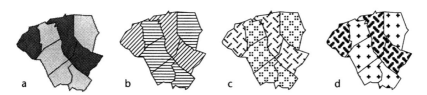

FIGURE 3.67. Discrimination of two map areas by value (a), by pattern arrangement (what cartographers would term orientation) (b), by local pattern geometry (c), and by the combination of all three (d).

primal sketch model, offers a sound basis from which patterns for use in interactive visualization might be devised (see Chapter 8 for discussion of one attempt to do just that).

Issues of texture discrimination have particular implications for design of area fills to be used in depicting qualitative information on maps. In this case, cartographic and perceptual logic suggests that patterns should avoid the use of differences in percent area inked because these will be seen as ordered. Without percent area inked, however, only pattern arrangement and local geometry are available.

Color Discrimination

In relation to color discrimination, Luria et al. (1986) point out that while color hue discriminability is "truly astronomical," the discriminable number of colors drops rapidly as their number in the scene goes up. These authors cite results of 98% correct discrimination among 10 colors, dropping to 72% for 17 colors. In spite of the fact that color discriminability may be orders of magnitude more limited than simple same–different experiments have suggested, there is considerable evidence that for visual search tasks, symbols that differ from others by hue are much more discriminable than those differing by either shape or size (Williams, 1967). In addition, for symbol conjunctions of color and shape or color and size, color seems to act as the dominant cue (Eriksen, 1952). Williams (1967) has provided evidence from eye movement studies that for these conjunction searches, subjects fixate on targets of the specified color to determine whether they are the correct size or shape rather than fixating on targets of a specified size or shape to check their color. Quinlan and Humphreys (1987), in a visual search experiment involving conjunction targets, came to a similar conclusion.

In graphic applications, Lewandowsky and Spence (1989) have demonstrated that discrimination of different variables on a multivariate scatterplot is higher for point symbols of three different colors than for three different geometric shapes or three different letters. When subjects were asked to estimate the correlation of variable pairs on the graphs, this difference in discriminability resulted in novices having more accurate correlation estimates with scatterplots using color than experts had for scatterplots using three letters (that were not individually very discriminable).

In relation to maps, Forrest and Castner (1985) cite an unpublished study by DeBrailes confirming the dominance of color in discrimination of point symbols for visual search tasks on maps. Although Forrest and Castner, along with other cartographers, argue that varying hue of point symbols will be a particularly good idea on maps where visual search is expected (e.g., travel maps, navigation charts, etc.), none have considered the fact that there is a pronounced male–female difference in color acuity, and that for both males and females that acuity drops with age.[13]

Neurophysiological evidence suggests that discrimination on the basis of color or luminance contrast is a lower level visual process than is estimation of luminance. Shapley et al. (1990) argue that early vision computes contrast (not reflectance as Land and McCann, 1971, had predicted). Experimentation with cats (a species whose early visual processes are considered quite similar to those of humans) has indicated that "the response of retinal, lateral geniculate and some primary cortical neurons is proportional to contrast over a low-to-medium contrast range, and then may saturate at high contrast" (Shapley et al., 1990, p. 435). This emphasis of early vision on contrast rather than reflectance helps explain phenomenon such as color constancy (that we see a color as the same under various lighting conditions), simultaneous contrast (that perception of a color or shade will change due to the background it is on), and assimilation (the additive effect on brightness of an object produced by the brightness of its background). The emphasis on contrast over reflectance may also relate to the important role contrast appears to have in segregation of figure from ground and the equivocal results that have been obtained when attempts are made to determine whether light or dark areas are most likely to be seen as figures (see discussion of figure–ground above).

Motion Discrimination

As we would predict based on the idea of indispensable variables discussed earlier in this chapter, human vision is very sensitive to motion (for which both location and time are changing). In relation to motion, the Gestalt principles suggest that the common fate of objects moving together will allow a viewer to visually group those objects and discriminate them from their background. Just as we can visually group objects moving together, however, evidence suggests that humans are very sensitive to constancy of spatial distance between moving edges (Mowafy et al., 1990). Their results indicate that discrimination between coherent and uncorrelated motion can be achieved with similar levels of accuracy to the ability for detecting any motion at all (i.e., changes of a few seconds of arc). Mowafy et al. (1990, p. 591) contend that "processing relative movements in the environment is a fundamental characteristic of human motion perception." Therefore the evidence for this ability to discriminate coherent motion has good evolutionary support. Ability to detect motion and discriminate between coherent and noncoherent motion has obvious implications for the design of animated maps. For example, on a map depicting flows, we might expect viewers to be attracted to even small deviations (in speed or direction) of a single moving arrow from the flow of a group. No cartographic research has been directed to this or related issues.

Judging Order

According to Shapley et al. (1990), one of several fundamental "facts" of perception is that if objects or patterns can be discriminated, we can usually also assign an order. Another "fact" (that they admit they have little scientific evidence for) is that there are "natural continua" along which discriminations are easy (e.g., larger–smaller, left–right), and that we can expect discrimination to be hard along non-natural continua (e.g., alphabetical order).[14]

For maps and other graphics, Bertin (1967/1983) contends that humans find inherent order in spatial location, size, color value, and texture. DiBiase et al. (1992) point out that time is inherently ordered as well, and that for animation temporal order provides one of three dynamic variables, all of which are intuitively ordered (the other two being duration and rate of change).[15] In *Some Truth with Maps* (MacEachren, 1994a), I suggest that color saturation and focus are ordered graphic variables that Bertin omitted from consideration and that color hue and orientation are marginally ordered. Few of these contentions, however, have been examined experimentally.

Most of the attention to whether various graphic variables used on maps are intuitively ordered has been directed to color hue, color saturation, and color value for area fills. The cartographic goal of the research has been to determine color sequences appropriate for choropleth and other quantitative maps. Experimental tasks used in this research have not allowed preattentive processes to be segregated from attentive ones. Whether the process of seeing the order of various hue-value sets is a logical–cognitive one or a purely visual one, therefore, cannot be determined from the evidence. Results do support the contention that color value and color saturation are ordered and that hue is not (Cuff, 1973; Gilmartin, 1988). In contrast to Cuff's empirical results, the phenomenon of advance and retreat is expected to cause red to appear to be located on a visual plane in front of blue. Travis (1990) offers three physiological explanations for this effect: (1) that because of chromatic aberration the eye's lens causes short wavelengths to have a shorter focus than long wavelengths, (2) that the visual and optical axes of the eyes do not coincide, and (3) that the apparent brightness of light depends on the point of entry through the pupil. He contends that using saturated red and blue will make objects literally "stand out" from the display.

A recent study, by students of mine, seems to favor the advance and retreat hypothesis and contradict Cuff's findings (Bemis and Bates, 1989). For hypothetical temperature maps, subjects were found to consistently see an order in shaded isotherm maps using a bipolar range of colors (with blues at one end and reds at the other). They proposed an interesting explanation for the contradiction between their results and Cuff's. Since 1973 when Cuff collected his data, a blue–red range has become much more common for temperature maps (e.g., on television news and in many newspapers). Bemis and Bates (1989) contend that the logic of the order has been learned, an explanation that is intuitively appealing.[16] A further test of this hypothesis would be to assess the relative order seen in value versus spectral series using reaction-time methods. If an identification of order for color value is a preattentive process while identification of order in a color hue sequence is a cognitive process, reaction times to judge "higher" values should be faster than reaction times to judge "higher" hues, and the presentation time threshold at which order can first be judged should be much less for value than for hue.

Although several studies have found value ranges to be judged as ordered (with dark values usually seen as the high end of the scale), this ordering is not perfect even for simple maps. McGranaghan (1989), for example, had subjects judge which of two states on a map of the western United States had the higher data value. While the darker of the states was selected as "more" in a majority of cases, only 30% of the subjects consistently saw darker as "more." Most of the inconsistent ordering occurred when the map's background was gray or black rather than white, with the black background resulting in about twice as many intransitivities as the white and the gray resulting in about four times as many.

Judging Relative Magnitude

Early perceptual research in cartography was devoted almost exclusively to attempts at deriving functional relationships between physical magnitude of different aspects of map symbols and psychological magnitude. McCleary (1970) reviewed this research and suggested that the one generalization that seemed to apply across the board was that map readers underestimate differences between map symbols. The precise functional relationship seemed, at first, to depend primarily upon the particular stimuli being tested and the questions asked (e.g., Olson, 1976). It gradually became clear that there is also substantial individual (subject) variation in magnitude judgments (McCleary, 1975; Griffin, 1985) and that map context created further problems (Gilmartin, 1981).

Based on the extensive testing Flannery did in 1956 and repeated in 1972, Robinson et al, (1984) and other authors of cartographic texts adopted the guideline of adjusting scaling on graduated circle maps to account for underestimation of differences. Others (e.g., Cox, 1976) have suggested that we might be better off simply providing more anchors (in the form of legend circles). The among-experiment and among-subject variations together with context effects have made many practitioners suspicious of the empirically derived guidelines for perceptual scaling, and today it is doubtful whether many cartographers actually use them.

In relation to gray tones for quantitative maps, there seems to be a bit more consensus. Kimerling (1985) was able to demonstrate a correspondence among what were apparently divergent results and showed that usable gray scales could be devised. The two most significant issues he considers are the interaction between area fill texture and perception of value and the interaction between judgment task and value perception. In terms of texture, Kimerling found that the finer the texture, the more curvilinear the relationship between perceived and actual gray tone. The implications of this finding are that a different set of gray tones is required for maximum discriminability if a map is produced on a laser printer (with dots spaced at about 60 lines per inch) versus on a film recorder (with dots at 100 or 120 lines per inch) (Figure 3.68). Judgment was also found to be dependent upon the visual task, with a different actual–perceived gray tone function for judgment of percent black versus a partitioning task or tasks leading to a set of maximally discriminable gray tones.[17]

PERCEIVING DEPTH FROM A TWO-DIMENSIONAL SCENE

Closely related to concepts of judging order and magnitude (as well as to the visual levels discussed above) is the simulation of depth in two-di-

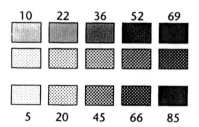

FIGURE 3.68. A gray scale designed for maximum between-category contrast with production on a 100 lines/inch image setter (top) compared to the same grays produced at 45 lines/inch resolution (typical of laser printers) (middle), and to grays adjusted in color value to achieve maximum contrast at the coarser resolution (bottom).

mensional displays.[18] Vision is designed to deal with a three-dimensional world. Interpreting depth in a visual scene is a complex process that appears to be facilitated by a large number of interdependent cues. For good evolutionary reasons, vision does not require all depth cues to be present in order to interpret features of a scene as being at varied distance from the observer. This makes it possible to trick vision into interpreting a map or other display as three-dimensional by combining some of these cues in appropriate ways. How vision interprets depth cues is relevant to cartography because cartographers are often faced with the problem of simulating three-dimensional information on a two-dimensional display (when depicting terrain, but also for more abstract multivariate information).

A Taxonomy of Depth Cues

Kraak (1988) provides a taxonomy of cues for depth perception and a review of the cartographic literature relevant to each. His taxonomy distinguishes between "physiological" and "psychological" depth cues. Some authors have called the latter "pictorial." Since this latter term puts emphasis on characteristics of the display rather than of the cognitive processing of that display, it is adopted here. The physiological depth cues have to do with the physical processes of vision as it reacts to the real three-dimensional environment. Pictorial cues, in contrast, are those related to the object's structure and the way that structure organizes visual input. In the context of computer graphics, Wanger et al., (1992) provide a list of depth cues similar to those cited by Kraak. Each of these sources includes pictorial cues (or subcategories of cues) omitted by the other and disagree on whether motion parallax should be considered a physiological or a pictorial cue (with Kraak opting for the former, and Wanger et al. for the latter). If we look to the art literature, we find additional depth cues not included in either the cartographic or the computer graphic taxonomies (along with some differences in terms for cues in common) (Metzger, 1992). A composite of these sources results in the following taxonomies of depth cues that may be relevant to maps:

> Physiological
>
> *Accommodation*: A change in thickness of the eye's lens as it focuses on an object.
> *Convergence*: The difference in angle of gaze by the two eyes focused on the same object.
> *Retinal disparity*: The difference in image (visual array) derived by each eye (which has a slightly different point of view).

Pictorial

Perspective: Kraak (1988) subdivides perspective into four components, and we will follow this subdivision here.

Oblique projection: Representation of a scene from a viewpoint that is not an elevation (profile) or plan view (overhead) suggests a three-dimensional solid, thus depth.

Linear perspective: Lines that are parallel in reality seem to converge with distance (e.g., a pair of railroad tracks).

Retinal image size: Objects appear smaller the farther away they are.

Texture gradient: Texture appears to decrease with distance.

Motion: Movement (actual or simulated) of the observer's point of observation produces changes in the relative retinal displacement of objects at different distances. Successive presentation of static images in which objects are displaced relative to one another can (particularly in the presence of other cues) also result in a sensation of depth.

Interposition: Using Gestalt principles of good continuation, vision will assume that whole objects juxtaposed with what appear to be part objects are really whole objects blocking our view of other whole objects farther away.

Shadow: A cue to obstruction or overlap, indicating that one object is blocking light from falling on another object.

Shading: Illumination gradient can indicate the shape and orientation of a surface.

Color:

Chromostereopsis (also called color stereoptic effect or, more commonly, advance-and-retreat): The differences in wavelength of colors are thought to result in apparent differences in distance (with reds appearing closer than blues at the same true distance).

Aerial perspective: With distance colors become less distinct (less saturated) and lighter (higher value), often with a bluish tint due to atmospheric scattering.

Detail: With distance detail becomes less visible and edges become blurred.

Reference frame: In order to judge relative size, vision must match retinal size to some frame of reference—apparent distance will therefore vary with what an object is compared to.

Not surprisingly, the bulk of cartographic attention to depth perception and how specific cues might prompt this perception is related to terrain mapping. Terrain is three-dimensional and cartographers have struggled with collapsing those three dimensions onto a two-dimensional page since the earliest maps were made. Although contour lines involve no depth cues, virtually all other methods of depicting relief rely on one or more of the cues listed above. Simulation of three dimensions on maps can be grouped into techniques that involve physiological cues, that rely on perspective, that use static nonperspective pictorial cues, and that include motion. For motion to cue depth, the user must assume a perspective view (but linear perspective is not essential). Since the possibility of motion as a depth cue requires a dynamic display, further discussion of these cues will be postponed until Chapter 8 in the context of geographic visualization environments (which, as they will be defined here, are dynamic).

Applying Depth Cues to Maps

Physiological Approaches

Computer technology has facilitated production of displays that make direct use of binocular parallax as the primary depth cue. Such displays consist of pairs of representations, usually perspective views, that depict the mapped area from slightly different points of view (simulating the different points of view resulting from the spacing of our eyes). Seeing depth in stereo pair maps usually requires that the observer's head does not change position while viewing, and/or that special glasses be worn. One technique, referred to as anaglyph plots, uses opponent colors of red and green to produce two overlapping views. When an observer wears glasses having one red and one green lens (if she has normal color vision) the two views will be separated with one seen by each eye. This technique was used for maps at least as early as 1970 in the Surface II package that could generate anaglyph fishnet maps.

Perspective Approaches

Included here are the four perspective cues of oblique projection, linear perspective, retinal size, and texture gradient. These cues are typically manipulated together on perspective view maps, with oblique projection common to all. Different representational techniques can put uneven emphasis on the remaining three perspective cues. The well-known fishnet plot (Figure 3.69), for example, emphasizes texture gradient. In contrast, layered contours (Figure 3.70) and block diagrams emphasize linear

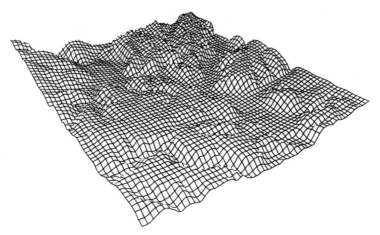

FIGURE 3.69. A typical fishnet plot depicting the terrain around Johnstown, Pennsylvania.

perspective and size disparity, and solid modeling (Figure 3.71) emphasizes linear perspective with shading and shadow as additional (nonperspective) depth cues. All of the methods mentioned make use of interposition as an additional cue (e.g., fishnet plots are rarely generated without hidden line removal). I have uncovered no empirical comparisons among the various styles of perspective map, but some attention has been given to perception of fishnet plots.

FIGURE 3.70. Layered contours applied to the same region as shown in Figure 3.69.

FIGURE 3.71. Solid rendering of the region from Figure 3.69.

That fishnet plots, with their strong texture gradient, do work was convincingly demonstrated by Rowles (1978). She found that subjects were able to judge relative height quite accurately, even when the point of view for the perspective was as high as 75° (nearly overhead) or as low as 15° (Figure 3.72). The view from 15°, however, results in considerable occlusion of map sections, something that probably helps to cue depth but can make the map much less useful (unless it can be dynamically oriented to allow hidden locations to be uncovered).

Nonperspective Approaches

Whether or not fishnet and other perspective view maps are effective, they all suffer from two problems. No matter what point of view is taken, there will be some hidden features and (if linear perspective is used) scale will change across the map. To avoid these issues, considerable attention has been given to use of nonperspective depth cues with the goal of an effective plan-view relief representation that suggests depth. Most cartographic attempts to create the illusion of depth without perspective use shading and/or color.

With shading, there is a long history of manual techniques using pencil, airbrush, and other tools. The procedures for what is termed "plastic relief"[19] borrow from principles of light and shadow in art and psychological principles of depth perception, but to be effective must also incorporate considerable knowledge of geomorphic structure of the terrain

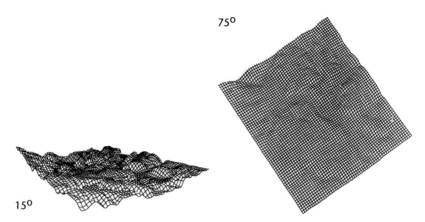

FIGURE 3.72. The Johnstown fishnet terrain map shown at elevations of 15° and 75°, viewpoints for which Rowles (1978) found no significant decrease in ability to estimate height. In Rowles's examples, however, relative relief was much greater.

being represented. Imhof (1965/1982) provides the most comprehensive account of these methods. One of the things that has been learned (primarily from long experience and years of marginal success at computerizing the process rather than from empirical research) is that perception is sensitive to what might be called the texture of shading as well as to its value. Humans can immediately recognize the difference between a perfect match of shading with slope–aspect values and shading that looks real. Not only do real surfaces not reflect light as perfectly as a virtual computer surface can, the real environment has complex interactions of direct with reflected light that our visual system has evolved to expect. For terrain shading to look real, it must incorporate at least some of the subtle variation from perfect reflectance that occurs in the real environment. Many theories have been proffered for the ideal reflectance model, but little empirical research has been done to determine their relative merits. In spite of the lack of empirical research, plastic shading has developed to the point in cartography that it has been successfully modeled with computer software (Figure 3.73). Perhaps the most impressive result thus far is Pike and Thelin's (1989) digital relief map of the United States, described by Lewis (1992) as a "cartographic masterpiece."

One issue that all disciplines interested in shading as a depth cue seem to agree upon is that the simulated light source needs to be from above the scene, and above-left is usually cited as best. This phenomena seems to be based upon a schema (or expectation) that light in the environment is from above. When applied to art (e.g., in the representation of a vase of flowers on a table or a figure in repose) this light-from-above rule is quite logical. On a map, the rule results in light from the north-

FIGURE 3.73. The Johnstown terrain map produced with computer-generated plastic shading (using ArcInfo).

west, a direction that is at odds with reality in the northern hemisphere. In spite of the physical impossibility of the scene, humans consistently treat terrain shading on maps in the same way that they treat shading on a painting. This reaction is so strong that a map produced with terrain illuminated from the south will appear inverted, with the hills looking like valleys and the valleys like hills.

In an effort to represent terrain aspect information clearly while also creating effective relief shading, Moellering and Kimerling (1990) developed a unique color-rendering process that has subsequently been labeled MKS-ASPECT™ (Moellering, 1993). They started with the assumption that aspect is a nominal (qualitative) phenomenon for which color hue differences provide a suitable representation.[20] They set out to devise a color-matching system that would allow observers to visually separate terrain regions with different aspects while also providing appropriate depth cues leading to interpretation as a three-dimensional surface. The system relies heavily on OPT (described above in the discussion of eye and brain). As noted above, OPT predicts four unique hues from which all others are derived. It also predicts that certain hue combinations are not possible: those across the diagonals of the square color space (red–green or blue–yellow). The four unique hues are considered to be the maximally discriminable hues (when at maximum saturation and medium lightness). One guideline that Moellering and Kimerling arrive at from OPT is that aspect should be grouped into four, eight, sixteen, and so on, classes using the four unique hues or these four plus their first order combinations, second order combinations, and the like. They argue that the resulting hues (for eight or more classes) should be seen as a circular progression of related colors.

Moellering and Kimerling (1990) had the primary goal of depicting aspect classes clearly. Initially they matched the four unique hues with cardinal directions. Although a discriminable map was obtained, the resulting representation prompted a number of inversions of features (e.g., ridges seen as valleys). Their technique (unlike true relief shading) does not take into account a light source or reflectance due to that light source. The impression of relief obtained is due entirely to slope aspect. Moellering and Kimerling were able to achieve a reasonable impression of depth by rotating the unique colors so that yellow (the highest value color) was aligned with the standard light source azimuth (315° or northwest), and the value of all other hues was adjusted to match the deviation of azimuth from northwest.[21] It is claimed that the MKS-ASPECT™ system eliminates one of the most severe problems with standard gray tone relief shading: that the visual interpretation of the scene will be highly dependent upon the exact angle of illumination for the hypothetical light source (Moellering, 1993). By not relying on color value alone, identification of ridge lines or valleys is not as dependent upon how their alignment matches with that of the illumination. No empirical test of Moellering's claims has yet been undertaken.

Recently Brewer (1993) has developed an alternative color scheme for mapping slope and aspect in conjunction. This scheme uses a hue range to represent aspect, with yellow as the anchor hue aligned with northwest. Other hues were selected so that a value progression was achieved in each direction from yellow, and each of the eight distinct aspect categories would have a sufficient saturation range for three saturation steps (plus unsaturated gray) to be discernable. Slope categories were depicted with these saturation steps; the higher the saturation, the steeper the slope. Its main advantage over Moellering and Kimerling's MKS-ASPECT™ system is that Brewer's color scheme results in a much more effective depiction of the terrain form, while still providing easily interpreted aspect information and adding three categories of slope.

An alternative use of color hue as a depth cue in terrain representation is found in Eyton's (1990) application of color chromostereopsis. As noted in the introduction to Part I, the idea of chromostereopsis (or more commonly advance-and-retreat) can be traced cartographically at least to Karl Peucker in 1898. The theory seems to have found at least partial support in research spanning the intervening decades (e.g., Eyton cites German publications on the topic as early as 1868 as well as Luckiesh, 1918; Kishto, 1965; etc.). A variety of explanations for the processes involved have been offered (see discussion of judging order above). It is uncertain, however, whether any standard layer tinting used on maps actually produces the effect (because there seem to have been no empirical cartographic tests). One problem, identified by Eyton (1990), in applying chromostereopsis to most paper maps is that the halftone processes of four-color lithographic printing will interfere with the effect. This interference is due to the fact that color appearance on lithographically printed maps results from the combination of overprinted inks and visual combination of adjacent dots.

Eyton (1990) experimented with several methods of producing the chromostereopic effect. He achieved limited success when a set of spectrally ordered colors were used on a layer tint map in which contours were created by adjacency of different colors. When he added black contours, the result (viewed as a color transparency) was said to have a "quite apparent" effect. Only an informal evaluation is offered, however, with 21 of 23 students in a cartography class claiming to see the effect. A map with black contours at double the contour interval was found to produce a weaker effect, leading Eyton to conclude that contour interval controlled the degree of depth seen. To explain the impact of the black contours, Eyton (1990, p. 23) argues that "the contour lines helped to create a rounding of the terrain form. Without contour lines the colors floated in planes; with the contour lines the display took on the appearance of a plastic surface with smooth, rounded features." An even more dramatic effect is cited for a continuous tone version of the map (in place of layer tinting). On this map, contour intervals of 200, 100, and 50 feet were compared, with the 50-foot interval producing the most dramatic effect. Eyton suggests that an explanation for the impact of contour lines on the perception of depth might be found in the fact that contour spacing is a cue to steepness of slope. This contour-enhancing effect remains to be empirically evaluated.

According to Eyton, the main problems involved in successful printing of maps using the chromostereopic effect are that standard printing changes the relative brightness of various hues (and brightness or color value interacts with the effect) and printed colors (particularly when inks are overprint or dithered) lack spectral purity. A solution proposed is to use fluorescent inks at full saturation with no overprinting. Fluorescent inks give the appearance of reflecting more light than is incident on the page. Again, Eyton provides anecdotal evidence, indicating that few students saw a fluorescent ink map as having depth, but the same map with black contours appeared to be three-dimensional to almost all the students. A final variation on the maps was obtained by adding hill shading (in gray) to the fluorescent ink maps. The added cue seemed to aid the perception of depth, but again the effect was strongest when contours were included as well.

A final depth-cue technique for static plan view maps worth noting is the "Tanaka method." Tanaka (1932) made use of shadow rather than shading (or color) to produce a sensation of depth in contour maps. The basis of the technique is to treat contours as if they represent a three-dimensional "layer-cake" model of terrain (which in a perspective view would result in a layered contour depiction of the sort discussed in Crawford and Marks, 1973). Tanaka's technique simulated the appearance of a layered contour map by putting white and black contours on a gray background. Contours toward the light source are in white with those away from the light source in black (Figure 3.74). Width of contours "varies

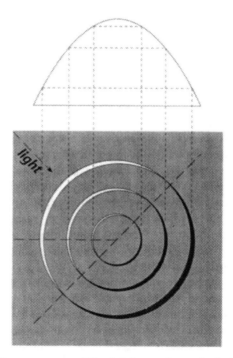

FIGURE 3.74. Representation of Tanaka's layer-cake method of terrain depiction. *After Japan Cartographers Association (1980, Fig. 4, p. 162).*

with the cosine of the angle θ between the horizontal direction of the incident ray and the normal to the contour at the point under consideration" (Japan Cartographers Association, 1980, pp. 162–163). Although the ability of Tanaka's method to provide a 3-D appearance is clear from examining a color version of a map using the Tanaka method (Japan Cartographers Association, 1980, f. 162), no empirical evaluation of the method exists nor any empirically derived guidelines on appropriate maximum widths for the variable contours.

SUMMARY

The goal of this chapter has been to provide an overview of a range of issues relevant to visual processing of maps. Perspectives from neurophysiology, psychology, cognitive science, human factors engineering, and cartography are woven together in an effort to build an understanding of how maps are seen that can serve as a framework for research on and guidelines for map symbolization and design. It is only by understanding what vision is for and its limits that we can hope to comprehend the complex process involved in "seeing" a map.

Vision has been treated as a complex information-processing system that generates a succession of "representations." At the lowest level are representations of the visual scene on the retina of the eye. These are processed by our neurological hardware through a series of stages leading toward an organization of input into a coherent description of the visual scene in a form that can be interrogated by higher level cognitive processes.

After briefly reviewing the neurophysiological hardware issues and the limits that they place on map displays, the bulk of the chapter emphasized perceptual organization and perceptual categorization and judgment—two areas that have received considerable attention in the cartographic literature of the past four decades. In terms of perceptual organization of map information, the topics of perceptual grouping, attention, visual search, and figure–ground are emphasized. Psychological, cartographic, and other research on these topics is considered in relation to Bertin's contentions about the fundamental graphic variables available for creating map symbols. Although a great deal of cartographic research of the 1960s and 1970s dealt with issues of magnitude estimation, it is now clear that other aspects of perception are more relevant to map design. In the section on perceptual categorization and judgment, the emphasis, therefore, was placed on what we know about discrimination of symbols and patterns and about the propensity of the visual system to distinguish differences in kind and differences in order, two topics that seem particularly relevant to design of interactive visualization tools. Finally, the chapter concludes with a section devoted to the simulation of three dimensions on flat two-dimensional maps. This is an area of cartography with a long history, but one in which the integration of psychological principles and cartographic practice has been minimally addressed. Again, the topic of simulating the third dimension has become more important than ever in the context of visualization.

The perceptual emphasis in this chapter sets the stage for discussion of how knowledge is linked to perceptual input in the interpretation and use of maps. We pick up this thread in the next chapter, where the emphasis is on cognitive processes of mental categorization and spatial knowledge representation. These topics are critical to design of interactive visualization tools intended to facilitate visual thinking as well as to the formalization of spatial knowledge that will be required by expert systems for map generalization, symbolization, and design.

NOTES

1. A fixation is a brief focus on a small section of the visual field where the item at the center of the fixation is "seen" by the foveal area of the retina and the cells connected to it.

2. The sections below are ordered because language requires order. The processes, however, are interdependent and are as likely to occur simultaneously as in the order presented or any other order. Some detection differences, for example, are necessary for grouping and attention. On the other hand, some grouping is required for objects to be isolated for discrimination and some attention to particular locations is needed to note differences between these locations.

3. Although Slocum did not specifically mention Gestalt psychology, he provides standard examples of the laws of similarity, proximity, and good continuation and directs the reader's attention to Arnheim (1974) and Woodworth (1938) as sources.

4. Eastman never cites the graphic organization variables he considered as Gestalt principles. To put his study in the context of the present discussion, however, I have made this link to Gestalt theory explicit in my review of Eastman's research.

5. This finding supports OPT, which would predict that wavelength averaging can happen for red–yellow and green–blue but cannot happen for red–green and blue–yellow mixtures. These mixtures are not possible (according to OPT) because there is no neurological mechanism for mixing across the diagonals of opponent color space.

6. The term "divided attention" is potentially misleading. Pomerantz and Schwaitzberg (1975) use it as an antonym for selective attention. Attention is not really divided (in time) between parts, as the term might imply, but is directed to the whole created by relationship of the parts. Thus "holistic" attention might be a more appropriate term.

7. Each of these hypotheses has implications for the development of visual-

ization tools designed to help analysts notice unexpected spatial patterns. These possibilities are considered in Chapter 8.

8. See Chapter 6 for details of focus as a graphic variable and Chapter 10 for consideration of uncertainty representation.

9. This use of textured patterns rather than solid fills is a function of printing technology for which large textured areas pose fewer printing problems than solids. In addition, textured fills allow the use of screens to obtain a range of values for a particular printing ink, a process that is much more efficient that using one ink for each value.

10. Aspects of Olson's (1989) research on color deficiency are described in Olson (1994).

11. It is this human propensity to notice differences from these "natural" categories that was the bane of student cartographers in the precomputer era. Cartography instructors seem to be particularly adept at noticing deviation from these natural categories, a possible indication that these "natural" categories are linked to experience (and that they can be taught).

12. For those who were not sure, there were four sizes on each map using point sizes of 6, 8, 10, and 12 points for map *a*, and 6, 8, 11, and 15 points for map *b*.

13. Most empirical research by both cartographers and psychologists has been done with college student subjects. Results related to color discrimination for those subjects may not extend to the broader (older) population.

14. As discussed in the next chapter, these natural continua may be related to what Lakoff (1987) terms "kinesthetic image schemata." These are embodied relationships that function at a preconceptual level and that we seem able to metaphorically map into abstract domains. For example, verticality or up–down of our bodies serves as an image schema to which quantity (e.g., ordered data such as temperatures on an isotherm map) is mapped. This results in a natural continuum of more to less.

15. In Chapter 6, I present an extention to this set of dynamic variables.

16. In terms of ideas presented in the next chapter, I would suggest that a new schema for interpreting colors on layer tint isoline maps has been developed.

17. As discussed in Chapter 2, one potential flaw in the gray tone research Kimerling (1985) discusses is that none of it has tested the guidelines in actual maps.

18. Since many of the most effective techiniques for simulating 3-D on maps involve color and/or motion (neither of which was available here), illustrations are used sparingly in this section. The reader is encouraged to consult Kraak (1988), Wanger et al. (1992), and Raper (1989), as well as the March 1993 special issue of *IEEE Computer Graphics and Applications* for visual examples of a variety of 3-D simulation techniques.

19. Plastic is used here in the sense of a *process* of molding or modeling— *not a material* from which things are made.

20. See Chapter 6 for a discussion of the map syntactics that lead to this matching of hue with qualitative differences.

21. This adjustment turned out to be a complex problem, but it is beyond the scope of our discussion here. Interested readers should refer to Moellering and Kimerling (1990) for details.

CHAPTER FOUR

How Maps Are Understood
VISUAL ARRAY → VISUAL DESCRIPTION
↔ KNOWLEDGE SCHEMATA
↔ COGNITIVE REPRESENTATION

To examine the next major stage in human–map interaction, how maps are understood, we must consider how visual descriptions of map displays (that result from how maps are seen) interact with existing knowledge and the form that knowledge takes. The perspective taken here is that existing knowledge, in the form of *propositional, analog,* and *procedural representations,* is brought to bear on the interpretation of visual scenes through knowledge schemata that serve as an interface between visual descriptions and knowledge representations. These schemata act to structure what we know about objects, concepts, relationships, and processes in the world. As a result, they also structure what we see by making certain groupings, categorizations, patterns, and so on, more likely than others. In a complementary fashion what we *see* (in the sense outlined in Chapter 3) provides input to generalized schemata. These complementary processes allow us to make sense out of the visual scene.

In order to understand the complex interactions among the visual array resulting from perception of a map, the visual description that provides structure to what is *seen* when studying the map, and the knowledge schemata that provide an interface with previous knowledge and a structure for conceptualizing about the input, we must consider the forms that knowledge schemata might take. There seem to be three possibilities that are probably used to some extent by all humans. These three are propositional, image, and event schemata. Each will be described in detail below.

Underlying our human ability to form and make use of knowledge schemata, of whatever form, is the human propensity to categorize. How we mentally categorize (what things are grouped together through what kinds of relationships) determines the elements and kinds of relationships available to knowledge schemata. Most entities in knowledge schemata are categories to which we "map" individuals that are encountered. Thus in order to understand how knowledge schemata link information acquired by visual (or other) senses to stored knowledge, we must first consider the nature of conceptual categories.

MENTAL CATEGORIES

"Without the ability to categorize, we could not function at all" (Lakoff, 1987, p. 6). If we accept this premise, the logical cartographic corollary is: Without categorization, maps would not be possible. Like Lewis Carroll's story of a map produced at larger and larger scale until it was the same scale as the environment (quoted in Muehrcke and Muehrcke, 1978, p. 22), a map without any categorization would be more cumbersome than using the environment as its own map. Maps depict categories rather than individuals.[1] As such, to understand how maps work, we must understand categorization. To make maps that work, we must depict categories using methods that match the structures of human mental categorization. Human mental categorization processes are, of course, relevant to map interpretation as well as to map creation. Maps depict information that has the potential to be categorized by viewers in a variety of ways that relate to their goals, interests, backgrounds, and so on. How stimuli in a visual scene are grouped into categories can be dependent upon relationships among the complete set of stimuli (see Handel et al., 1980).

The classical approach to mental categories assumes that categories exist in the world and that we can discover them. These natural categories are taken to be clearly defined and mutually exclusive. Cartographers have accepted this proposition (at least in part) and assumed that there are optimal (or correct) categories to be uncovered in any mappable information. Following from this perspective on categories, cartographers (and their geographic colleagues in climatology, soil science, etc.) have devoted considerable energy to developing optimal categorization (or optimization) methods for category delineation (e.g., Jenks, 1977). Map categories derived by such methods are typically defined with precision, resulting categories are mutually exclusive, and the category system is comprehensive (i.e., everything fits into a category). The failure to use "optimal" classification procedures (by noncartographers) has lead to critiques (by cartographers) of maps produced to represent disease and other topics.[2] The assumption that optimal categories for statistical maps are possible has led to conflicts between cartographers and their map author clients of the kind cited by Krygier (forthcoming).[3]

The classical approach to mental categorization that cartographers and others have accepted as fact has come under increasing attack in the linguistic, philosophical, and psychological literature. Championed by the empirical efforts of Rosch (1973), a convincing body of evidence has been produced to support a view that few mental categories fit the classical mold. It is now clear that many of the scientific categorization systems that had been taken as evidence of the classical view of categories do not, in fact, support that view. Lakoff's (1987) *Women, Fire, and Dangerous Things: What Categories Reveal about the Mind* provides perhaps the most compelling account of what a shift from classical categorization theory entails in relation to both mental categorization itself and, more generally, how the mind organizes knowledge. A few researchers in the GIS community have begun to grapple with what this new view of mental categorization means for natural language approaches to GIS user interfaces and to GIS information structures (e.g., Peuquet, 1988; Mark and Frank, 1991; Bjorklund, 1991; Usery, 1993). Also, from the perspective of databases for GIS, Suchan (1991) describes the implications of alternatives to classical categories in the context of land-use classification. It is time that cartographers also begin to question past approaches to categorization for maps and to consider what has been learned about mental categorization in relation to how information should be presented on, and is derived from, maps.[4]

To understand the implications of recent evidence about mental categorization, we must begin with the classical approach to categories. From Aristotle to the present, two basic tenets were taken for granted by most disciplines: (1) *Categories are like containers* (with things either inside or outside) and (2) *Individual things are assumed to be in the same category if and only if they have certain properties in common.* A third assumption that usually goes along with classical categories is that natural categories exist, and that if we are diligent and resourceful we can discover them. Biological taxonomies and color are often cited as evidence for this last proposition, but as we will see below, neither is independent of the human observers who developed the categories.

The classical approach to categories was not based on empirical study; it was assumed to be so obvious that a priori philosophical speculation was sufficient to consider the tenets as fact, rather than theory subject to verification. One implication of the acceptance of classical categories is that any element in a category must, by definition, be as representative of the category as any other element. This perspective is applied in cartography when we produce choropleth maps, soils maps, climate maps, and so on, in which all map units falling in a particular category are depicted with identical symbolization.[5] On a choropleth map not only are we implying homogeneity *within* geographic units, but homogeneity *among* units in the same category. Several spatially adjacent units that are grouped in the same numerical category, therefore, are taken to represent a homogeneous region that is assumed to differ from other regions on the map. The cartographic concept of a choropleth map depends upon the classical theory of categorization being "correct." The understanding derived from choropleth maps based on classical categories depends upon how the categories are interpreted by users (i.e., in terms of classical or alternative categorization systems).

Elenore Rosch (1973, 1975a, 1977, 1978; Rosch et al., 1976) has been the primary force behind a dramatic change in how we view human categorization, and by implication how we view the human mind and human reason. She was among the first to view classical theory as just that, *a* theory, and to seriously question its assumptions. Two of these assumptions were specifically addressed in research by Rosch and her collaborators: (1) If categories are defined by properties that all members share, no members should be more representative of the category than others; (2) if categories are defined by properties inherent in their members, then categories should be independent from characteristics of humans doing the categorizing (e.g., neurophysiology, perception, culture, etc.). What Rosch and others found was that neither assumption held in practice. Most categories investigated consisted of individual members judged to be more or less representative of the category, and the characteristics of the humans doing the categorizing were fundamental to the categories defined. The alternative to classical categories that is supported by empirical evidence collected by numerous researchers in varied disciplines over the past two decades has been termed *prototype theory* in reference to one of its basic premises, that category membership is determined not by a match to a fixed set of properties, but by similarity to a *prototype* representing the most *typical* category member.

Prototype Effects

The central feature of Rosch's theory of categorization is this concept of a prototype. Prototypes have been variously defined as abstract mental concepts of a typical (but not necessarily existing) member of a category, as exemplars (or "best") members of a category, or as collections of features (and relationships among them) that are most strongly associated with the concept. In differentiating her approach from classical category theory, Rosch (1975b) initially advocated a conception of prototypes as a kind of abstract mental composite of the most typical members of a category.[6] At least with common objects such as fruit, however, this mental composite idea of a prototype does not work well (e.g., it is difficult to conceive of what a composite of apples, oranges, and bananas might be like—other than as a fruit salad—or how this composite might be used to judge the "fruitness" of another potential, but less typical, category member such as a tomato). The idea that prototypes for a category are simply exemplars of the most typical category members that a person has encountered has similar problems explaining how the prototype could be used to judge whether new objects or concepts fit in the category. In addition, exemplars are likely to be highly individualistic, and therefore to be too idiosyncratic to be useful in interpersonal interaction.

A view of prototypes based on features of category members seems to be most plausible. This view differs from classical accounts of categories (also based on features) in that the features are not expected to be common to *all* category members. Rather, the feature lists of a prototype are assumed to represent those features that together represent the most typical and distinctive characteristics of the category (Roth and Frisby, 1986). Smith and Medin (1981) propose a "probabilistic" model as one mechanism for using features to establish the relative "goodness" of entities as category members. This model matches the probability that two entities are in the same category with the degree to which they share features. It discounts the notion of a common core of critical properties that are necessary for membership in the category. As Murphy and Medin (1985) point out, however, a limitation of the probabilistic approach is that it cannot determine which features form possible concepts (prototypes) and which form incoherent ones. By itself, therefore, a probabilistic model might be able to describe the structure of a particular category (in relation to entities that people tend to assign, or not assign, to the category), but we need something further to understand what qualifies as a category in the first place. Murphy and Medin argue that concepts are fundamentally linked to "theories" about the world, and claim that these theories are the "glue" that holds concepts together.[7]

In the realm of maps, Downs et al. (1988) found that the category "map" itself demonstrates prototype effects. Both children and adults shared a concept of a prototypic map (with the *Weekly Reader* World Political Map being a good example), but children chose a much narrower range of displays that they were willing to categorize as a map than did adolescents or adults. Vasiliev et. al. (1990), in a similar experiment, found that the category "map" was defined by a set of characteristics that in various combinations determine "mapness." This finding seems to agree with the feature-list approach to prototypes described above.

Family Resemblance

As early as the 1950s, evidence began to accumulate that classical categories having clear boundaries and common properties did not fit all situations. Wittgenstein (1953; cited in Lakoff, 1987) used the category "game" as an example of a category that does not conform to classical theory. As a category, games do not share a common set of properties. Some are based on competition and others are not, some depend on chance (e.g., dice), while others depend on skill (e.g., darts). Some games are for amusement (e.g., pinball), while others have an educational goal (e.g., *SimCity*). The main point made by Wittgenstein is that some categories are defined not by common properties, but by *family resemblance*. In people or animals, members of a family resemble each other in various ways (eye or hair color, shape of the face, build, height, etc.). To be recognized as part of the family does not require sharing all characteristics with other members. Closer members in a family tend to share more features, while distant members may share none but still be identified as in the same family by virtue of sharing separate features with a more central member (e.g., Joe has Uncle Harry's eyes while Sally has Uncle Harry's sense of humor).

Category membership defined by family resemblance is much more flexible than are classical categories. It allows individual entities to maintain category links to multiple categories. In addition, the concept of family resemblance provides a logical mechanism by which potential new category members can be added over time (as evolving knowledge and technology generates them). Historically, the category "numbers" exhibited this growth, beginning with integers and subsequently extending to include rational numbers, real numbers, complex numbers, and so on (Lakoff, 1987). As Stephen Hall's (1992) overview of mapping technology applications makes apparent, the category "map" is being extended into new domains as a result of computer technology producing spatial depictions (i.e., maps) of new *frontiers* in medicine, biology, physics, mathematics, and astronomy. This expansion of the category "map"—due to the flexibility of the concept of family resemblance—has the potential to expand the bounds of the category "cartography" as well. If maps are not restricted to depictions of the earth's surface or the surfaces of other celestial bodies (a restriction found in the 1973 International Cartographic Association [ICA] definition of cartography) the definition of cartography will have to be rethought. In 1987, the ICA actually began an attempt to redefine the field by redefining "cartography." The revised definition is most significant in its shift of emphasis from "making maps" to "organization and communication of geographically related information" (Board, 1989; cited in Taylor, 1991). Such a formal redefinition, however, is unlikely to prove satisfactory unless cartographers accept the idea that the object of their interest, the map, is a category in constant flux (rather than a static classical category).

One of the most important aspects of the concept of family resemblance for cartography is that it allows for category members that share *no* properties. This is exactly the approach recently taken in attempts to derive optimal categorization procedures for quantitative statistical maps (e.g., Jenks, 1977; Coulson, 1987). Jenks's statistically optimal classification procedures delineate categories in which all members are more similar (numerically) to some central prototype (a mean or median that may not actually exist in the data set) than to the central prototype of adjacent categories. Jenks's optimal categorization scheme does not (as would a classical categorization scheme) require that all elements of a category share some common property. If a map user misinterprets the resulting categories in classical terms, however, the map as a whole will be misinterpreted (Figure 4.1).

Fuzzy Categories

Some categories like "*large* cities" have no clear bounds. As city size decreases, the certainty that a city is large will probably decrease, but there is unlikely to be a distinct boundary above which an individual classifies cities as large and below which cities are classed as small. "Map" as a category, in addition to an apparent structure based on family resemblance,

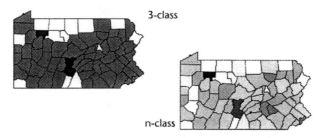

FIGURE 4.1. The potential for misinterpretation is apparent if we compare an optimally classed three-class choropleth map with its n-class counterpart (a map in which each data value is given a shade of gray corresponding to its position in the data range). The clear regions with sharp edges apparent in the classed map are not really there.

has been demonstrated to have *fuzzy* bounds (Downs et al., 1988; Vasiliev et al., 1990).[8] Downs and Liben (1987) specifically measured the fuzziness of "map" as a category by asking subjects to respond to a series of examples with one of three answers: Yes, I think it is a map; No, I think it is not a map; or I am unsure or can't decide whether this is or is not a map. Vasiliev et al. (1990) posed a similar task using a 5-point scale from "definitely a map" to "definitely not a map." Both experiments demonstrated fuzziness or uncertainty concerning category bounds. Both also point out that scale is a factor in "mapness," with national scale more typically a map than city scale. Downs and Liben (1987) found this scale factor to be particularly important for children, only slightly over half of whom accepted a "map" of Philadelphia as a map (while 100% of the adults did so).

The phenomenon that some categories appear to have fuzzy boundaries was initially investigated by Labov (1973). Using drawings of household objects as stimuli, he attempted to determine whether the assignment of an object to a category could be influenced by the context within which the object is presented. Such a context effect would, of course, be at odds with classical category theory that assumes all categories have clear-cut and stable boundaries. His drawings all depicted objects that could be considered vessels for holding things. These were presented in one of four contexts: neutral—where subjects were to imagine someone holding the object in their hand; coffee—where subjects were to imagine someone holding the object and drinking coffee from it; food—where subjects were to imagine the object filled with mashed potatoes and sitting on the dinner table; and flower—where subjects were to imagine the object on a shelf filled with cut flowers. As the appearance of the drawing was less and less cuplike (broader and shallower), responses that it was a cup declined in all four contexts. A clear context effect, however, was present. For the neutral and coffee cases, the identification of objects as a cup decreased much less markedly across the set of examples than it did for the food and flower cases (Figure 4.2). Labov's (1973) research, and subsequent findings by Barsalou and Sewell (cited in Barsalou, 1985) among others, suggest that people's perception and structuring of categories does not reflect invariant properties of the categories. Category structuring appears to be highly dynamic and context-dependent.

Stephen Hall's (1992) *Mapping in the Next Millennium* may provide cartographic examples of context effects. Because this book is purportedly about mapping, and he uses the term "map" when referring to visual depictions of the brain, DNA, atoms, galaxies, and fractals, readers undoubtedly accept them as such. The same readers, however, in a different context, might be more likely to categorize the depictions presented as medical images, star charts, diagrams, or perhaps abstract art.

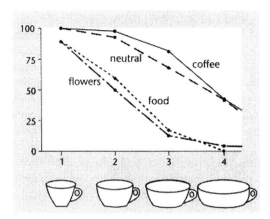

FIGURE 4.2. The percentage of "cup" responses for four of Labov's drawings in four different context conditions. *Derived from Roth and Frisby (1986, Fig. 1.7, p. 50).*

A mechanism for conceptualizing and working with fuzzy categories such as those Labov found was developed nearly a decade earlier by Zadeh (1965). With this method, known as *fuzzy set theory*, membership in a category is not a binary decision of in or out. Values between 0 and 1 are allowed. At its most basic level, fuzzy set theory allows operations on sets that are generalized from the analogous operations on ordinary sets. Within the area of spatial representation, versions of fuzzy set theory have been used in an effort to simulate the uncertainty in category assignment within a GIS.

Typicality Effects

One of the key initial findings underpinning prototype theory as an alternative to classical category theory was the identification of typicality effects in category structure. Some categories, although they may have relatively clear bounds (e.g., trees), have a graded internal structure resulting in differences in terms of how typical an individual example is of the category (e.g., oak vs. mulberry). Rosch (1973; Rosch and Lloyd, 1975) demonstrated typicality effects within categories convincingly in a series of experiments in which subjects rated potential category members according to their goodness as an example of a specified category. Categories tested were furniture, vehicle, weapon, carpenter's tool, toy, sport, fruit, vegetable, and bird. In all cases, subjects found the task easy and there was substantial agreement on the ratings given. Results provide clear evidence that, at least for these common categories, membership is not a simple binary choice but involves a grading from typical to atypical.

A chair or sofa, for example, represent highly typical members of the category "furniture," while lamp and stool are clearly within the category, but much less typical. Other items, such as vase and telephone, are atypical.

If we consider a standard cartographic topic such as land use or land cover, it seems likely that we will find a similar graded range of members at all levels of classification. As an example, an avocado orchard is probably less typical of "agricultural land" for most people than is a wheat field and a 100-acre cattle ranch is probably more typical than a 2-acre biointensive herb farm (although both may generate similar income identified as "farm" by the IRS). As noted above, the category "map" itself exhibits prototype effects, and these seem to include typicality. Some maps while clearly within the category are less suitable as exemplars of the category than others.

Among the best known empirical demonstrations of typicality effects (and fuzzy boundaries) is Berlin and Kay's (1969) research on color categories, a topic of increasing cartographic relevance in an age of virtual color CRT maps. They conducted a series of empirical tests of the generally accepted notion that different languages can partition color in arbitrary ways. What they found was that a set of basic color terms exist across languages (names consisting of a single morpheme: yellow vs. sun-colored, not contained in another color—yellow vs. gold, applicable to a broad context—yellow vs. blond, and generally known—yellow vs. saffron). Not all languages examined had the same number of basic terms, but when languages had a specified number, they were typically the same (e.g., those languages having only three basic color terms use terms for white, black, and red; yellow or green are added in four- and five-color languages; blue is added in six-color languages; and brown is added in seven-color languages). The complete set of what Berlin and Kay identified as basic color terms is black, white, red, yellow, green, blue, brown, purple, pink, orange, and gray. Obviously, since I was able to list all colors in single English words, English contains all eleven. A key feature of the basic color terms is that when individuals were asked to select (from a set of color chips) the *best* example of any color included as a basic color for their language, they were very consistent in selecting the same color. In addition, this color choice corresponded across languages, even when languages had different numbers of basic colors (e.g., for all languages having blue as a basic color, the same blue was picked).

For each basic color, therefore, there is a central member considered to be the best example of the category and the category grades away from this central color. While Berlin and Kay found consistency in the prototypic color representing each basic color category, they did not find similar consistency in identifying category bounds. Colors seem to be fuzzy categories with the fuzzy set size for specific basic colors varying from language to language, resulting in lack of consistency for color boundaries. This cultural difference suggests potential problems in trying to select colors for maps used in cross-cultural contexts or default colors in mapping systems for a multinational market (e.g., tourist maps).

In an attempt to explain the phenomenon of basic color categories, Kay and McDaniel (1978) looked to neurophysiological evidence of opponent color processes in vision (see Chapter 3) combined with an adaptation of fuzzy set theory. They cite DeValois and Jacobs (1968) as evidence for opponent processes based on blue, yellow, red, and green, plus light and dark. According to an opponent process model, human vision is expected to react to these six distinctions in a direct way. Kay and McDaniel (1978) point out that these six neurophysiologically basic colors are the first six basic colors that appear in language. That these colors are basic, they contend, is due primarily to human neurophysiology. This view has subsequently been disputed by Ratner (1989) whose results suggest that cognitive as well as neurophysiological mechanisms may play a role even for these most common focal colors. The remaining basic colors are thought to come from combinations of the first six based on cognitive mechanisms making use of something akin to fuzzy set theory. Orange is derived from a fuzzy set intersection of red and yellow, purple from blue and red, pink from red and white, brown from black and yellow, and gray from white and black. As Lakoff (1987, p. 29) rightly notes, "Color categories result from the world plus human biology plus a cognitive mechanism that has some of the characteristics of fuzzy set theory plus a culture-specific choice of which basic color categories there are."[9]

Map as a Radial Category

Lakoff (1987), in combining the ideas of prototypes, family resemblance, fuzzy bounds, and typicality effects, suggests the concept of a *radial category*. Radial categories have a clearly defined center or prototype. The center of the category is predictable from family resemblance to prototypic members. Noncentral members are not predictable from prototypes, but, according to Lakoff, are *motivated* by family resemblances to them. By "motivated," Lakoff seems to mean that there is a cognitive economy in recognizing similarity on some criteria (e.g., appearance, function, etc.) between the potential category member and the central prototype. For maps, we are motivated to consider a contour plot of brain cell activity (see Figure 3.13) to be a map because this allows us to interpret the display as we would a topographic surface, thus immediately "recognizing" peaks and troughs in density of active cells.[10]

The category "map" seems to be a clear case of a radial category. Its category space can be defined by two orthogonal axes. One is the grading of mapness due to scale identified by both Downs et al. (1988) and Vasiliev et al. (1990). The other involves map abstraction. John Ganter and I have argued that spatial displays form an abstractness continuum from images to graphics (or diagrams) (MacEachren and Ganter, 1990). Typical maps occupy a middle ground along this continuum. Combining the ideas of a prototypic scale and abstractness for maps gives us a map category space within which a variety of spatial depictions can exist (Figure 4.3).

The fact that "map" is a fuzzy and radial, rather than a precisely defined, category is important because what a viewer interprets a display to be will influence her expectations about the display and how she interacts with it.[11] Prototypic maps, being at the midpoint of the scale axis of this space, are probably more readily interpreted and more readily accepted as objective depictions (than are maps at the atomic or astronomic extremes) because they depict reality at a scale closer to human experience. Considering the other axis, graphs and diagrams are often viewed with some suspicion due to their abstract nature, while images (i.e., photos) are generally accepted as unbiased depictions of what can be seen from a particular vantage point. The prototypic map, lying between the image

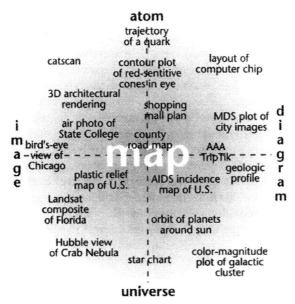

FIGURE 4.3. The radial category *map*, illustrating two possible axes of the category space.

and diagram extremes, clearly involves some processing, and therefore some potential for bias (but also becomes functional due to this processing). The idea that function is often a defining feature of categories is stressed by Lakoff (1987) and others. A depiction is considered a map if it can function like one (e.g., be used to plan a trip).

Basic-Level Categories

Lakoff (1987) credits research by Brown (1958) as the first to note that categories are organized hierarchically. Certain labels within the hierarchy of category names have a higher status, and these categories are correlated with actions (e.g., "flower"—can be sniffed—compared to the more generic "plant"; "ball"—can be bounced—compared to "sports equipment"). This distinctive level of categorization has come to be termed the *basic level*. It was the anthropological research by Berlin (1972) that first explored basic-level categories in detail and served as the impetus for Rosch et al. (1976), Tversky and Hemenway (1984), and others to extend the idea of basic-level categories to all kinds of mental categorization (including categories of events as well as objects).

Berlin examined, in considerable detail, the plant names used by the Tzeltal Indians of Mexico. The categories that they routinely identified were found to correspond with categories at a particular level of scientific taxonomy for plants, the level of genus. Berlin argued that there was cognitive economy in grouping plants at this level because the genus is the highest level at which many visually distinctive attributes are held in common by most members of the category—plants that look alike therefore are likely to be in the same category when categories are defined at the level of the genus. Lakoff (1987) points out that the correlation between the folk classification at this basic level and the scientific level of the genus was not a coincidence. Linnaeus built his taxonomic system around the level of genus, and similarity in some readily apparent characteristic (e.g., shape of fruit) was his key criterion for grouping at this level. As Lakoff makes clear, therefore, the scientific classification system (at what Linnaeus considered the most important level) was grounded in folk categorization emphasizing visual similarity.

Tversky and Hemenway (1984) argue, using evidence from psychology, linguistics, and anthropology, that for all category systems (not just biological ones), there will be a basic level at which a variety of aspects of perception, behavior, and communication converge. These basic-level categories are the most general categories for which we can form a single image. They contend that the basic level is distinctive primarily because there are relatively large numbers of attributes (parts) associated with it and because these attributes share both perceptual and functional features. "Car" and "table," for example, are cited as more basic categories than "vehicle" or "tool" because most people can define more attributes of the former than of the latter and because many of those attributes have both form and function (e.g., tires on a car). This important role for the function of attributes shared by category members is echoed in Lakoff's (1987, p. 12) contentions that properties of certain categories depend upon human capacities and experience of functioning in the environment, and that concepts that are "used" are more fundamental than those that are simply "understood intellectually."

Lakoff (1987) reviews much of the evidence for basic-level categories. From this review he contends that basic-level categories are basic in at least four respects:

1. *Perception*: Basic-level categories are the highest level categories having similar overall perceived shape, a single mental image, and fast identification.
2. *Function*: They are the highest level categories for which a person uses similar motor activities to interact with them (e.g., sitting on chairs vs. ? on furniture).
3. *Communication*: Basic category labels are the shortest, most commonly used, and most contextually neutral words; they are the first learned by children; and they are the first to enter the lexicon.
4. *Knowledge organization*: The basic level is the level at which most of our knowledge is organized and for which the most attributes are stored.

The search for a basic level of categorization is related to cartographic questions surrounding quantitative data classification for choropleth maps and the issue of when, or if, classification is appropriate (vs. use of unclassed maps).[12] As mentioned above, Jenks's so-called optimal classification makes use of one of the aspects of prototype theory, that categories can be delineated on the basis of family resemblance such that variation internal to the categories is minimized and that variation among categories is maximized. Jenks and Caspall (1971) seem also to have anticipated the idea of basic-level categories in arguing that comparing variance for a hierarchical series of classifications is a good way to identify the most efficient position in that hierarchy. Specifically, they advocated computing a "goodness-of-variance-fit" for two through n categories, graphing the results, and determining the number of categories at which the graph levels out (Figure 4.4). This number of categories should be the most cognitively efficient for the map reader. With a more detailed cate-

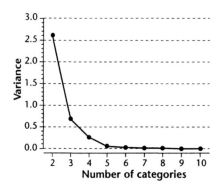

FIGURE 4.4. A plot of decreasing classification variance with increasing numbers of categories. Such plots typically suggest a decreasing accuracy return on the cognitive investment of processing more and more differences.

gorization, cognitive effort to process the increasing number of categories would continue to increase, while within-category variance would improve only marginally.[13]

Building from the extensive research on basic-level categorization in a variety of contexts, then, we arrive at an argument in support of classed choropleth maps over unclassed maps based on features that have been identified for basic-level categorization, such as cognitive efficiency and enhanced memorability. My research on memorability of choropleth and isopleth map patterns supports this contention by demonstrating that even if we give map readers more detailed maps (with up to 11 categories), they seem to "remember" at what may be the basic level of three categories (high, medium, and low) (MacEachren, 1982). The alternative argument is, of course, that if readers are capable of *seeing* basic-level categories in a map display having subordinate categories, we can provide more detail to the readers without interfering with their ability to ferret out the basic-level structure. This argument is supported to some extent by Muller's (1979) demonstration that subjects asked to regionalize n-class choropleth maps grouped enumeration units into roughly the same groups as did Jenks's statistically optimal data classification procedure (when both subjects and Jenks's algorithm were asked for three classes) (Figure 4.5). As we will discuss below, however, different map readers bring different schemata to the map interpretation task and these differences may result in differences in what is identified as a basic-level categorization.

Cartographically and geographically, we often categorize the world on the basis of spatially contained regions. This process of geographic regionalization differs from standard choropleth categorization and much of the natural object categorization that Tversky and Hemenway (1984) and others studying basic-level categories have addressed. The primary difference is the interaction of spatial location with nonspatial attributes. Tversky and Hemenway (1984) offer an extension to the idea of basic-

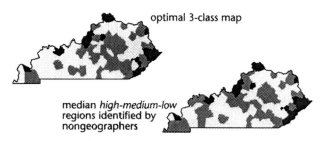

FIGURE 4.5. Muller's composite group map compared to a three-class optimal map (derived using Jenks's classification algorithm). *After Muller (1979, Fig. 9, p. 247). Adapted by permission of Blackwell Publishers.*

level categories that may be relevant to geographic regionalizations. In their emphasis on the role of parts in basic-level categories, Tversky and Hemenway distinguish between what they term *taxonomies* and *partonomies*. Taxonomies involve organization by kinds (i.e., grouping things that are similar), while partonomies involve organization by parts (i.e., grouping things containing similar parts or attributes). In relation to geographic regions, a taxonomy could group locations that were similar in spatial position and overall appearance (e.g., sparsely populated rolling hills dominated by livestock pastures), or some composite measure of characteristics (e.g., a politically conservative, agrarian economy, with an aging population). A spatial taxonomy, then, might serve to divide the world into spatially bounded categories that are organized hierarchically such that inferences about the largest regions can be applied to subregions within them. A partonomy, in contrast, would focus on region parts (or attributes) and identification of region cores based on locations with the greatest number of shared attributes. Regionalization by partonomy, in contrast to the spatial taxonomic regionalization described above, may be spatially quite fragmented with ill-defined (fuzzy and perhaps overlapping) spatial bounds.

My own research on the role of map compilation exercises in influencing regional images supports the spatial taxonomy–partonomy distinction (MacEachren, 1991a). I found that spatial and attribute components of regional images were at least somewhat independent. Subjects in an initial survey exhibited considerable location mismatch between the location they associated with a major national region (the Midwest) and where the attributes they associated with the region actually are. The task of compiling maps of these attributes led to changes in the location component of regional images to correspond more closely with attribute location (Figure 4.6). The category "Midwest" was therefore shown to exist in two separate but interrelated cognitive models that can conflict and can influence one another when those conflicts are brought to the indi-

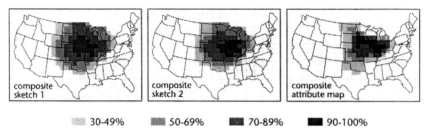

FIGURE 4.6. A composite Midwest location based on responses from 32 students (left), a composite map of the location of Midwest attributes identified and compiled from existing maps by those same 32 students (right), and a composite map of Midwest location as specified by these 32 students 10 weeks after completing the Midwest attribute map compilation exercise (middle). Shading on all maps corresponds to the percentage of students for whom a cell was identified as in the Midwest (for the two location maps) or for whom a cell contained two or more Midwest attributes (for the attribute map). *After MacEachren (1991a, Fig. 1, p. 153). Adapted by permission of The Cartographic Journal.*

vidual's attention. That the locationally based categorization of the "Midwest" was altered to correspond more closely to the attribute-based categorization seen on cartographic maps is evidence for the importance assigned to "map" as a category. The depictions classed as "real" cartographic maps were accepted as more valid than the subject's own mental maps of regional image location.

Regional images represent a "folk" classification in that they are derived by nonspecialists for use in everyday spatial decision making and behavior. Like plant taxonomies, many presumably objective geographic regionalizations (e.g., climate zones) are based in part on folk categorization, presumably at the basic level. Climate categories, for example, in order to be functional, must relate to issues of human comfort and/or productivity (e.g., the ideal temperature and moisture conditions for agricultural crops). This structuring of regionalizations around a functional basic level of categorization has proved effective, even if we were not conscious of using a fundamental process of human mental categorization when we devised the categories.

In relation to socioeconomic factors, folk classifications (of high-income neighborhoods, ghettos, etc.) combine space and attributes in ways similar to the physical regionalizations of climate and soils. Why, then, do we ignore space when classifying data for a choropleth map? Jenks's statistical classification procedures seem to have (if only coincidentally) taken into account the principle of basic-level categories. The next step in our evolution of quantitative data classification, then, should be to add space to the equation. Jenks, in fact, attempted to do this in a "contiguity-biased" data-classification procedure that he developed in collaboration with John Davis; the method, however, was never published. Monmonier (1972) also suggested a procedure for contiguity-biased classification, but few cartographers seem to have accepted the idea. It may be that cartographers, ignoring the implications of prototype theory and still accepting classical category theory, believe that contiguity-biased classification is "biased" (read "not objective"). As we are beginning to realize, however, "objective" is a relative term and can only be interpreted in a limited context. The question that must be addressed is whether it is better to offer map readers an artificial aspatial classification system that is statistically objective (and matches at least the family-resemblance ideas of prototype theory) or a classification that may be viewed by some as less objective, but that comes closer to folk categorization, and therefore comes closer to matching the probable knowledge structures of map readers.

Cartographers need to take several aspects of nonclassical categories into account in evaluating their strategies for deriving map categories. These include the idea that fuzzy categories are the norm for mental processing, that no human category system can exist independently of mental processing and human experience/interaction with the world, and that basic-level categories combining location and attribute information are the most functional because they are grounded in space (outside of which interaction with attributes is not possible).

Natural versus Cultural Category Structures

Classical categorization assumes that the world is divided into categories that it is our task to discover. Anthropologists, who were among the first to find flaws in classical category theory, came to the conclusion that structure does not exist in the world: it is imposed by categorical processes of the human mind (Leach, 1964). They came to this conclusion by observing the variety of ways different cultures categorized the natural world (e.g., languages that had from two to 11 basic color terms; those that group plants on the basis of appearance, kind of propagation mechanism, or medicinal use, etc.).

Rosch, the most influential critic of classical categorization theory and proponent of prototype theory, took a middle ground. Her initial view (Rosch et al., 1976) was that the world has structure and that this structure influences how the mind divides things into categories. She contended that there are *natural correlations* of properties of things in the world that tend to go together (e.g., wings with feathers). These natural correlations influence, but do not determine, the groupings we observe and find useful. Characteristics of human perception and cognition were seen to interact with these natural structures to highlight some and to bypass others. Kay and McDaniel's (1978) explanation for the existence of a structured but variable set of basic colors in the world's languages fits this conception. In subsequent research Rosch (1978) saw flaws with this *realistic* view. In particular, it became clear to her that many categories could not exist independently of the observers doing the categorization. Certain attributes of categories (e.g., "seat" for a chair) could only be applied after knowledge of the object as a particular category was established, others (e.g., "large" for a piano) were meaningful only in terms of a superordinate category (furniture vs. built object), and yet others were functional attributes that depended upon how humans interact with the world (e.g., humans eat, therefore we have developed the category of "food").

Lakoff (1987) places particular emphasis on this *interactional* nature of many category properties. He contends that properties are "the result of our interactions as part of our physical and cultural environments given our bodies and our cognitive apparatus" (p. 51). He suggests that interactional properties often seem objective or natural when we consider basic-level categories because humans perceive certain aspects of the environment very accurately at the basic level. Lakoff contends that cognitive models that structure thought and are used in forming categories are either directly *embodied* or linked directly to embodied models. By "directly embodied," he means that the models are based upon human biological capacities and experiences of functioning in a physical and social environment. Part of our cognitive model for understanding a road map, for example, includes a basic understanding of front and back in relation to the human body, and origins, paths, and destinations in relation to everyday movement.

Multiple Representations

Classical categorization assumes that there is always a single correct way to categorize any phenomenon. Prototype theory with its typicality grading, fuzzy category bounds, family resemblance, and hierarchical category structure presents a more flexible view that allows for multiple representations of individual concepts.

Dual Representations: Common and Scientific

It is easy to demonstrate that there are differences between scientific and common categorizations of the world. As Roth and Frisby (1986) point out, common categories are often clear-cut (e.g., male–female, life–death). Scientifically (and legally) there is often considerable debate about the bounds of such categories. Nonexperts tend to focus on the core of categories and shared characteristics while experts may focus on category divisions and distinctive features. In addition to this expert–nonexpert distinction, individuals often hold more than one kind of representation of a concept suited to different applications. A cartographer, for example, may accept digital databases as being "maps" at a conceptual level, but when she goes to the local bookstore looking for the map section in preparation for a trip abroad, she would probably be surprised (and not very happy) to be led to a rack of floppy disks containing Digital Data Bank of the World files.

An important issue that cartographers must begin to grapple with is the choice of category representation. Up to now, we have taken a view that there is one "objective" way to categorize any data set and have tended to rely more and more heavily on quantifiable attributes of categories to achieve this objectiveness. As Krygier (forthcoming) points out, this model for what maps are and how they should depict is at odds with the philosophical perspective of much of modern human geography. There is a need to explore the possibility of varying levels of categorization for different goals, applications, and perspectives, and to explore how our maps might incorporate some of the less precisely defined (but no less truthful) ways of categorizing the world. The distinction is not just one of common versus scientific categories, but also occurs among scientific categories based upon differing philosophical foundations (e.g., see Yapa, 1992, on the category GNP).

Fuzzy Representations of Well-Defined Concepts

Many kinds of categories can be demonstrated to be ill-defined or fuzzy categories. Categories like "map" are fuzzy categories at both common and scientific levels because there is no precise definition (either an embodied functional one or an arbitrary formal one) that allows us to clearly identify whether an object is in or out of the category. Other domains can be quite well defined. Roth and Frisby (1986) offer the triangle (a category defined quite precisely as an object having three and only three sides) as an example of a particularly well-defined domain. Cartographically, a "railroad" on a map is a similar unambiguously defined concept (although its position may not be well defined). Armstrong et al., (1983) have provided evidence that even for well-defined categories people often have a fuzzy representation. This happens even when they know, intellectually, that the category is not a matter of degree. Roth and Frisby (1986) suggest that the tendency to have fuzzy representations of well-defined domains is cognitively efficient because these fuzzy representations provide a means for fast judgments. This notion of fuzzy representation relates to the pattern recognition model for cartographic visualization that John Ganter and I have devised (see Chapters 2 and 10) (MacEachren and Ganter, 1990). In that model, we suggest that use of maps as visualization tools is effective because of a human tendency to see (or notice) patterns quickly that are usually correct or significant, then reason about them to determine how accurate this initial noticing actually was. Recognizing that a display has some similarity to a likely prototype is often enough to result in an initial categorization of an object or pattern noticed on a map. This initial categorization will remain accepted until evidence to the contrary is encountered, or until a reasoning-why process determines that an alternative categorization has a higher probability. These fuzzy categories also allow objects or concepts that have some similarity to the prototype for a category to be linked with that prototype even without meeting some formal criterion for category membership. Figure 4.7, for example, can be classed as triangular even though it is not geometrically a triangle.

KNOWLEDGE REPRESENTATION

Mental categories occupy a fundamental level of human knowledge. Without categories, we would be quickly overwhelmed with specific unique elements of information (for color alone, as an example, humans are capable of discriminating more than 7 million colors). At the level of mental categories and above, we are faced with the issue of how knowledge in general is organized. Such knowledge includes (in addition to category structures) spatial and aspatial relationships among categories or category members, procedures for action, and more. A number of theories of knowledge organization or representation exist. The one I will draw

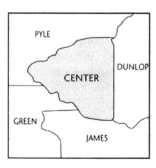

FIGURE 4.7. A map region that could be interpreted as in the fuzzy category of triangles, but not the classical category of triangles.

upon here is that offered by Rumelhart and Norman (1985) in which they present a typology that divides knowledge representations into three types: propositional, analogical, and procedural.

Kinds of Knowledge Representation

There has been considerable debate among psychologists concerning the form that cognitive spatial representations take. The competing theoretical positions were described by Lloyd (1982) as ranging from radical image theory (see Kosslyn, 1981) through dual-coding theory (see Paivio, 1969) to conceptual-proposition theory (see Pylyshyn, 1981). The distinction among the theories arises from the role that imagery is thought to play in the coding, storage, and retrieval of spatial information. Conceptual-proposition theory contends that all knowledge exists as conceptual propositions and that images, if they are accepted as existing at all, are an epiphenomenon with no real function in thought. Radical-image theory, at the other extreme, posits that images exist, are used in thought, and are stored in an analog form. Although there is little hard evidence to support the idea of this "pictures-in-the-head" view, there is considerable evidence that images are a real phenomenon that can be mentally manipulated and used in abstract thought (see the discussion of "Processing of Imagery" in Chapter 2).[14] In addition, imagery seems to use some of the same processes as seeing because imagery cues have been demonstrated to help detection (Peterson and Graham, 1974). While most psychologists and neurophysiologists scoff at the *picture*-in-the-head idea, a more schematic *map*-in-the-head view has growing support. The maps of activity in the brain presented in Chapter 3 (Figure 3.13) provide one piece of evidence for this view. Neurobiologist John Allman (1990; cited in Hall, 1992, p. 16) contends that one of the most distinctive differences between human brains and those of other animals is the size of our neocortex and that one form of learning, which he calls "perceptual memory," is stored within "maps" in the neocortex. Stephen Hall (1992, p. 17) in reference to work by Allman and others speculates that "it may be that the human brain not only perceives but stores the essentials of a visual scene using the same geometrical, quasi-symbolic, minimalist vocabulary found in maps."[15]

Both propositional and analogical representations appear to be applicable to encoding concepts or static scenes. In relation to maps and the geography they depict, it seems that propositional knowledge representation might be most suited to the organization of what Golledge and Stimson (1987) refer to as "declarative knowledge," knowledge about objects, attributes, and places. Analog representations seem suited to organization of configurational knowledge about space—knowledge of spatial relationships among entities in space. In relation to environmental learning, Golledge and Stimson contend that declarative knowledge is at a lower level of cognitive processing with configurational knowledge achieved through a developmental-like process with increasing experience in an environment (Figure 4.8). I have found that learning can be facilitated by simulating this developmental sequence by presenting a map as a hierarchically structured set of route segments that gradually build toward a configural representation (MacEachren, 1992a). As Lloyd (1989) has shown, however, when a map is the source of information, configural knowledge can be acquired quickly (10 minutes of map study resulted in more accurate distance and direction knowledge than 10 years of living in an environment). In the study of sequenced presentation of map components (cited above) I found evidence that different kinds of knowledge representations can be stimulated by different knowledge-presentation procedures (MacEachren, 1992a). It is likely that map reading involves reliance upon knowledge structures dealing with both declarative and configurational knowledge and that the process of map reading can generate or alter both propositional and analogical knowledge representations.

What the bipartite approach to knowledge representation suggested by the proposition–imagery debate leaves out is what Golledge and Stimson (1987), in an environmental learning context, refer to as procedural knowledge, knowledge of the sequence of steps needed to get from one place to another. More broadly, procedural knowledge is considered simply knowledge of how to do something.[16] Rumelhart and Norman (1985) contend that procedural representations are a distinct category of knowledge representation; that propositional, analog, and procedural representations all play a role in human knowledge organization; and that they in-

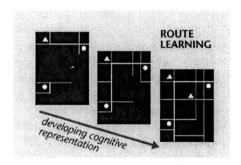

FIGURE 4.8. A schematic depiction of the hypothesized developmental sequence of environmental learning. *Derived from MacEachren (1992a, Fig. 1, p. 252).*

teract in complex ways. Evidence from research by Golledge et al. (1992) suggests that procedural knowledge obtained during route learning is difficult to transform into a configurational (analog) representation. Frequent reports of traveler frustration with maps (that probably induce an analog representation) suggest that transformation in the opposite direction (analog to procedural) is equally difficult. This implies that alternative navigation aids that provide procedural rather than configurational information may have an advantage over traditional maps for some travel needs, particularly for following a route once it has been selected (McGranaghan et al., 1987).

That cognitive representations, exhibiting both propositional and analog characteristics, can be generated from maps has been demonstrated by a number of authors (Garling et al., 1983; Peterson, 1985; MacEachren, 1992a). It has been argued by some that map-derived representations are typically analog (image) in form and it has even been suggested that these can be picturelike (Levine et al., 1982). Corresponding to the likelihood that maps generate analog representations, research supports a view that mental imagery plays some role when information learned from maps must be recalled and applied to solving specific map-use tasks (e.g., Steinke and Lloyd, 1983a; Lloyd and Steinke, 1985; Rice, 1990). Steinke and Lloyd (1983a), for example, demonstrated that images can be formed of maps and that these images can be transformed (i.e., mentally rotated). Two colleagues and I (Goldberg et al., 1992), however, found that when dealing with perspective terrain maps, only a small portion of subjects seemed to use mental rotation as a strategy for comparing map views. Our results suggest that a variety of problem-solving strategies can be used in map comparison tasks and that some of these strategies involve interrogation of a propositional rather than an analog representation. My research on the role of maps in environmental knowledge acquisition supports a hypothesis that individuals differ in the tendency to organize map-derived knowledge in an analog versus a propositional (and/or maybe a procedural) form (or in the tendency to retrieve knowledge in that form) (MacEachren, 1992a).

As a whole, cartographic evidence supports Rumelhart and Norman's (1985) contention that there are separate propositional and image representations. At this point, whether or not maps can stimulate procedural representations remains an open question. It seems likely, however, that dynamic real-time navigation aids based on moving-window map displays are likely to do so. That usable cognitive representations of some kind can be generated from such displays has been demonstrated by Ford (1985). Whether or not the representations generated are procedural, it is clear that map-derived knowledge must interface with procedural representations in the process of wayfinding (e.g., Crampton, 1992a).

Kinds of Knowledge Schemata

Regardless of how knowledge is represented in long-term memory, whether in propositional, analog, or procedural form, we must have some way to interface between visual descriptions (and other temporary sensory input stores, such as those for acoustic input) and these knowledge representations. The dominant view in the psychological literature is that some structuring mechanism or mechanisms exist that provide a common format for organizing sensory input and information retrieved from long-term knowledge representations. These structuring mechanisms have been given a variety of labels depending upon the kind of knowledge they are hypothesized to deal with and the proclivities of the author suggesting them (e.g., schemata, frames, scripts, mental models, idealized cognitive models, etc.). Although there are some fundamental differences implied by these labels (and among authors who use the same label), for simplicity in discussing how map understanding may depend upon these cognitive structures I will adopt a single term, *schemata*, to refer to them.

Following from the introduction of schemata in Chapter 2, schemata will be defined here in rather general terms as structures for representing and organizing concepts. These structures can be conceived of as models containing slots (or nodes) and links among them. The slots represent categories or attributes of categories and the links specify the possible relationships that can hold among the categories or attributes. Mental categorization of elements in a visual description is influenced by potential relationships specified by a schema that an observer brings to bear on a situation—such as understanding a map. Conversely, human mechanisms for categorization, identified as aspects of prototype theory, will control how unknowns of knowledge schemata can be filled in (i.e., instantiated). An important feature attributed to schemata by most theorists is that they are active devices—"plans for finding out about objects and events" to quote Neisser (1976) again.

In one of the few attempts thus far to consider schema theory in a cartographic context, Eastman (1985a) emphasizes the hierarchical, embedded nature of schemata. He equates schemata with cognitive structures that define prototype cases and points out that each schema consists of concepts or entities (each of which is a category that may be described by a prototype structure of its own) connected by relations. Eastman uses this argument of hierarchical schemata to contend that there is really no difference between schemata and "chunks" other than the typical use of the latter term to label the more primitive structures contained in a schema at a particular level of description.

Schemata are hypothesized (by various authors) to exist for dealing with all three kinds of information suggested by the typology of propositional, analog, and procedural representations. In the psychological literature schemata are typically assumed to be propositional models or organizing structures, even when they are applied to procedures or images (Rumelhart and Norman, 1985). I suggest, however, that different schemata are likely to exist for interfacing with different knowledge representations. A categorization of schemata as propositional schemata, image schemata, and event schemata seems appropriate. The idea that schemata serve as a structure to interface between sensory representations and long-term memory representations is derived from work by a number of cognitive psychologists, cognitive scientists, and linguists. I believe, however, that the proposal offered here is unique in suggesting three categories of schemata as cognitive structures matched with the three categories of knowledge representation delineated above.

As organizing structures, schemata can be embedded one within the other, representing knowledge at all levels of abstraction. These embedded complexes of schemata can include combinations of all three kinds identified. For most maps, we can expect at least propositional and image schemata to interact in map understanding. According to Neisser (1976, p. 55), an important function of schemata in seeing "is to direct exploratory movements of the head and eyes." In visual search of maps for items such as placenames or point symbols, propositional schemata may guide search strategy and image schemata may control precognitive decisions about whether a name or a symbol is likely enough to be the one searched for to warrant closer examination. In exploratory analysis of complex map patterns, propositional and image schemata in combination may determine what spatial relationships we initially see. Whenever a map depicts a dynamic event (particularly with dynamic or animated symbols) or is used dynamically as a decision-making tool (e.g., in wayfinding), procedural schemata are likely to play a role.

This typology of schemata and the contention that each higher level schema can include subordinate schemata of more than one type parallels (to some extent) Lakoff's conception of an *idealized cognitive model* (ICM). An ICM as presented by Lakoff (1987, p. 126) offers "a conventionalized way of comprehending experience in an oversimplified manner. It may fit real experience well or it may not."[17] ICMs can use any of four structuring principles: propositional, image-schematic, metaphoric, and/or metonymic. In Lakoff's view, ICMs that deal directly with automatic, normal functioning (functionally embodied concepts) have primacy. Lakoff devotes most of his attention to concepts related to "things." Procedures are given minimal consideration and no separate procedural category is proposed. Although procedures are omitted from Lakoff's view, the concepts of metaphor and metonymy are added as structuring "models." These concepts seem to be at a different level than propositional or image-schematic models (or than my typology of schemata). Both metaphor and metonymy deal with *how* one thing or concept is able to *stand for* another and, as Lakoff presents them, are structures internal to ICMs or that serve as links among elements of different ICMs.

Like Lakoff's ICMs, schemata as presented here are conceived of as simplified structures that can include imagistic and propositional concepts (as well as procedural ones). Lakoff's concept of an ICM (or schema) as a structure that is prototypic of a concept or set of relations, rather than being a template that must be matched exactly, parallels the perspective that John Ganter and I took in our pattern identification model for cartographic visualization. In that context we contended that when scientists apply a pattern identification schemata to visual analysis of a cartographic display, "There is seldom an exact match, instead we do a sort of curve-fitting to get an approximate answer—very quickly" (MacEachren and Ganter, 1990). A similar position is also found in Uttal (1988, p. 29) who states that "humans generate approximate solutions to geometry of the stimulus on the basis of soft rules of approximate inference."

The most completely developed formulation of schemata applied to interpretation of visual information displays is Pinker's (1990) conceptualization of a graph schema. Although Pinker's approach to graph schemata, highlighted in Chapter 2, offers an attractive starting point for conceiving of how people understand graphs (or maps), there is something intuitively uncomfortable about trying to explain graph or map understanding while relying exclusively upon propositional structures. Pinker's propositionally based approach may work quite well for computational models in which all information must ultimately be processed linearly. For human map understanding, however, Arnheim's (1985) contention that humans can reason imagistically as well as verbally (propositionally) seems undeniable. Basic concepts involved with all three kinds of schemata are outlined below. This outline will then be built upon in a proposal for how map schemata might be formed and used.

Propositional Schemata

Some cognitive theorists contend that all knowledge is fundamentally propositional in nature. The major proponents of schema theory seem to have held this perspective because most attempts to formalize the idea of schemata for thinking and knowledge acquisition have modeled schemata in propositional terms. Rumelhart and Norman (1985), for example, discuss schemata and frames as subcategories of organizing structures for propositional representations. They contend that schemata exist for concepts underlying "objects, situations, events, sequences of events, action and sequences of action" (Rumelhart and Norman, 1985, p. 36). They go on to describe schemata as "packets of information that contain *variables*." In their view, each schema has a "fixed part" corresponding to characteristics that are (usually) true of exemplars and a "variable part" that deals with those characteristics that are likely to be unique to individuals. The variable parts are assumed to have default values (information about what to assume when incoming information is left unspecified). They offer a schema for the category "dog" as an example (Figure 4.9). Among the fixed parts of the schema would be "four-legged animal." Variable parts would include color, size, and so on (with defaults possibly set to match an individual dog with which the person is familiar).

A similar example schema for a map symbol interpretation might be associated with activity location symbols on a National Park Service (NPS) map (Figure 4.10). The fixed part of the schema might involve the concept of a small, black, roughly square symbol. The variable part would involve shape of the symbol interior with possible subschemata for categories that have similar appearance (e.g., skiing). The default value for the variable might be the pup tent or picnic table. The concept that schemata are active devices (proposed by Neisser, 1976, and by Rumelhart and Ortony, 1977) is significant here. According to this view of schemata, they are not only a structure for interfacing with visual descriptions but have the embedded goal of determining their own goodness of fit to the visual description. Seeing an NPS-style camping symbol used on

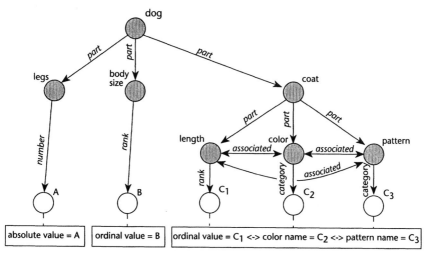

FIGURE 4.9. A schema diagram for knowledge about dogs.

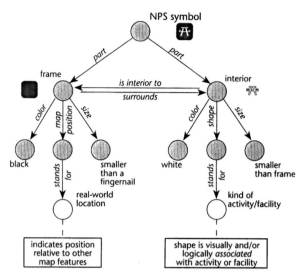

FIGURE 4.10. A possible schema diagram for interpretation of National Park Service point symbols.

a non-NPS map, for example, might activate an NPS symbol schema resulting in an expectation that other symbols encountered on the map will also correspond to the NPS symbol set. This, in turn, would lead to attempts to match other symbols to this schema, a strategy that might lead to confusion if all point symbols on the map did not come from the NPS set. Following this logic, we could predict map-reading problems in any situation for which map symbols associated with a particular map type are mixed with other symbols.

Of the attempts to formalize the concept of propositional schemata, Pinker's graph schema introduced in Chapter 2 comes closest to the kind of schema that might underlie at least some components of map understanding. I will, therefore, use it as an exemplar of propositional schemata and fill in some of the details omitted from the discussion in Chapter 2.

Pinker introduces his conception of a graph schema by diagramming the potential interface role (see Figure 2.12). The graph schema as depicted in this diagram serves as a model that has three roles: (1) to determine the kind of interpretation possible from the visual description, (2) to structure conceptual questions (queries) in a format that is compatible with the structure of the visual description, and (3) to allow different visual descriptions to be categorized (as particular graph types). It seems, then, that the first stage of graph understanding (and map understanding if it is analogous) is to draw upon aspects of a schema that process the visual description at a global level, allowing particular categories of the object to be recognized. This recognition (e.g., that the visual description is of a histogram or a choropleth map) allows the system to access lower level schemata that provide for more precise interpretation. At both global and local levels, the schemata are active in that they include a goal of assigning elements of what is seen to conceptual categories. These, in turn, allow links to be made with knowledge organized according to these categories.

A nonspatial but map-related example may be useful to present the formalisms incorporated in Pinker's propositional schemata. Consider the numbering system for interstate highways in the United States and how we can interpret those numbers to derive more than an arbitrary label. A schema is diagrammed in Figure 4.11 in a manner similar to that used by Pinker for a telephone number. The schema contains slots or parameters

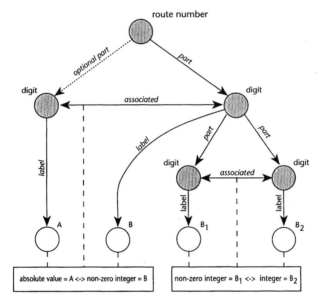

FIGURE 4.11. A schema for interpreting U.S. Interstate Highway numbers. Pinker terms the rectangles in schema diagrams such as this "message flags" because they contain the definitions and relational information needed to derive knowledge, which he terms "conceptual messages." When instantiated with values from the visual depiction, the message flag on the left is interpreted as follows: When A is present, a connecting or bypass route is indicated and B then represents the primary route it is linked to. The right-hand message flag has the following possible interpretations when instantiated: (a) If B_2 is an even number, the route is a dominantly east–west one and the magnitude of B_1 represents the position north–south in the United States (the higher the number, the farther north); (b) if B_2 is an odd number, the route is a dominantly north–south one and the magnitude of B_1 represents the position east–west in the United States (the higher the number the farther east).

for as yet unknown information to be matched against a visual description containing constants in place of the parameters. What the interstate numbering system allows us to determine (if we have the appropriate schema) is (1) whether the highway labeled is a primary route, a regional connector, or an urban beltway; (2) if a through route or connector, whether the route is predominantly east–west or north–south; and (3) if an urban route, whether it is a circular beltway or a crosstown expressway. The schema works because it includes fixed knowledge (that interstate numbers have two or three digits, that two-digit numbers are primary routes, etc.) and variable knowledge that can be filled in from a limited number of possibilities. With this schema, you can recognize that a route labeled M1 is not a U.S. interstate (it is the British equivalent). The recognition is accomplished at the level of individual parameters (in this case representing slots that can be occupied by single alphanumeric characters), with M1 being rejected because the interstate numbering schema limits each parameter to the set of one-digit integers. The number 270, however, passes the recognition test as a possible interstate number because each of the three schema parameters could be replaced by one of the three digits (or numerical predicates).

Pinker goes on to develop a rather elaborate schema for interpretation of bar graphs. The schema contains a fundamental separation of the overall graph framework (L-shaped) from the pictorial content (bars) (Figure 4.12). These elements, however, are linked in two places with the vertical part of the framework joined to the bar heights depicting data values and the horizontal component of the framework joined with the categorical information that provides information on what each bar represents. Because Pinker has devised an explicit formalism for representing graph schemata, I will quote at length from his description of the realization of this schema so that the notation used can be interpreted.

> The height and horizontal position of each bar are specified with respect to coordinate systems centered on the respective axes of the framework, and each bar is linked to a node representing its nearby text. An asterisk followed by a letter inside a node indicates that the node, together with its connection to other nodes, can be duplicated any number of times in the visual description. The letter itself indicates that each duplication of the node is to be assigned a distinct number, which will appear within the message flags attached to that instance of the node.
>
> The message flags specify the conceptual information that is to be "read off" the instantiated graph schema. They specify that each bar will contribute an entry to the conceptual message. Each entry will equate the ratio value of the first variable (referred to in the description as "IV," for Independent Variable) with the horizontal posi-

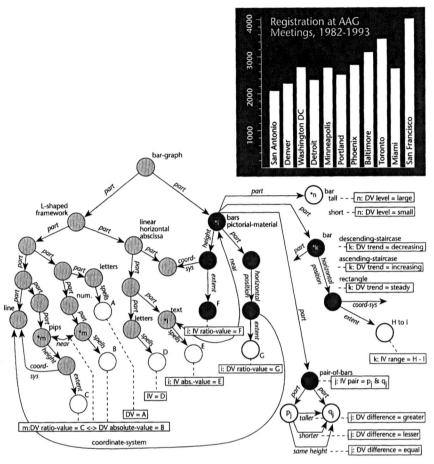

FIGURE 4.2. Pinker's bar graph schema diagram and the kind of bar graph it is intended to deal with. *After Pinker (1990, Fig. 4.17, p. 98, and Fig. 4.18, p. 100). Adapted by permission of Ablex Publishing Corporation.*

tion of the bar with respect to the abscissa, and will equate the ratio value of the second variable (the "DV" or Dependent Variable) with the bar's height with respect to the ordinate. In addition, the absolute value of the independent variable for an entry will be equated with the meaning of whatever label is printed below it along the abscissa. Finally, the referents of each variable will be equated with the meaning of the text printed alongside its respective axis. (Pinker, 1990, p. 97)

Pinker (1990) hypothesizes that the process of applying the graph schema to a particular graph understanding context involves four procedures that can repeat in an iterative cycle as appropriate structures representing the graphic information are accessed. The first procedure is termed the *match* process and involves comparing the visual description (in parallel) with every memory schemata for a visual scene. A goodness-of-fit is calculated for each schema and the one with the highest goodness-of-fit is chosen. This schema is instantiated (by replacing parameters with appropriate constants from the visual description). The main function of the match process is to recognize a graph as being of a particular type.

The second procedure that Pinker describes is *message assembly*. This procedure examines the instantiated graph schema for message flags and when found the message contained is added to the conceptual message (i.e., knowledge representation). It is expected that capacity limits will prevent all message flags from being converted to a conceptual message. Message flags are most likely to be added to the conceptual message if Pinker's third process, *interrogation*, comes into play. Interrogation takes place when the reader needs some specific piece of information not already in the conceptual message. The generation of a conceptual question causes the message flag in the graph schema corresponding to the question to be specifically examined. If the node or relation of interest is instantiated, the message flag "equation" is simply added to the conceptual message. If an uninstantiated parameter is found, the relevant part of the visual description is checked for the information. If the visual description is also missing the information, lower level visual processes will be activated to fill in this portion of the visual description. If graph (and other visual display) reading does proceed as Pinker hypothesizes, it would explain why we often seem only able to see what we know to look for. What will be "seen" will be message flags with the highest encoding likelihood and these will be the message flags that we actively seek to acquire or ones that are instantiated automatically "at a glance" on encountering the display. Bertin's (1967/1983, 1977/1981) semiology of graphics (as discussed above in Chapters 2 and 3 and below in Part II) seems to be grounded in a belief that he has identified the uses of visual variables that will lead to this automatic instantiation of map and graph schemata (although he does not present it in these terms).

Pinker (1990) goes on to propose a final stage that he calls *inferential processes*. This stage includes a wide range of human abilities to draw conclusions from what we encounter. Some examples that Pinker offers include the ability to perform arithmetic operations (that allow a graph reader to note the rate of increase for a set of bars on a graph), the ability to infer from the graph's context what kind of information might be obtained, and the ability to form conclusions based upon domain knowledge. Together, this and the other three procedures are summarized by

Pinker in an elaboration of his initial information-processing diagram (Figure 4.13).

In addition to proposing schemata that provide the structure necessary to interpret specific graphs, Pinker (1990) also advances the concept of a *general schema* (Figure 4.14). A general schema is a more abstract, less detailed schema that might be considered a prototypic schema. It includes those parameters and relations found in typical graphs but does not necessarily represent knowledge about any particular graph type. The main purposes of general schemata, according to Pinker, are to allow entities never before encountered (a new graph type) to be recognized as belonging to a particular category (a kind of graph) and to provide an initial structure that can be modified and added to in order to develop a specific schema for understanding this new entity. The hierarchical structure of Marr's 3-D model representation (described in Chapter 2) allows us to recognize categories without recognizing individuals because information about arrangement of parts of one size is segregated from information about internal organization of those parts. This same ability at a higher level of processing is what Pinker seems to be suggesting with his general schema. The notion of a general graph schema translates easily to the context of maps. In Figure 4.15, the main attributes of Pinker's general graph schema are listed along with the corresponding attributes that I would propose for a *general map schema*.[18]

Pinker's propositionally based schema for graph understanding explains many facets of how graph readers can link their knowledge with sensory input to interpret this abstract form of representation. Many of the key relations specified in Pinker's graph schema, however, are spatial. The kinds of spatial relationships Pinker encodes in propositional form (e.g., relative height, proximity, etc.) are just the kinds of relations that Lakoff argues are more logically represented via image schemata. In dis-

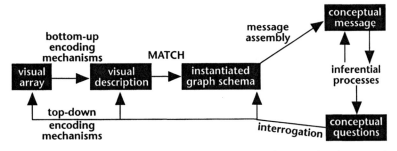

FIGURE 4.13. Pinker's elaborated model of the process of graph schema application. *After Pinker (1990, Fig. 4.19, p. 104). Adapted by permission of Ablex Publishing Corporation.*

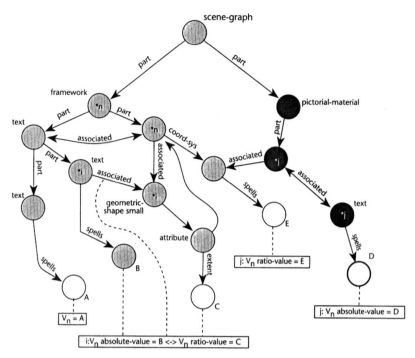

FIGURE 4.14. A depiction of Pinker's general graph schema. *After Pinker (1990, Fig. 4.20, p. 106). Adapted by permission of Ablex Publishing Corporation.*

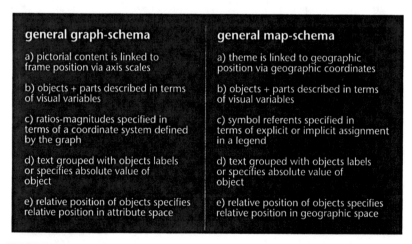

FIGURE 4.15. A chart indicating the main features that Pinker identified for a general graph schema and their proposed complements for a general map schema.

cussing where graph schemata come from, in fact, Pinker presents what might be interpreted as a strong case for image schemata as the core feature of his graph schema. He cites Lakoff and Johnson's (1980) work when he states that "a great many abstract concepts seem to be mentally represented by structures originally dedicated to the representation of space and the movement of objects within it, a phenomena that manifests itself in language in many ways. . . . In particular, abstract quantities seem to be treated mentally as if they were locations on a spatially extended scale" (Pinker 1990, pp. 105–107). It is to this issue of the indispensability of space and the potential role of image schemata that we now turn our attention.

Image Schemata

Lakoff (1987) suggests that image schemata provide a format for encoding information from vision and language simultaneously. If Lakoff (1987, p. 440) is correct when he says, "I believe that they [image schemata] structure our perceptions and that their structure is made use of in reason," image schemata must occupy a more fundamental level than propositional schemata in how we interact with concrete visual representations such as maps. This perspective is complemented by John Allman's contention that the brain extracts information from a noisy environment "by making maps that accentuate the useful information" (quoted in Hall, 1992, p. 16). He goes on to suggest that these maplike structures are not only generated by the brain, but retained in the neocortex for repeated use. Thus, there seems to be not only a functional–evolutionary argument for image schemata, but a neurophysiological one as well.

Our capacity for basic-level Gestalt perception is at the core of our everyday interaction with the world. Lakoff (1987, p. 270) asserts that research on basic-level categorization demonstrates that our experience is preconceptually structured at the basic level and that "basic-level *concepts* correspond to that preconceptual structure and are understood directly in terms of it." Basic-level preconceptual structures arise due to our capacities for dealing with part–whole structure in real-world objects via Gestalt perception, motor movement, and the formation of mental images. Complementary to basic-level preconceptual structures, Lakoff (1987) hypothesizes preconceptual image-schematic structures that can be independent of any concepts. Johnson (1987) makes a strong case for embodied kinesthetic image schemata as the basis of this alternative preconceptual structuring, and Lakoff (1987) elaborates upon this contention. He points out that the logic of image schemata can be represented in formal logic using predicate calculus notation, but the flaw in doing so is that this formalism produces a set of meaning postulates composed of symbols having no inherent meaning but which would be "given meaning" by the set theoretic models they are specified in. According to Lakoff, however, image schemata are inherently meaningful. Lakoff's point seems to be that some concepts that can be described with abstract propositionally based formalisms emanate from *inherently* meaningful schemata or relations that derive their meaning directly from human experience with the environment—what he and Johnson refer to as "embodied image schema."

Lakoff (1987) identifies a *container schema* as one of the best examples of an embodied image schema. The schema consists of a boundary distinguishing interior from exterior; these distinctions can be understood directly in relation to our own bodies which take in air and breath it out, ingest food and excrete, and so on. The container schema is applied to a vast array of experiences in both a literal and a metaphorical sense. Johnson (1987), for example, cites 21 cases in which an in-out schema plays a role in the first few minutes of a typical person's day (e.g., waking *out* of a deep sleep, going *out* of the bedroom *into* the bath, taking the toothpaste *out* of the medicine cabinet and putting the toothbrush *into* your mouth, etc.). Lakoff goes on to describe how the container schema is transferred metaphorically to a variety of contexts. Among the metaphorical translations relevant to maps is the understanding of the visual field in terms of a container with things coming into and going out of view. Beyond this, the container schema plays a direct role in many aspects of mapping: most maps include bounded areas that define an inside and an outside (e.g., lakes, forests, counties, environmental impact zones), individual symbols are interpreted as being in or out of particular regions, and so on.

While these are all fairly literal applications of the container schema, it is also applied in a metaphorical sense to all information depicted on a map. A given population density or soil type, for example, is considered either in or out of a particular map category. The fact that such map categories are prototypic rather than classical categories has been recently recognized by those concerned with visualizing data quality (e.g., Fisher, 1994; MacEachren, 1992b). In an effort to communicate the probabilistic nature of being *in* a soils category and typicality effects within these categories, Fisher (1994) has devised a dynamic soils map that directly depicts the ambiguity of the categories. On this map, individual cells or pixels are shown as in a category (by having a particular hue) for lengths of time proportional to the likelihood that the categorization is correct. Such a dynamic display seems more effective than written documentation of category ambiguity in overcoming the inflexible in–out container schema that we normally bring to interpretation of map categories.

Kinesthetic image schemata are distinguished by Lakoff (1987) from *context-bound specific conscious effortful rich images* of the sort investigated by psychologists (Kosslyn, 1980; Shepard and Cooper, 1982) and cartographers (Steinke and Lloyd, 1983b; MacEachren, 1992a). Instead of representing particular objects, kinesthetic image schemata are models or prototypes dealing with fundamental spatial relations (in–out, up–down, etc.). Like other schemata described here, kinesthetic image schemata can be considered interfaces for matching sensory input to knowledge representations or for structuring thought processes that make use of information retrieved from knowledge representations. Conscious mental images of the sort studied by most cartographers thus far are specific instantiations of a cognitive representation in working memory, consciously generated to solve a particular task. It could be hypothesized, however, that mental images of the sort that can be consciously scanned or rotated rely upon a complex of schemata to generate them from propositional, analog, and procedural representations, with kinesthetic image schemata likely to dominate. In addition, image schemata seem to derive from repeated use of a particular imagistic structure. Over time with repeated use, therefore, we might expect experts to transform context bound rich images into image schemata. In addition to this conversion of mental images to image schemata through repeated use, Kosslyn and Koenig (1992) contend that we often used imagery directly as an aid in symbolic reasoning. In this case, Kosslyn and Koenig contend that reasoning makes use of a mapping of nonspatial problems onto an image (e.g., locating your assessment of people's intelligence along an imagined yardstick). This "mapping" of nonspatial concepts to a spatial framework seems analogous to Lakoff's ideas about metaphorical transfer of spatial relationships into nonspatial situations.

Beyond the container schema, Lakoff (1987) details a number of other image schemata that he contends operate at a preconceptual level as structuring devices for processing sensory input as well as for general processes of reasoning. Among these are *part–whole, link, center–periphery, source–path–goal,* and *linear order* schemata. All of these schemata seem to have direct application to maps. The link schema, for example, is based on the fact that human interaction is structured via physical and metaphorical links (from the umbilical cord and holding a parent's hand to establishing social *ties* or a desire for freedom because we object to being *tied* down). On maps, physical links between places are depicted with symbols representing highways, air routes, and so on, and metaphorical links extend these basic symbol conventions to flow maps depicting topics such as fiscal transfers or fishnet plots of AIDS incidence (that suggest a continuous surface linking isolated locations where the disease is present). Any locations on a map with a continuous symbol stretching out between them are assumed to be linked in some way.

The *linear order* schema suggested by Lakoff (1987) is particularly important in relation to how people grapple with abstractions of space, such as occur with strip travel maps (Figure 4.16). In this case, the map is reinterpreted from the prototypic map schema in which two-dimensional map space represents two-dimensional environmental space. With a strip map, the map space represents a linear sequence or order of decision points along a route between an origin and a destination. An even more abstract metaphorical extension of the linear order schema occurs for quantities depicted on maps. Quantitative information is interpreted as ordered and cartographers attempt to use a variety of graphic means to prompt viewers to recognize the order (to prompt them to apply a linear order schema). Legends on graduated circle and choropleth maps, for example, appear as isolated object groups in which the categories are arrayed in numerical order and symbols are assigned in what is expected to be an intuitively logical order (e.g., by size or value as found in Figure 3.38). A common objection by cartographers to maps of quantities produced by noncartographers is that these maps often ignore the importance of the linear order schema and employ a set of eye-catching (but randomly ordered) hues. Sometimes the hues are ordered, but according to wavelength of the hue. Wavelength ordering is not immediately recognized by our visual system, and therefore is unlikely to prompt the appropriate linear order schema on the part of the viewer.

Many of the image schemata identified by Lakoff (1987) seem to underlie propositional schemata. For example, container schemata are the basis of both classical and common notions of conceptual categories. The prototype theory of categorization presented in this chapter, however, does not "map" to container schema perfectly.

At least one image schemata, the *source–path–goal* schema, may underlie what are identified below as event schemata. The source–path–goal schema is based on the fact that whenever we move anywhere, there is a starting location (a source), a destination (the goal), and a sequence

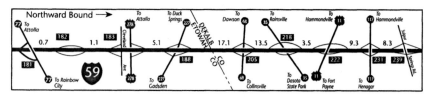

FIGURE 4.16. An example of a schematic strip format travel map that emphasizes the order in which decision points will be reached.

of contiguous locations that we traverse in achieving the destination. The strip map illustrated above is one of the most obvious cartographic examples of a map designed to match this schema. In keeping with Lakoff's view that image schema are embodied, a more appropriate label for the underlying schema might be *origin–path–destination*. The source–path–goal schema is actually the metaphorical extension of this basic spatial movement schema to purposes of any kind.

All of the image schemata described by Lakoff are thought to be based on a structuring of bodily experience preconceptually. These embodied structures have a basic logic and they "motivate metaphors that map logic into abstract domains" (Lakoff, 1987, p. 278). This ability to map image-schematic structures (which are taken to be inherently meaningful) into abstract domains is, according to Lakoff, at the root of human conceptualizing capacities. He goes as far as contending that, "image schemata define most of what we commonly mean by the term 'structure' when we talk about abstract domains" (1987, p. 283).

Of direct relevance to cartography and to geographic visualization is Lakoff's contention that much of our technology is aimed at facilitating this mapping of embodied schemata into abstract domains in order to make those abstract domains seem more like the environment we are used to interacting with (e.g., through the use of tools that can enlarge or reduce objects to human dimensions or can depict nonspatial relationships in terms of familiar spaces).

Elaborating on the role of image schemata in human reasoning abilities, Lakoff (1987, p. 283) offers what he terms a *spatialization of form hypothesis*. This hypothesis includes the following points:

- Categories (in general) are understood in terms of *container* schemata.
- Hierarchical structure is understood in terms of *part–whole* schemata and *up–down* schemata.
- Relational structure is understood in terms of *link* schemata.
- Radial structure in categories is understood in terms of *center–periphery* schemata.
- Foreground–background structure is understood in terms of *front–back* schemata.
- Linear quantity scales are understood in terms of *up–down* schemata and *linear order* schemata.

Overall, then, image schemata, in addition to having directly understood structures of their own, are used metaphorically to structure most abstract concepts. As indicated above, for example, image schemata probably form the basis for our apparent ability to consciously form context-bound rich mental images. Such image-schemata-based mental images seem to underlie some aspects of wayfinding, scientific problem solving, and other endeavors in which geographic space must be conceptualized or for which space serves as a convenient media within which to grapple with abstract concepts.

Lakoff (1987) contends that science has come to rely on tools that extend basic-level perception in ways that allow image schemata to be applied in the manner detailed in his spatialization of form hypothesis. Technology provides "ways of extending basic-level categorization by extending the means for gestalt perception and for the manipulation of objects" (Lakoff, 1987, p. 302). Extensions of basic-level perception and manipulation "makes us confident that science provides us with real knowledge" (Lakoff, 1987, p. 298). This may be one explanation for the early development of maps in the history of civilization and the important role they have played (and continue to play) in attempts to understand the environment. It may also explain why recent advances in computer graphics that made *Scientific Visualization* possible were so rapidly accepted as viable tools for doing science.[19]

Lakoff cites chemistry's use of plots from nuclear magnetic resonance (NMR) spectrometers to demonstrate how concrete representations of abstract concepts can be such complete metaphorical mappings of embodied schemata that they come to be treated as the phenomenon itself rather than a representation of it. The NMR spectrometer produces a spectrogram, in the form of a curve, to represent a substance. Chemists using NMR have come to accept the resulting curve as a *property* of the substance on a par with its crystal configuration and more revealing than the substance's color. This extended basic-level perception has come to be regarded as primary data.

Maps are a means to extend basic-level perception to places too distant, too large, or too complex to be experienced directly. That maps functioning as an extension of basic-level perception are often also treated as primary data is one of the main arguments behind Harley's (1989) call for "deconstructing the map" (an issue to be dealt with more directly in Chapters 7 and 10). Essentially, his argument is that cartographers have (*perhaps* unconsciously) conspired to present maps as a direct window on the world rather than a highly processed representation of selected aspects of it.

Event Schemata (Scripts and Plans)

As we made a case above for distinguishing procedural representations from both propositional and analog representations, it is useful to consider event schemata as a separate category. In practice (when reading a map or walking from home to work) it is likely that we employ multiple schemata to build hierarchical structures for interfacing between "the world out there" and our various kinds of cognitive representation. Following from the contentions that schemata are hierarchically organized and different kinds of schemata can link in complex ways within the hierarchies, it is probable that event schemata include subschemata of image or propositional form. The point of describing event schemata separately is that they are schemata with separate goals and with primary links to a different kind of knowledge representation. In contrast to propositional schemata (which, as described here, structure knowledge about objects, concepts, and categories) and image schemata (that deal exclusively with fundamental spatial relationships), event schemata are structures that place primary emphasis on time, sequence, and process. In understanding the role of the three kinds of schemata proposed here, an analogy to spatial databases is helpful. In spatial-database terms, we would associate propositional schemata with database attributes, image schemata with specification of location in space, and event schemata with specification of location in time.[20] While both propositional and image schemata seem to have a role in all map understanding, application of event schemata to map understanding is probably limited to maps that depict dynamic processes, maps used to initiate or guide a process (e.g., wayfinding), or maps that use dynamic symbols or interaction as a tool to reveal features and relations in the data.

Among the first to direct attention to what I have labeled event schemata were Schank and Abelson (1977) with their concept of *scripts*. Like other schemata, scripts are structures that include variables (slots, parameters, or placeholders for categories of things), requirements about what can fill the variables, and relations among them. In Schank and Abelson's formulations, a distinction is made between variables that must be filled by people (called *roles*) and those filled by objects (called *props*). When we consider the propensity of humans for metaphor, however, this distinction is quickly blurred. In abstract conceptualizations, objects are often assigned characteristics of humans and could fill a "role" rather than a "prop" position in a script. For example, consider a forest-succession simulation and the "role" of a non-native tree species introduced in a forest as it gradually transforms the competitive balance of species and takes over most of the territory from a native species. With many map-reading situations for which event schemata are applicable, then, roles will be filled metaphorically by an element of the map theme.

Schank and Abelson (1977) direct their attention to scripts in a less abstract context: those that deal directly with human events. They focused particularly on describing potential scripts that might guide common kinds of social events. A restaurant script, for example, was devised to explain how the event of going to a restaurant might be structured.

The script includes props (e.g., table, food, bill, etc.), roles (customer, waiter, cook, etc.), entry conditions (a hungry customer with money), and results (a transfer of money and cessation of hunger). As we might find with a script for a play, Schank and Abelson's scripts are divided into scenes that represent sequences of linked events (entering the restaurant, ordering, eating, and exiting).

A key point that Schank and Abelson (1977) make about scripts is that the structure of a script is an interconnected whole with the contents of any variable affecting all others. In addition to being a whole (or a gestalt), scripts are considered to be stylized or stereotypes representing a sequence of actions defining some well-known situation. As with other schemata, then, scripts will have approximate fits to particular events and the script applied will be the one that has the highest goodness-of-fit to the situation. One of the advantages (and sometimes disadvantages) of scripts (or event schemata) serving as prototypic models of events is that certain variables and actions will be assumed even when sensory input does not indicate them directly (or instantiate them in the schemata). For example, a script to handle driving an automobile will include the role of a driver for any other automobile on the highway and assumptions about the driver's behavior. The script's assumptions would probably include an expectation that both the individual with the script and all other drivers (using similar scripts) will stop at stop signs. This commonly held script allows traffic to flow in an efficient manner down through streets having no stop signs, but can lead to accidents when an unlicensed teenager takes a car for a joyride and ignores the script (or uses an alternative based upon other assumptions).

While the standard driving schema (and potential flaws in it) may inform efforts to create real-time cartographic navigation aids, it is in more abstract domains that event schemata are likely to be most important in cartography. On an animated weather map, for example, change in location of the jet stream shortly before initiation of a midlatitude cyclone will probably be interpreted in cause-and-effect terms. This interpretation is likely because standard event schemata probably include an assumption that the order of scenes in an event corresponds to an underlying process. In this case, the schema is likely to lead to a reasonable conclusion, but often it will not.

There are at least two kinds of cartographic applications where event schemata have a potential role. The most obvious, perhaps, is in relation to interpretation of maps that depict dynamic events. Szegö (1987) has provided an elaborate account of temporal cartographic events and the key features of these events that we attempt to map. He structures his analysis around a theater metaphor with stage = geographic setting, play = spatiotemporal event, and actors = objects of the map theme. Beyond consideration of how dynamic events are depicted on maps, however, attention to event schemata is relevant cartographically in relation to human–map interaction with virtual maps and hypertext/hypergraphic environments. Monmonier (1989b) has borrowed directly from Schank and Abelson (1977) in an attempt to develop prototypic scripts for *narrations* that lead a viewer through an exploration of data—what Monmonier calls an *Atlas Tour*. Both Szegö and Monmonier's work will be considered in more detail in Chapter 6 when we look at how formalized cartographic systems for structuring maps can be matched to human perceptual–cognitive structuring systems.

DEVELOPMENT AND APPLICATION OF COGNITIVE MAP SCHEMATA

As noted in Chapter 2, cartographers have directed only limited attention to the concept of knowledge schemata and how they may serve to mediate between sensory input and long-term knowledge representations. Therefore, the ideas presented below, are speculative with a view to identifying the questions we should pose rather than to supplying answers. The appeal of exploring the concept of *map schemata* is that the approach offers a way to bring together ideas about perception of map symbols, cognitive processing of map-derived information, and the roles of knowledge, experience, practice, and training on the part of map readers. I conclude our look at "How Meaning Is Derived from Maps," then, with a brief consideration of how map schemata might develop, how a particular map schema is selected to match individual map-reading situations, and how map schemata may be used.

It would be impossible to delineate all potential sources and uses of map schemata. This section will instead focus on a prototypic case: schemata for interpreting isarithmic representations of terrain. Terrain representation provides a good example of how map schemata might work because the representational techniques devised for terrain depiction have been adapted (through metaphorical extension) to so many topics (from atmospheric chemical concentrations, to population density, to neurophysiology of the brain, to molecular structure, to abstract mathematical concepts).

How Map Schemata Are Developed

It is Lakoff's (1987) belief, and I share it, that human abilities to understand and conceptualize about the world around us (including the ability to make concrete visual depictions of space in the form of maps and to interpret those made by others) are grounded in fundamental principles of human–environment interaction. He contends that "meaningful thought and reason make use of symbolic structures *which are meaningful to begin with*" (Lakoff, 1987, p. 372). These symbolic structures underlie the view of map schemata presented here. Map schemata, like graph schemata or any other schemata described above, then, are assumed to derive essentially from preconceptual structures associated with basic-level concepts and kinesthetic image schemata. These, in turn, are derived in part from characteristics of human physiology and in part from learned behaviors required to adapt to the physical and social world. It is likely that general schemata that have evolved to deal with everyday visual experiences will be adapted for map understanding (and may be the cause of misunderstanding in those situations where decisions of map authors do not follow them).

Physiological Bases for Map Schemata

One potential physiological basis for a map schema can be found in the advance and retreat theory mentioned in the introduction to this section. Cartographers advocating a red-to-blue range of color for elevation maps have argued that this color sequence matches the prediction that long wavelength reds will advance toward a viewer while short wavelength blues will retreat away. If this neurophysiological response does occur, we might expect an image schema to have evolved that makes use of it. If such a red = near–blue = far image schema does exist, adapting it to the map-specific case would be quite easy.

A related potential physiological source for image schemata that may be incorporated in map schemata can be found in opponent process theory for color vision. Eastman (1986) (as noted in Chapter 3) speculates that opponent processes of color might lead to tendencies to *see* certain color combinations as ordered mixtures of two variables (what he called balance scales) with others *seen* as bipolar scales. This speculation suggests two possible extensions of linear order schemata, one a metaphorical extension to ordered mixtures, the other a more direct application to order in two directions (as locations along two paths from the same physical place might be ordered). An alternative schema that could underlie bipolar scales is a front–back schema, which deals directly with position in two directions from some central point. The fact that Eastman's hypotheses about opponent color processes was at least partly confirmed in empirical testing implies one of two possibilities. The first is that Eastman's subjects did have image (or propositional) schemata that linked the physiology underlying opponent process theory to a front–back or linear order schema. Alternatively, repeated exposure to the kinds of color combinations used may have produced the more specific map schema that specifies these relationships as a default when an area fill map is encountered. In either case, Eastman's finding that value differences dominated a hue sequence supports a contention that a light–dark schema is more likely to exist.

In addition to explaining aspects of hue perception, OPT takes into account the presence of light–dark opponent cells. Our eyes are, in fact, dominated by cells that react only to light–dark differences. This physiological propensity to recognize light versus dark (and a range of grays in between) combines in our everyday lives with experiencing a daily order from dark to light to dark. The light–dark schema implied may, of course, itself be derived from the more general linear order schema. Evidence for metaphorical extension of the light–dark image schema to maps is relatively strong, with Cuff's (1973) and Gilmartin and Shelton's (1989) findings that value ranges suggest an order even when no legend is provided. Whether application of the light–dark schema to quantitative maps is preconceptual or conscious is not clear at this point. Only in cases where the map's background is intermediate to the light–dark range, however, is a light–dark schema not consistently applied (McGranaghan, 1989).

The potential sources of schemata noted above are just a sample of those that are possible. In relation to use of layer tints on terrain maps, we have at least three (but probably many more) potential physiologically based image schemata that might be applied separately or in combination. A color range crossing a unique color might activate an linear order image schema extending in two directions from the perceptually unique color in the center of the range (which might be used to indicate sea level); a color range from red to blue might activate a front–back image schema (which could be used, as many cartographers have, to indicate mountains to valleys); and a value range might activate both a linear order image schema and a more = up image schema (with the value range "mapped" to the elevation range and more ink mapped to more elevation). Evidence seems to suggest that a value range is most consistent in prompting application of the appropriate image schema.

Developmental Bases for Map Schemata

There are several developmental issues that underlie the ability to generate the kind of general map schema described above. First, an individual must grasp the idea that a two-dimensional graphic display can *stand for* some portion of the world, what Downs et al. (1988) term a *holistic stand-for relationship*. The holistic stand-for relationship requires, at a general level, an ability to achieve metaphoric mappings between domains (the environment and the map). This ability seems to be achieved between 3 and 6 years of age. In an empirical examination of what constitutes a map, Downs et. al. (1988) found that children of kindergarten age and older recognized certain prototypic maps as place representations, but that younger children frequently misidentified some maps as object representations of various kinds (e.g., a Washington, DC, tourist map was seen as a cage or spaceship by some children). This inconsistency indicates that even when an ability to grasp holistic stand-for relationships is achieved by children, appropriate cues must be present in a particular display before a map schema is selected by the child as the appropriate vehicle for understanding. The ability to recognize certain objects in a display as map cues appears to develop with age and probably relates to the second category of stand-for relationships cited by Downs et al. (1988), *componential stand-for relationships*. A conception of these relationships is required to recognize map symbols as something other than what they appear.

Beyond stand-for relationships, general and specific map schemata depend upon the development of image schemata and metaphorical mappings that deal with space in a variety of ways. To cope with plan views, an individual must be able to mentally project themselves to a vantage point and understand space in a Euclidean sense such that distance and direction can be inferred from relative position. In addition, to deal with geographic hierarchy, the individual must have developed container, linear order, up–down, and part–whole schemata, have the capacity to metaphorically extend these schemata to abstract situations, and have the ability to relate them to one another in an integrated structure. Based on most theories of development, all of the above abilities evolve gradually. We can, therefore, expect that children have quite different map schemata than do adults and that their map schemata will be continually adapting as their ability to cope with space, scale, and representation increases.

The prototypic map, for both children and adults, seems to be a plan view of relatively small scale—for example, a political map of the world showing countries as bounded places identified by varying colored fills (Downs and Liben, 1987). Thus, if we assume that it matches the prototype reasonably well and that it allows maps to be understood rather than misunderstood, the general map schema must include an expectation of a plan view and transformations that allow the world to be split open and flattened (in terms of both sphere to plane and 3-D to 2-D). It is, of course, possible that these elements are absent from some individuals' map schemata. When they are missing, we would expect that individual to make some characteristic misinterpretations of the map (e.g., assuming that the distance from the United States to Japan is much farther than it really is because, on a Euro-centered world map, they are at opposite ends of the world). That young children accept a world map as prototypic of maps does not, of course, prove that they have a map schema that includes a full understanding of plan views and map projection. It seems, on the basis of developmental theory, that the ability to incorporate these ideas in a general map schema must wait until the child develops projective and Euclidean spatial concepts, an ability that preoperational children do not have and one that develops only gradually with topological spatial concepts coming first. Children without projective and Euclidean spatial concepts who recognize a world political map as a typical map must do so through application of a very different general map schema than that assumed to operate for adults.

Developing projective spatial concepts are suggested by Wood (1977) to underlie the gradual evolution of children's depictions of topographic relief. His investigation offers support for the notion that as children progress toward adulthood, an ability develops to deal with projective spatial concepts and that this ability allows for inclusion of plan views as a basic feature of the developing map schema. Downs and Liben (1987), however, provide a variety of evidence to indicate that even many college students do not have all of the spatial abilities that Piagetian theory suggests that they should have. In particular, projective spatial concepts are not universally developed. If projective spatial concepts underlie the inclusion of plan views as a component of general map schemata (as hypothesized here), we must assume that many young adults will be unable to make use of maps effectively.

In addition to plan views, understanding of geographic hierarchy is critical to a fully developed general map schema. Geographic hierarchy, according to Downs et al. (1988, p. 694), "refers to the structural arrangements of places which result from the recursive spatial partitioning of larger areal units into smaller ones, creating a series of nested spaces." To understand that objects can simultaneously exist as independent objects and be *in* another object clearly depends upon development of a container schema and the ability to map it into abstract domains. In addition, however, geographic hierarchy requires a part–whole schema, a linear order schema, and an up–down schema to recognize the part–whole relationships of embedded and superordinate regions and the ordering and dominance implied by the hierarchy. Piagetian theory contends that the operations of class inclusion (knowing that if A is in B, B cannot be in A) and transitivity (knowing that if A is in B and B is in C, then A is in C) are fundamental to comprehension of geographic hierarchy, and that these operations do not emerge until children reach the concrete operational stage of development, at about 7 years (Downs et al., 1988). We should anticipate, then, that children in kindergarten and first grade will not include concepts of geographic hierarchy in their general map schema, and as a result will typically have trouble dealing with maps in which embedded geographic regions appear.

The conventional wisdom, derived from Piagetian theories of development, is that by the time children reach a level of *formal operations* (beginning at about age 12 and continuing through adolescence) all of the basic abilities described above should be present. From this perspective, then, we would assume that all adult map readers will have developed the same set of core image schemata and abilities to metaphorically and metonymically extend them to abstract domains. Differences among adults in map schemata should be due to different causes than differences between children and adults (e.g., the extent to which particular schemata are used on a routine basis or specific training in concepts underlying general or specific map schemata). As noted above, however, conventional wisdom might be wrong. There may be a segment of college-age people who have not yet achieved full development of projective spatial concepts.

General-to-Specific Map Schemata

I propose that humans possess a general map schema comparable to the general graph schema hypothesized by Pinker. Such a schema would draw upon noncartographic schemata derived for and from everyday experience. A general map schema is, in fact, more likely for most humans than a general graph schema because, as Hall (1992, p. 17) points out, "geography is essential to survival—to finding food, to migration, to reproduction and nurturing" and mapping behaviors have been evident in all cultures for the past two millennia. Wood (1992) contends that even cultures that do not make many tangible maps exhibit active mapping behavior. General map schemata will also be closer to real-world experience than graph schemata are because map schemata rely upon space to represent space.

A general map schema, as I envision it, must include at least the following principles (which extend upon those outlined in Figure 4.15): (1) position on a map is linked to position in space via some coordinate system; (2) the map space represents a geographic space within some size range (probably city to globe); (3) the space depicted is continuous and hierarchically structured such that features within the bounds of an area on the map are within that area in the world, features adjacent on the map are adjacent in the world, features connected on the map are connected in the world, and relative distances are consistent throughout the space; (4) point, line, area (and maybe volume) objects exist in this space and are represented in the schema as generic point, line, area, or volume variables or slots; (5) graphic primitives linked to these variables represent the category of point, line, area, or volume object; (6) relationships exist among the graphic primitives such that locations whose symbols look alike are expected to be alike in some recognizable way, locations that look different are expected to be different, and graphic primitive differences are expected to be meaningfully related to actual differences (e.g., differences can be in kind or in order/amount); and (7) scale of map marks (objects + graphic primitives) is independent of the scale at which the geographic space is depicted.

A subordinate schema that might be termed a *general notational schema* underlies (or is embedded within) the last two features of the map schema as defined here. This notational schema is derived from Kosslyn's (1989) interpretation of Goodman's (1976) theory of symbols. The notational schema makes use of a container schema at two levels, that of the symbols and of the phenomenon, what Kosslyn identifies as the syntactic and the semantic requirements of notational systems. The two syntactic requirements are that a given map mark (an object + graphic primitive combination) should map into only one symbol category and that discriminable differences will exist between these map marks such that it is possible to decide into which symbol category that map mark falls. At the semantic level, the notational schema includes the rule that each symbol category will map into one and only one phenomenon category (what Goodman called "a compliance class" and Kosslyn termed "the referent") and that it will be possible to identify the phenomenon category into which each symbol category should be placed.[21]

The general map schema (and notational subschema) outlined is presumed to be held in varying levels of detail by most adults. It requires an ability to understand both holistic and componential stand-for relationships—an ability that, as Downs et. al. (1988) point out, develops gradually in children. Particularly below the second- or third-grade level, then, children are anticipated to hold a very different general map schema if they have one at all. An initial general map schema for a preoperational child, for example, might be "lines on a flat surface" with lines seen as defining the shape of objects on an unspecified surface (or areas to be colored with a crayon).

For adults, specific map schemata can be develop by using a general map schema to recognize that the object of interest is a map, then to note those features that do not match or are missing from the general schema. Developing a specific map schema, then, will be a process of modifying, expanding, and filling in the details of the general schema (or possibly modifying another specific schema that is a reasonable but not perfect fit to the situation at hand). This process can be achieved by being told how, observing map use and inferring how, applying other schemata from related domains that seem appropriate to the situation (e.g., red on washroom faucets means hot and blue means cold; therefore red on a climate map indicates hot and blue indicates cold), or simply by trial and error (e.g., assuming that a blue line on a highway map is a connecting link between two towns and that this link is a road—until it proves to be a river at which point "rivers are blue" might be added to the schema).

Certain specific map schemata—for reading an area cartogram or an isochrone map—require metaphorical mapping of spatial aspects of the general schema into an abstract domain. For a map that depicts categorization uncertainty directly, on the other hand, the general notational schema described above would need to be modified to include the possibility of fuzzy assignment of symbolic categories to phenomenon categories (see the discussion of visualizing uncertainty in Chapter 10).

While most adults probably have a general map schema including at least some of the components cited above, most will have only a few specific map schemata (perhaps for road maps, weather maps, and geopolitical [atlas] maps). Schemata for understanding less common map forms are likely to be accumulated only by experts in some field that makes use of those map types on a routine basis (e.g., geologists develop a geologic map schema that allows them to interpret individual symbols as well as complex color and pattern combinations in an automatic or nearly automatic way, seismic sounding map schema that allow them to translate contours into a three-dimensional image of the density of subsurface layers, etc.). The point is that although they may be based on basic-level categories and/or image schemata (some of which may be innate), specific schemata are learned. Cartographers need to consider the fact that not all potential map users will have the appropriate schema with which to interpret a particular category of map. As a result, attention to training geared toward developing appropriate schemata may lead to more significant improvements in how maps work than our more frequent attempts to improve the maps.

Chang et al. (1985) examined the differences in strategies used by experts and novices for interpretation of topographic maps. They point out that "a common method of teaching topographic map reading is through illustrations of contour patterns made by specific landform features. The idea is that, when map readers become familiar with those contour patterns, the process of map reading would be just a matter of detection and identification" (Chang et al., 1985, p. 88). This suggests a procedure in which there is development of prototypes for landform categories together with a structuring of relations among these prototypes to form a schema. This schema can be used to guide map examination and the categorization of what is encountered. Chang et al. (1985) go on to draw an analogy between schemata that are used to guide map interpretation and patterns stored by chess experts and used to guide decision making concerning potential moves. The authors speculate that the ability to form the initial patterns is related to general abilities for pattern recognition. A study by McGuigan (1957) is cited in which contour interpretation ability was found to be highly correlated with general pattern recognition aptitude.

To make some of the ideas presented about schemata more concrete in relation to map understanding, I describe an idealized specific map schema for layer tint terrain map understanding—referred to for convenience as the *hypsometric map* schema.[22] The schema builds from the general map schema described above to include categories and relationships necessary to deal with both the abstract features of this particular map symbolization and with the nature of the phenomenon underlying the symbolization. At the level of the phenomenon, the schema takes into account the three-dimensional nature of terrain by including extensions of up–down image schema to directly indicate height above a base level. In addition, a source–path–goal schema is used to deal with the potentially circuitous route needed to travel between places while avoiding major terrain features. In relation to the symbolization method itself, linear order and light–dark schemata are linked to the up–down schema in order to map the value range onto the elevation range. In addition, a center–periphery schema will be invoked to interpret closed contours and their indication of a peak or pit at the center of an area with gradually differing elevations.

One of the notable points about the hypsometric map schema suggested is that it makes fairly direct use of several of the kinesthetic image schemata that Lakoff identified as preconceptual embodied schemata, and it does so with only modest metaphorical extension of these concepts. The up = more schema, for example, is interpreted directly and the symbolization indicates it through a straightforward mapping of other image schemata onto the up–down schema. This interpretation has interesting implications for statistical maps of enumeration unit data. A shaded isopleth map, for example, is a minor modification from a layer tint elevation map. In terms of schemata, the only difference is the acceptance of a metaphorical extension of up–down and link schemata to abstract derived quantities (e.g., population density). For readers using the appropriate schema, then, isopleth maps may be remembered better than choropleth maps, not, as I once suggested (MacEachren, 1982), because they are inherently simpler in a geometric sense, but because they are interpreted with an inherently simpler cognitive map schema.

As we move farther from general map schemata toward schemata for specific categories of map, we can expect more complex mapping of embodied schemata into increasingly abstract domains. Schemata for specialized maps typically depend upon sophisticated domain knowledge that has become so ingrained that these schemata begin to take on properties of embodied image schemata. As with the chemists' use of NMR spectrometer plots cited above, map users can come to accept quite abstract symbolic representations as properties of the environment represented because they develop schemata (derived from embodied image schemata) that allow for preconceptual processing of what is seen.

How Map Schemata Are Selected

We can see in an illustration only what we know to look for (i.e., what the schema allows us to see). Cues in the sensory input will prompt a choice (or adaptation) of schemata having the highest goodness-of-fit. This choice can only be from among schemata held by the individual experiencing the sensory input and will be influenced by elements of the display that generates the visual array and subsequent visual description to which our visual system tries to find an appropriate match.

To some extent, the selection and possible modification of a schema to match sensory input will be an individualistic act. The knowledge that a person holds along with experience in dealing with a particular kind of sensory input will determine, in part, the choice of schema that a goodness-of-fit can be evaluated for. Lakoff (1987), however, argues that at the most fundamental level, humans share a common image-schematic structuring of bodily experience. This is essentially an argument that we all possess linear order, container, source–path–goal, and other basic image schemata and that these are the most likely schemata to be initially applied to sensory input. As our visual and cognitive system processes input, a series of increasingly more specific schemata will be applied and it is at this stage that significant differences among individuals in the schemata selected for a particular situation will arise.

Judging from anthropological research on differences in basic-level categories across cultures, we can expect both schema development and schemata matching (to particular sensory input) to be influenced by cultural as well as individual factors. The fact that different cultures divide color hues into a different number of named categories, for example, suggests that people from these cultures would bring different schemata to the act of map reading. A culture whose language includes only the distinction between light and dark, for example, is unlikely to find it easy to develop a schema to deal with a full-spectral, layer tint elevation map.

Beyond the cultural influences and individual limits on the store of schemata available, some evidence suggests that what might be called "global cognitive styles" can influence visual input–schemata matching. For example, an interesting difference in apparent schemata applied to dot maps was identified by McCleary (1975) in his dot map regionalization experiment cited in Chapter 3 (although he does not discuss it in terms of schemata). McCleary had subjects identify regions on dot maps by outlining groups of dots forming those regions. He noted two quite different approaches (i.e., cognitive styles), the identification of many small tight clusters of dots versus a few large general regions (see Figure 3.30). He labeled individuals following the two approaches atomists and generalists, respectively. In relation to issues of schemata, McCleary's results could be interpreted as two different extensions of a container schema. In both cases, regions are viewed as areas in which elements are either in or out. The difference seems to be in how this container schema is adapted to the dot map context and how it interacts with Gestalt grouping principles. Atomists seem able to extend the container schema to situations in which a boundary between in and out is clearly defined by presence or absence of dots. Generalists, on the other hand, seem able to extend this image schema farther. They are able to see regions of similar dot density as units, with the container boundary identified as a change in dot density. Whether the atomist–generalist difference might be due to individual differences at a genetic or environmental level, or to broader cultural differences, is not clear. That these differences exist, however, is certain to influence the impact of interactive visualization tools for different users (see Part III).

No matter how individual or group differences might influence schemata selection, the schemata applied to a particular sensory input also depend upon how the information is presented to us and at what scale (Antes and Mann, 1984). It is these issues, of course, that are of the most concern to cartographers because we have more control over how information is presented than over the schemata available to a reader.

Eastman's (1981) research on symbolic indications of map scale suggests that adults make assumptions about the abstractness of metaphorical mapping used that do not always match those intended by map authors. He found that detailed depictions of base information lead to assumptions that the map is at large scale with relatively unaggregated data. Generalized base information, on the other hand, suggests generalization of, or uncertainty about, the theme. Eastman proposes that we might take advantage of the tendency to assume a similarity in the abstractness of metaphorical mapping throughout the map. By adopting this schema as a map design guideline, for example, we might help viewers grapple with scale change in dynamic interactive data exploration environments that computers are allowing us to create.

In addition to making use of likely schemata, we can also manipulate our map designs to act as a catalyst for specific schemata. There has been one particularly interesting investigation that addresses the issue of how we might prompt appropriate schemata for terrain map reading (although, again, the authors did not present their investigation in these terms). This experiment by DeLucia and Hiller (1982) focused on the role of legends in understanding layer tint elevation maps. Specifically, they devised what they termed a *natural legend* and evaluated it against a standard legend. With the standard legend, sample area fills are arrayed in labeled boxes next to the map. The natural legend, in contrast, depicted (in a somewhat generalized form) a section of the map in which all area fills were present (Figure 4.17). DeLucia and Hiller tested two groups of subjects on a series of map interpretation tasks selected to represent typical data acquisition and terrain visualization tasks. For the data acquisition tasks, their subjects performed slightly better with standard legends. For visualization, however, subjects with the natural legends had

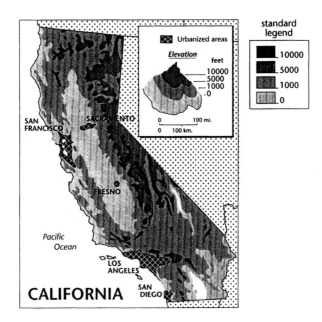

FIGURE 4.17. An exmple of a layer tint elevation map with a natural legend, as it compares to a standard legend.

substantially better performance, leading the authors to conclude that their natural legend noticeably enhanced subjects' ability to visualize a landscape.

In relation to schemata as interfaces between visual descriptions and knowledge representations, DeLucia and Hiller's natural legends have one substantial advantage over the standard box legends. While the box legends can help invoke a linear order schema by arraying categories in order and an up = more schema by putting the symbol for the highest elevation at the top of a column of boxes, a natural legend adds a cue for link and center–periphery schemata as well. These additional schemata are necessary to understand a layer tint terrain map completely because it is essential to recognize that enclosed contours are peaks (or pits) at the center of a region that either rises (or falls) from this central point and that the elevation categories on the map are always linked in the order depicted by the legend (i.e., you cannot move from 100 feet to 300 feet without passing 200 feet). link and center–periphery schemata would not be required in order to use the map effectively for comparing spot elevations, but would be necessary to accomplish visualization tasks such as determining intervisibility between locations. DeLucia and Hiller's results are precisely what this application of schema theory would predict: there was no between-group difference in accuracy for spot height comparisons, but for terrain visualization tasks, the group using natural legends performed significantly better.

How Map Schemata Are Used

Although I do not share Pinker's (1990) opinion that graph (or map) schemata will be exclusively propositional, I do agree with him that schemata (in whatever form they take) will be part of an iterative process of seeing, organizing, interpreting, and interrogating a visual description. The number of iterations required to extract particular information from a visual description will be a function of both the visual description itself and the appropriateness and completeness of the schemata brought to the viewing process.

In detailing his concept of a graph schema, Pinker considers some of the factors that may cause graph comprehension difficulty and how aspects of the graph schema predict these difficulties. In addition, he speculates on how idealized graph schemata can provide clues to training procedures that might reduce graph-reading difficulty in certain cases. He bases his argument on a *Graph Difficulty Principle* that he defines as follows (Pinker, 1990, p. 108): "A particular type of information will be harder to extract from a given graph to the extent that inferential processes and top-down encoding processes, as opposed to conceptual message look-up, must be used." This is essentially the equivalent of Bertin's (1983) distinction between maps to be seen and maps to be read. A map to be seen allows immediate apprehension of particular relations, while a map to be read requires conscious data extraction.

Pinker argues that two factors will influence whether a particular kind of information will be present in a conceptual message on initial interaction with a graph. The information can only be present if "the visual system encodes a single visual predicate that corresponds to the quantitative information" (1990, p. 108). Complementary to this low-level perceptual control on what information makes it into the visual description and becomes available to be matched with a graph schema, is the encoding likelihood that particular predicates will be attached to the graph schema. Those patterns that are highly practiced (or that the reader is prompted to look for) are more likely to be specified in the graph schema, and are therefore more easily extracted from the scene. Together, these two factors allow a prediction that certain graphic constructions will be recognized in an almost effortless way, without the need for top-down queries to prompt the visual system to extract more details.

To support his contentions, Pinker (1990) cites his own empirical work and that by Simcox (1983) related to interpretation of bar graphs, line graphs, and a novel graph form (using length and angular change of linked line segments). Simcox examined the hypothesis that line graphs facilitate encoding of overall trend information while bar graphs facilitate encoding of individual magnitudes or comparisons between magnitudes. Simcox found, in a sorting experiment, that height of individual bars (in a pair) could be selectivity attended to regardless of slope between the two bars. Slope between the bars, however, could not be attended to without interference from relative bar height. For lines, the opposite was true. Sorting by height of a line segment endpoint was slowed by variation in height of the other endpoint, but sorting by slope of the segment was not affected by the overall height of the line. This difference at the level of perceptual organization transferred to tasks in which subjects were asked to extract specific kinds of information from bar and line graphs. Subjects were faster at extracting slope information from line graphs than bar graphs, but faster at magnitude judgments with the bar graphs. The explanation for these results is that schemata for different graph types have different default encoding likelihoods for particular kinds of information.

Cleveland (1991) provides a simple, but dramatic, example that gives further support to this view. His example focuses on the issue of whether comparison of two data sets can be accomplished most efficiently using overlay of the two variables or through merging the data into a single depiction. In his example, the data are depicted with overlaid line graphs, making relative magnitude estimates for various points in the time series hard to judge (Figure 4.18). Pinker (1990) suggests that line graphs are ill-suited to extraction of specific values (or value comparisons between lines or between places on the line) because Gestalt grouping principles cause us to process each line as a whole unit. The schema used by the typical analyst interpreting these bivariate line graphs is missing a key constraint. To interpret the relationship between lines correctly, the schema must include the constraint that comparison is only valid perpendicular to the dependent variable (i.e., parallel to the Y-axis) (Figure 4.19). Using a bar graph instead of a line graph minimizes incorrect comparison because it emphasizes differences in the Y-direction, thus prompting the appropriate schema component. With a bar graph, however, the advantage for judging relative difference between variables is balanced against making the individual slopes hard to extract.

Pinker (1990, p. 121) summarizes the hypothesis that is the basis of his overall theory of graph understanding as follows: "Graphs will be easy to comprehend when the visual system naturally encodes the geometric features of the graph with visual predicates that stand in one-to-one correspondence (via the graph schema) with the conceptual message that the reader is seeking." Graphic constructions seem to work well for par-

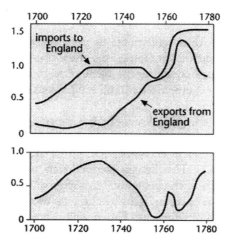

FIGURE 4.18. Example of the difficulty of extracting relative value information from line graphs that depict two variables indendently. On the top graph, the impression obtained by most viewers is that the two variables become quite similar by 1755 and remain so. If, however, we look at the plot of differences between the variables (bottom graph), it is apparent that the decrease in difference reverses itself about 1760, reaching a peak in 1762. *After Cleveland (1991, Fig. 4). Adapted by permission of the author.*

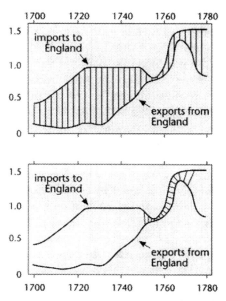

FIGURE 4.19. The bivariate line graph with an indication of the correct interpretation schema in which separation of the lines in the Y-direction is attended to (top) versus the more typical (but inappropriate schema) in which separation between the lines on the basis of shortest distance in any direction is attended to (bottom). *After Cleveland (1991, Fig. 5). Adapted by permission of the author.*

ticular tasks when those constructions parse information in a manner that facilitates perceptual grouping into visual units that match the conceptual units demanded by the problem context (e.g., discrete objects when specific values are required and continuous features when trends are the focus).

The overall theoretical perspective presented here links visual processes involved in generating a visual description with schemata that provide logical interfaces between these visual descriptions and knowledge representations. This dual approach offers a theoretical structure within which much of the perceptual task efficiency work done thus far by cartographers and psychologists can be organized and interpreted. For example, the kinds of differences Pinker and others have found between graph types on different information acquisition tasks is reminiscent of research by Phillips and his colleagues (1984) concerning suitability of different terrain representation methods for a range of map-reading tasks. We can interpret their results in relation to the theory of visual information processing and map schemata detailed here. Phillips's finding that specific values are derived most efficiently from spot heights, second best from contours (which put visual emphasis on discrete lines on the maps), and least well from layer tints (that put emphasis on trend direction) seems to match Pinker's results for tables compared to bar and line graphs. The reverse pattern for a visualization task (that focuses attention on slope and trends) also corresponds to Pinker's graph results.

In extending his theory from quantitative graphs to qualitative charts and diagrams, Pinker (1990, pp. 122–123) contends that the ease with which charts will be processed will depend upon whether the visual system parses them into "important chunks of conceptual information." This argument is similar to the one made by Eastman (1985b) in relation to how people learn map information. Eastman examined the impact of various design changes on how information from a typical reference map was chunked in memory. He found that design differences did influence chunking, but had little effect on subsequent recall of information (see Figure 3.29). He did not, however, examine whether the variations in visual chunking of map information induced by design differences matches with any hypothesized map understanding schema or whether these chunking differences made it more or less easy for map readers to solve different categories of map use task (e.g., route planning vs. relative distance estimation vs. intervisibility estimates).

CONCLUSION

The processes and relationships underlying mental categories interact with knowledge schema complexes to allow humans to interpret the world. When dealing with a map or other graphic display, the human propensity to categorize and to apply knowledge structures to sort out what is seen leads to both the great advantages and (often hidden or overlooked) disadvantages of visual tools as a prompt to thinking.

Graphic displays can be designed to take advantage of human perceptual organization tendencies leading to nearly automatic identification of certain relationships via a mental structuring ability (defined here as schemata). This contention echoes that of a number of other writers who have examined why graphics are effective at communicating relationships and prompting visual thinking. Larkin and Simons (1987, p. 92), for example, propose that "diagrams and the human visual system provide, at essentially zero cost, all the inferences we have called 'perceptual.'" Bertin (1967/1983), in relation to diagrams, networks, and maps, contends that there are specific uses of his set of graphic variables that are immediately apparent to viewers. He distinguishes between *maps to be seen* and *maps to be read*. The distinction translates to a difference in viewing *efficiency*, or the *mental cost* that the viewer must expend to extract information, with maps to be seen extracting a low cost and maps to be read a high cost. Bertin has gone as far as to devise what he terms "standard schemata" to guide the information designer in producing efficient graphics.

Merging Pinker's and Bertin's perspectives would lead to the hypothesis that information displays will be most effective when the information designer uses a logical schema to organize the display and the viewer employs an identical schema when viewing the display. Improvements in information design (for maps and/or other graphics) could be expected to result through user training in the schemata employed by information designers and by information designers developing design schemata that match the general schemata of potential viewers in intuitive ways so that they find it easy to adapt their general schemata to the particular case at hand. In Part III of this book these issues will be addressed from the perspective of the cartographic information designer faced with questions of design for interactive geographic visualization. Before doing so, however, Part II will focus specifically on the rule structures (schemata) of information designers from the perspective of semiotics.

NOTES

1. Virtually everything on a map is a representation of a category, not of an individual entity. The categorization may be one imposed by the cartographer or by the context within which the map is being produced (e.g., industrial, residential, agricultural land use). Categorization may also simply be that imposed by language (e.g., tree, house, road, etc.).

2. Some of the most interesting critiques were never published. These involved reactions by cartographers to medical maps presented at the 1976 Workshop on Automated Cartography and Epidemiology and the 1980 Auto-Carto IV meeting (with its theme of Cartography and Computing: Applications in Health and Environment). In both cases, cartographers were adamant in condemning choropleth maps with quantile, equal interval, and arbitrary data classes, along with those in which raw totals were depicted.

3. Krygier (forthcoming) recounts an experience in which a client (a historian) objected to "optimal" data classification because it did not reveal the pattern that she "knew" was there.

4. To begin addressing the latter issue, we can look to evidence from psychological literature concerning the impact on mental categorization of interaction between characteristics of the scene and of the individual (Barsalou, 1983, 1985; Handel et al., 1980).

5. Jenks's (1967) "data model" concept, as applied to choropleth maps, is founded upon this view of categories.

6. Rosch later backed away from the view that her research supported a

theory of mental representation, emphasizing prototypes as "goodness judgments" people make about category membership (Rosch, 1978).

7. They use "theory" broadly to include both formal scientific models and common sense knowledge of how the world works—both of which can at times be wrong.

8. Within the category of maps, specification of large versus small scale, thematic versus reference, and so on, are also fuzzy categories.

9. Eco (1985a) presents a semiotic argument for culture as a major variable in colors that are actually defined by various groups.

10. Lakoff's (1987) "motivation" can be considered analogous to Murphy and Medin's (1985) "theories." Both provide explanations for the choice of criteria on which similarity is judged, and for which similarities are actually noticed.

11. In the terminology to be developed more fully in the next section, categorization of a display as a map is likely to bring various aspects of our cognitive map schemata to bear on interpretation of the display.

12. See Dobson (1973), Muller (1979), and Peterson (1979) for the debate.

13. Cartographers have frequently used the well known 7±2 limits of short-term memory to argue for a maximum of about six or seven classes on choropleth maps, regardless of where the curve levels out. This argument assumes, however, that for a person to remember a map pattern long enough to compare it to another one, she must convert it to some kind of propositional representation. Research by Phillips (1983), Baddeley (1988), and others suggests that this conversion might be unnecessary. What a viewer sees on a map can be retained in the short-term visual store (or the visuospatial scratch pad) for which the 7±2 limits do not apply.

14. Philosopher Mark Rollins (1989) contends that although evidence for image storage as pictures may not exist, neither does evidence that images are not stored as pictures.

15. Kosslyn and Koenig (1992) provide a compelling account of the neurophysiological basis of imagery.

16. The use of the term "procedural" knowledge representation is somewhat broader in the context of spatial cognition/environmental psychology than in that of skill acquisition. In the latter, "procedural" is reserved for fully acquired skills in which knowledge of procedures reaches a level at which it can be accessed without conscious effort (see Anderson, 1982). From this perspective, simply knowing the sequence of steps required to accomplish a wayfinding task would be considered declarative knowledge. It would be considered procedural at the point where a person can negotiate a trip without paying direct attention to the route (as happens when someone travels the same route to work everyday).

17. The idea that a map itself is an ICM will be taken up in Part II, where we will consider how cartographers and cartographic practice provide maps with meaning.

18. Although Pinker does not discuss category theory, his view of dynamic active schemata that can adapt to new input and evolve with experience seems

to correspond with Barsalou's (1985) contention that categories are not invariant. Pinker's systematic approach to schemata also suggests a mechanism by which Murphy and Medin's (1985) "theories" can be used to structure concepts.

19. See Part III for elaboration of these and other aspects of schemata in scientific visualization.

20. The perspective is thus analogous to Peuquet's (1994) tripartite conceptual model for space–time data structures.

21. See the discussion of "unambiguous" and "vague" signs in Chapter 5.

22. I originally intended to diagram this schema following Pinker's model for a bar graph schema. As I began to do so, however, it became clear that such a diagram would imply a propositional basis for this map interpretation schema—an implication that is far from my view. Ultimately, I decided that the sense of interlinked image schemata, procedural schemata, and propositional schemata intended would be best conveyed through description rather than illustration. The reader is encouraged to refer to the map in Figure 4.17, however, as the suggested schema is contemplated.

PART TWO

How Maps Are Imbued with Meaning

> A systematic study of the universal properties and of the structure of cartographical language is still at an elementary stage.... The fundamental basis for such a study is seen in semiotics.
> —SALICHTCHEV (1983, p. 17)

Part II takes as its point of departure the position suggested in the epigraph from Salichtchev: a semiotic approach to map representation provides a framework for exploring how maps structure knowledge. The contention made is that *maps are imbued with meaning by virtue of semiotic relationships*.

Semiotics can be defined as "the science of signs," with "sign" considered to be a relationship between an expression (the sign-vehicle) and its referent (content). According to Peirce (reprinted in Innis, 1985[1]), who many view as the founder of semiotics, this relationship determines how "something stands to somebody for something else in some respect or capacity," or how one thing represents another. Semiotic relationships pervade map representation from the explicit joining of individual map marks with referents to the proclivity of maps to "stand for" points of view, claims to and about territory, and power over space.

Developments in semiotics have been closely linked to those in linguistics due to a recognition that the fundamental feature of natural language is the stand-for relationship. Some semioticians have treated natural language as a prototypic subfield of semiotics, with an implicit assumption that understanding the semiotic basis of natural language will provide insights into all semiotic systems. For their part, some linguists consider semiotics as the application of linguistic concepts to communication outside of natural language. Following the lead of this latter group, several researchers have attempted to devise a model of cartography based on "natural language" (see Dacey, 1970; Head, 1984; Saushkin and Il'yina, 1985; Lyutyy, 1985). The language analogy, however, has proved to be a weak one both for semiotics as a whole (Hervey, 1982) and for cartography in particular (Schlichtmann, 1984, 1985). In both cases, although some similarities between natural language and other semiotic systems can be pointed to, there are nearly as many differences.[2]

Although certain fundamental semiotic concepts have their origin in the application of semiotics to linguistics, emphasis here is placed on how maps represent via signs—not on how maps converse. Below I outline a semiotics of map representation that does not assume cartographic representation to be a form of natural language. Cartographic inquiry can profit from a semiotic (as opposed to a natural language) approach for two reasons. First, semiotics provides a conceptual framework for developing a cartographic representation logic that can take advantage of what we know about cognitive representations, mental categories, and knowledge schemata. Second, aspects of semiotics that deal with meaning offer a way to integrate approaches to map representation that emphasize both explicit and implicit meaning, logical and expressive meaning, denotation and connotation, and more. We gain an opportunity to draw strength from the diversity of views on the role of maps in society rather than becoming entangled in arguments about map "objectivity."

As a complement to the private, cognitive perspective on representation taken in Part I, here I use a semiotic perspective to consider the public functional and lexical perspectives that together establish the relationships that underpin the "meaning" with which maps are imbued. A semiotic theory of map signs appears to be analogous to the theory of "schemata" discussed in the preceding section, but considered from a philosophical–logical–sociological rather than a cognitive point of view. Howard (1980) hints at a link between schemata and semiotics when he comments that "some cognitive psychologists divide their 'inner representations' or 'schemas' into enactions, icons, and symbols." As will be detailed below, a "sign" is generally considered to be a mediating relationship between a sign-vehicle (e.g., map symbol) and a referent (category of geographic feature to which the symbol refers). As presented in Part I, a schema is a mediating relationship (or set of them) between a visual description (a set of sign-vehicles generated by vision) and a cognitive representation (a set of referents). Frutiger (1989), in discussing the historical cross-cultural development of graphic symbols, goes so far as to suggest that "schemata" for viewing and thinking about the world around us may be (to some extent) inherited and provide a propensity to recognize (or create) certain archetypical forms. Whether innate or learned, the point is that knowledge schemata are intertwined with human use of graphic symbols.

A semiotic approach to functional and lexical properties of map representation is presented here in three parts. First, Chapter 5 provides a brief primer on semiotic concepts relevant to mapping. Two main applications of the semiotic approach to cartography are then detailed. Chapter 6 focuses on efforts to devise logical systems for cartographic representation (with the goal of clarifying meaning and reducing ambiguity). Chapter 7 addresses the mechanisms by which cartographic entities (including map "symbols," patterns, and entire maps) are explicitly and implicitly assigned (or acquire) meaning. The distinction, then, is between the structure of semiotic relationships and the process of establishing them.

NOTES

1. This citation of Peirce's work, and those similarly labeled below, refers to an excerpt from *The Collected Papers of Charles Sanders Peirce* (Hartshorne and Weiss, 1931). The essay reprinted from the above source involves a compilation of pieces from several manuscripts and two contributions to Baldwin's 1902 *Dictionary of Philosophy and Psychology*.

2. See Head (1984) for the most elaborate attempt to depict maps as natural language.

CHAPTER FIVE

A Primer on Semiotics for Understanding Map Representation

Two dominant semiotic traditions can be identified, one with roots in C. S. Peirce's "semeiotic" and the other in Saussure's "semiology" (associated with North America and Europe, respectively). Peirce (1839–1914) approached semeiotic as a science of signs, taking the perspective of a scientist (with training in chemistry) interested in the "logic of science." He achieved induction into the U.S. National Academy of Sciences in recognition of his contributions to logic. Saussure (1857–1913), in contrast, was trained in France as a linguist. He envisioned the topic of semiology as a science which "studies the life of signs in society" and attempted to link it with the developing field of social psychology, a field that he also saw as focusing on the most important components of language (Hervey, 1982). Saussure's semiology presented language as the analytical paradigm for all other sign systems. The traditions traced to these founding scholars remained, for more than half a century, surprisingly separate. According to Nöth (1990), it was not until 1969 that the term "semiotics" was generally agreed upon as the label for the discipline (when it was selected by the initiators of what eventually became the International Association of Semiotic Studies).[1] Although many "schools" of semiotics (with both theoretical and applied emphases) exist today, their links to either the Peircean or the Saussurean tradition remain evident. The Peircean tradition has provided the most elaborate analysis of the typology of signs and how they "stand-for" their referents, while the Saussurean tradition has had a decisive influence on the semiotic theory of codes (i.e., the study of sign systems).

The above distinction suggests two fundamental issues of semiotic inquiry relevant to map representation: *the nature of map signs* as relationships between map marks and referents (and associated typologies of signs) and *the nature of map sign systems* as relationships among map signs. Each includes functional components (i.e., related to the mechanism of representation) and lexical components (i.e., related to kinds of meaning and how it is achieved).

THE NATURE OF SIGNS

Among the initial issues that we must address is a precise definition of the topics of discussion: the sign and its components. The terminology of semiotics can be particularly confusing because of the interdisciplinary nature of the field (with terms contributed from linguistics, philosophy, anthropology, logic, psychology, and sociology). To make matters worse, individual scholars are often inconsistent in their own use of terminology. Part of the difficulty seems to be related to issues of dual-category representation (common and scientific), discussed in Chapter 4. Many terms used by semioticians (including the term "symbol") had common meanings before they were usurped for scientific use. Often the common and scientific uses become intermixed, even within the same essay. I preface our semiotic primer, therefore, with a discussion of the terminology to be adopted.[2]

At the broadest level, semiotics considers the relationship between an "expression" and the "concept" to which that expression refers. Not all semiotic theories include reference to the "real" world. In those that do, however, sign is expanded to include the "object of reference." Following Nöth (1990), I adopt the convention of using *sign* to specify this overall relationship: the "entity" encompassing an expression, the concept it stands for, and the object of reference. Thus, a sign (as defined here) is not a "symbol" in the common sense, nor any other kind of mark that carries meaning. Neither is it a physical device used to inform or dictate behavior (as used in the term *traffic sign*).

The "carrier of meaning" will be referred to here as a *sign-vehicle*. While this term is perhaps clumsier than "symbol" or "expression," it does not suffer from the multiple meanings of "symbol" nor the implied links to natural language of "expression." In addition, as will become clear below, the term "symbol" has come to have a very narrow definition in the semiotic literature. A physical traffic "sign" is an example of a sign-vehicle, as is a drawing of a pair of crossed pickaxes on a topographic map.

The "meaning" (or concept) to which the sign-vehicle refers is termed the *interpretant*. This term (borrowed from Peirce) has been selected over the many alternatives (meaning, sense, idea, content, signatum, notion, significatum) because it is unlikely to be confused with an actual object in the real world (as "content" or "signatum" might be) and because it suggests an act of interpretation (making clear that the sign relationship is more than one of simple definition).

Finally, the object of reference to which the sign-vehicle is linked via the sign (in those theories where such an object is included as part of semiotic inquiry) will be labeled the *referent*. This term does not imply that all signs represent physical entities (as Peirce's use of "object" does), nor does it limit consideration to explicit relationships (as Morris's [1946/1971] use of "denotatum" seems to).

Models of the Sign

The two semiotic traditions referred to above (i.e., those traced to Peirce and to Saussure) are linked with two general models of "sign" as a relation. These models are referred to as *dyadic* and *triadic* models, alluding to the number of elements identified in their sign relationship.

For Saussure in 1916 (1959 translation reprinted in Innis, 1985) a sign was the relationship between a sign-vehicle (what he called a *signifier*) and an interpretant (what he called a *signified*). In his linguistic application of the idea, these become a "sound image" and a "concept," respectively (Figure 5.1). This dyadic model for Saussure's "sign" explicitly omits the referent. For Saussure, semiology (i.e., semiotics) operated within the sign system which was, in his view, completely arbitrary (Nöth, 1990). Saussure's theory of the sign, then, had nothing to do with how sign-vehicles refer to real-world entities, only with how they refer to mental concepts. As Nöth (1990, p. 61) notes, "According to Saussure's structuralist view of semantics, meaning is the *value* of a concept within

FIGURE 5.1. A depiction of the sign *tree* as proposed by Saussure (1959 translation reprinted in Innis, 1985) and a similar relationship as it might be applied to the sign *campground* as depicted on a map. Note that the referent (or the signified) in the map case could be a mental image of a campground, a propositional representation of a campground, or the word "campground," which in turn could have its own sign relationship with an image or proposition. *Derived from Saussure (1916/1986, p. 99).*

the whole semiological system.... These semantic values form a network of structural relations, in which not the semantic concepts as such, but only the differences or oppositions between them are semiotically relevant."

Applying this view to mapping, as at least one recent critique of cartography has done (see Woods and Fels, 1986), we arrive at the conclusion that maps do not refer to the real world, but to concepts about the world. This perspective on map representation seems counterintuitive if considered in relation to a general map schema that has topographic maps as a prototype. Most cartographers would probably argue that the real-world referent is a critical part of the signifying relationship for a topographic map.[3] The idea that map signs do not refer directly back to the real world is most plausible when applied to maps of something like global-climate-model predictions of temperature change due to increased CO_2. Questions of what a map's referents are (and whether there are referents corresponding to all sign-vehicle—interpretant pairs) will be taken up below (in Chapter 7).

In Peirce's theory of signs, the referent (his "object") plays a critical role.[4] The sign, according to Peirce (Innis, 1985, p. 5), "is something which stands to somebody for something in some respect or capacity. It addresses somebody, that is, creates in the mind of that person an equivalent sign, [an interpretant].... The sign [the interpretant] stands for something, its *object*." Elsewhere, Peirce (quoted in Hervey, 1982, p. 27) states that "a sign [sign-vehicle] mediates between the *interpretant* ... and its object." Hervey uses this statement to propose a graphic model of the sign relation (Figure 5.2). He describes the implications of this interpretation of Peirce's triadic model as follows:

> In this triadic correlation, the role of a sign is to establish a habit or general rule determining both the way the sign is to be "understood" on the occasions of its use, and the kind of perceptible, or at least "imaginable," features of experience to which the sign may be applied. Thus we may take it that the way a sign is to be "understood" implies some kind of mental activity or state, whereas the features to which a sign can be applied presuppose something perceptual or experiential. (Hervey, 1982, p. 28)

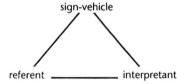

FIGURE 5.2. Hervey's graphic interpretation of Pierce's triadic correlation between object (referent), sign (sign-vehicle), and interpretant. *Derived from Hervey (1982, Fig. 1.6, p. 28)*.

Cartographically, this tradic view of signs suggests attention to the way in which map "symbols" are simultaneously linked to actual or possible referents and to concepts about those referents on the part of the map user. From this perspective, map "symbols" might be evaluated on dual grounds: on the basis of the concepts they prompt (or the knowledge schemata they cue) and on the basis of the manner in which they correspond to the real or the imagined world.

Although a triadic interpretation of the theory of signs dominates North American semiotic literature, the relative position of the three elements in the relationship has varied. The primary alternative to Peirce's view of sign-vehicles as the mediator between referent and interpretant was offered by Ogden and Richards (1923). They also seem to have been the first to provide a graphic depiction of the sign relationships in the form that has become known as the "semiotic triangle" (Figure 5.3). The Ogden–Richards triangle depicts an interpretant (which they call the "thought or reference") as mediator between the sign-vehicle (labeled "symbol") and the referent. Their immediate application was to language. Their diagram, then, was meant to suggest that a word (as a sign-vehicle or symbol) has a causal relationship to a thought (interpretant), which in turn refers to a thing (or referent). The "stand-for" relationship between the word and the thing is depicted as less direct than that between interpretant and either sign-vehicle or referent. The word is thus portrayed as linking the thing, primarily through a thought or concept (rather than the concept linking to the thing through the word).

The Ogden–Richards triangle has somewhat different implications for the analysis of cartographic signs than does the initial Peirce triadic model. In the Ogden–Richards approach, emphasis is placed on the nature of interpretants as links between map "symbols" and referents. Attention, for example, might be directed to alternative interpretations of the sign-vehicle–referent relationship. As suggested in Part I, these alternative interpretations can be modeled in terms of knowledge schemata as the mediator between what is seen and what is known. As discussed below, one application of the Ogden–Richards semiotic triangle to visual representation has reinterpreted the connection between sign-vehicle and referent to suggest a connection that can vary in strength (i.e., re-

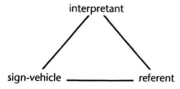

FIGURE 5.3. A depiction of the semiotic triangle with the interpretant (rather than the sign-vehicle) as mediator. *Derived from Ogden and Richards (1923, p.11).*

flecting the degree of similarity between sign-vehicle and referent) (Knowlton, 1966) (Figure 5.4).

Typology of Signs

Signs, whether they are treated as dyadic or triadic relationships, can be categorized on a variety of criteria. Nöth (1990), for example, cites Eco's (1973/1977) proposal of ten criteria.[5] Peirce (Innis, 1985) initially offered three trichotomies of signs, from the point of view of the sign-vehicle, the referent, and the interpretant. From the sign-vehicle perspective, Peirce proposed *qualisign* (a quality that is a sign-vehicle), *sinsign* (a thing or event that is a sign-vehicle), and *legisign* (a law that is a sign-vehicle). From the referent perspective. Peirce proposed *rheme* (a sign of qualitative possibility, it represents "such and such a kind of possible object"—a name is a rheme), *dicent* (a sign that represents in terms of or asserts the actual existence of something), and *argument* (a sign that asserts the truth of something). Perhaps the most important sign categorization criteria in relation to cartographic applications (and certainly the one that has attracted the most attention from both semioticians and cartographers) is the kind of relationship that exists between the sign-vehicle and the referent (i.e., from the point of view of the interpretant). This was the criterion selected by Peirce in devising his well-known typology of *icon*, *index*, and *symbol*.

For Peirce (Innis, 1985, p. 7), the icon is a sign-vehicle that refers "merely by virtue of characters of its own." Through a rather convoluted argument, Peirce ends up deciding that "a possibility alone is an Icon purely by virtue of its quality," thus essentially eliminating the category of icon as a visible sign-vehicle. While no true icons exist (at least ones that can be used on maps), Peirce (Innis, 1985, p. 9) contends that other sign-

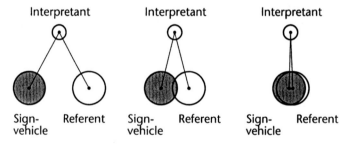

FIGURE 5.4. A depiction of Knowlton's variable strength semiotic triangle. *After Knowlton (1966, Fig. 5, p. 171). Adapted by permission of the Association for Educational Communications and Technology, Copyright 1966, Washington, DC.*

vehicles can be *iconic*, "that is, may represent its object [referent] mainly by its similarity, no matter what its mode of being." The term *hypoicon* is coined to specify these iconic sign-vehicles. Three forms are identified: (1) *images*—those that represent through similar visual qualities (e.g., a drawing of a ranger station on a National Park Service map); (2) *diagrams*—those that represent the relations of parts through analogous relations of parts (e.g., a network diagram that topologically represents stream order for streams in a particular drainage basin); and (3) *metaphors*—those that represent through a parallelism in something else (e.g., the use of up–level–down line orientation to represent increasing–stable–decreasing pollutant indices—as on the map in Figure 3.39). More so than the other hypoicons, metaphorical iconicity is a kind of similarity that generally depends on cultural codes (Lakoff and Johnson, 1980), although the particular line-orientation example cited could be argued to depend upon universal kinesthetic image schemata.

Peirce's definition of an index was as a sign-vehicle that refers to its referent "by virtue of being really affected" by it (Innis, 1985, p. 12). The reference is due "not so much because of any similarity or analogy with it, nor because it is associated with general characters which that object [referent] happens to possess, as because it is in dynamical (including spatial) connection both with the individual object [referent], on the one hand, and with the senses of memory of the person for whom it serves as a sign, on the other hand" (Innis, 1985, p. 12). Examples he cites include a yardstick, a photograph, and a pointing finger.[6] Three distinguishing features of indices are noted: (1) they have no significant resemblance to their referents; (2) they refer to individuals or individual units, collections or continua; and (3) they direct attention by "blind compulsion." With index, Peirce clearly had in mind a sign-vehicle property rather than a kind of sign, and even comments that it would be "difficult, if not impossible," to identify a pure index. Also, Peirce suggests that most sign-vehicles will have some level of indexical quality. Among the most clearly indexical sign-vehicles on maps are the graticule lines or tick marks used to "indicate" latitude and longitude.

Although map graticule provides an example of a map sign-vehicle that might be considered primarily indexical, any map symbol with fixed position has the property of spatial indexicality, regardless of the other sign aspects it may possess. The possibility that a sign can be indexical in relation to location while at the same time signifying some attribute of the place (perhaps iconically) accords with Keates's (1982) dichotomy of locational and substantive information and Schlichtmann's (1985) dichotomy of spatial and nonspatial characteristics. As Schlichtmann points out, this distinction can be separately applied to sign-vehicles and to interpretants (his sign expression and content).

The symbol, in Peirce's typology, represents by virtue of a "law" or "rule" or "convention." The choice of the term "symbol" was perhaps the most unfortunate one made by Peirce. This is because the term is defined in a multitude of ways by other semioticians, some equating it with "sign" and others with its Peircean opposite, the icon.[7]

Various authors have drawn on Peirce's trichotomy of iconic, indexical, and symbolic sign types. Most of these have identified limitations in scope. As a result, several alternative typologies have been advanced that contain more than three sign types. Two will be noted here because they deal with issues relevant to cartographic signs.

Hervey (1982) outlines a typology that he attributes to Martinet (1973) and associates with the branch of semiotics called "functional semiotics." The key differences between this typology and that of Peirce is an apparent limitation of index to "natural indices" (e.g., smoke indicates fire) and the addition (between Peirce's icon and symbol) of a category that has partial similarity (or "motivation"). The functional semiotic typology, then, includes the following types: (1) (natural) index; (2) icon (limited to strong similarity between sign-vehicles and referents—neither of which exist expressly for semiological purposes; a portrait of Queen Elizabeth is one example given); (3) symbol (for which the referent is not arbitrary—it exists—and the sign-vehicle to referent link is partly motivated and partly arbitrary or conventional); and (4) sign (for which both the referent and the sign-vehicle are arbitrary—the example provided is the sign-vehicle "pig" as it refers to the category "species of pig"). Although the addition of a partially motivated (nonconventional) category is an important addition that will be considered in more detail in the next chapter, the use of "sign" to refer to a category of sign and of "symbol" to refer to what Peirce might have termed a hypoicon are problematic.

Sebeok (1976) presents a different kind of variation on the Peircean typology of signs. He begins by stating that he is not attempting to classify signs, but only "aspects of signs." The distinction is an important one because it emphasizes the point (echoed by Eco, 1985b) that signs are seldom of one clearly defined type; instead, they have varying degrees of a range of properties. As noted above, this interpretation of sign typology was implicit in Peirce's approach, but it was ignored by many subsequent authors. These authors then criticized Peirce's categories by citing examples of sign-vehicles that did not fit unambiguously into one category or another. In Sebeok's system, Peirce's original aspects of signs (icon, index, and symbol) are retained with virtually the same meaning as delineated by Peirce. Three additional categories are added. They seem to deal with the intended impact of a sign as much as with the relationship between sign-vehicle and referent. Sebeok's (1976) three additions are: (1) signal—"*When a sign token mechanically or conventionally triggers some reaction on the part of a receiver, it is said to function as a signal*" (p. 121); (2) symptom—"*A symptom is a compulsive, automatic, nonarbitrary sign, such that the signifier is coupled with the signified in the manner of a natural link*" (p. 124); and (3) name—"*A sign which has an extensional class for its designatum is classed a name*" (p. 138). Names have no common property other than a shared label.

Clearly the "name" category is relevant to maps. While I can think of no use of "symptom" as an explicit map sign, it seems that symptoms are related to the use of map schemata for pattern analysis. A map schema that allows a person to recognize a relationship between homeless shelters and wealth of residents, as a *New York Times* map (Ahmad-Taylor and Montesino, 1992) juxtaposing income areas and facility locations allows, may function due to implicit signs that have the "aspect" of a symptom, an apparent natural link. The map can be considered to work to the extent that certain implicit signs (created by attribute–location combinations) are seen as a "symptom" of a particular public policy (protecting the rich from contact with the homeless). In contrast to symptoms, signals seem to be an explicit aspect of signs in a limited range of mapping contexts. An example is the AAA map I picked up to use on a trip to the Association of American Geographers meeting in Atlanta. The yellow highlight drawn by the AAA travel counselor is intended to "signal" me to turn or continue my current direction at various interchanges along the route. The role of dynamic dashboard-mounted maps in wayfinding is increasing and the advent of hand-held personal navigation assistants is predicted (Rhind, 1993). As a consequence, we are likely to see development of a set of dynamic map sign-vehicles intended to act as signals for travel behavior.

Typology of Discourse

Discussion of "signal" brings us neatly to our next topic, Morris's typology of discourse. Signals have a clearly behavioral goal. Morris (1901–1979) had roots in Peircean semiotics but approached all aspects of the field from a behavioral perspective. For Morris (1964, p. 6) the interpretant is defined as "a disposition to react in a certain kind of way because of a sign." Although issues of sign-vehicle to referent relationships were still of interest, an additional avenue of inquiry became dominant—how signs influence (or are intended to influence) behavior. Morris's goal in this effort was to develop a typology of the major kinds of discourse in everyday life. He set out to accomplish this task by delineating several "modes of signifying" that relate to purposes of sign use. Together these modes and purposes define a matrix of discourse types.

Morris initially proposed five modes of signifying which he reduced to four in his later writing (Hervey, 1982). These five are:

Designative: The sign directs attention to a referent by signifying "observable" properties—properties are "designed." A map example is a choropleth map of population density in which the sign-vehicle (shade or color) designates the density range for that country.

Appraisive: The sign signifies the "consummatory" properties of a referent, it directs attention to preferential treatment of the referent, or it assigns a value judgment. Signs on a highway map such as yellow = scenic route, dashed = unimproved road, etc., constitute appraisive signs.

Prescriptive: The sign signifies how a situation should be reacted to, it directs attention to performing a response (a signal as defined above would, typically be prescriptive). An arrow on a city street map specifying a one-way street can be considered prescriptive.

Indentificative: The sign directs attention to a certain spatial–temporal region (an index will use this mode of signifying). All map signs have indentificative properties in relation to geographic position. Some signs, like simple dots to indicate city location, may be primarily indentificative in mode. Morris dropped the indentificative mode from his later writing.

Formative: The sign signifies in a "logical," "grammatical," or "structural" way. Conjunctions ("and," "or") are considered formators. They perform operations on other signs. On maps, signs can use formative mode to suggest links between places. A double-ended hook symbol (common on tax maps) uses formative mode by indi-

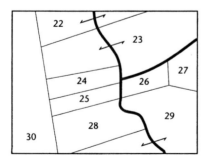

FIGURE 5.5. An example of the formative mode of signifying in which the sign's interpretant (for parcels 22, 23, and 29) is "these two map regions are part of the same land parcel."

cating an "and" relationship between two parcels of land (Figure 5.5).

Complementary to the modes of signifying, Morris (1946/1971, p. 172) offered four dimensions of use that address the "question of the purpose for which an organism produces the sign which it or other organisms interpret." These dimensions of use include:

Informative: The sign is intended to inform about something.
Valuative: The sign is intended to aid in preferential selection.
Incitive: The sign is intended to incite response-sequences.
Systemic: The sign is intended to organize sign-produced behavior into a determinate whole.

Morris (1964) contends that the modes of signifying and dimensions of sign use have no necessary links. In fact, he provides a matrix indicating all possible combinations and types of discourse to which they might apply (e.g., the designative–informative combination is typified by scientific discourse and the prescriptive–valuative combination is typified by political discourse). He does note, however, that "in general, designative signs are used informatively, appraisive signs are used valuatively, prescriptive signs are used incitively, and formative signs are used systemically" (p. 15). Although we might expect most map signs to match signifying modes and dimensions of sign use in this way, it is those cases that deviate from expectations that provide the most food for thought and that have attracted interest among recent critics of cartographic practice (e.g., Harley, 1989; Wood, 1992). I will come back to this issue in Chapter 7.

Morris's typology of discourse is closely paralleled by Guiraud (1975) who worked from a more Saussurean linguistic base in developing an analysis of "the functions of communication."[8] Since Guiraud contends that the function of signs is to communicate, his analysis can also be considered a typology of discourse. Guiraud's typology includes the following functions: referential, emotive, connotative or injunctive, poetic or aesthetic, phatic, metalinguistic, understanding and feeling, meaning and information, and attention and participation. Even without a detailed account of these functions, it is clear from their labels that correspondence exists with Morris's categories. The main difference seems to be in Morris's separation of modes of signifying from dimensions of sign use. Guiraud's one-dimensional typology does not allow the potential mismatches between how signs signify and how they are used to be identified.

Guiraud's (1975) poetic–aesthetic category has the least clear complement in Morris's typology. Defined as "the relation between the message and itself," this category suggests semiotic analysis of maps as an object of expression in and of themselves. Keates's (1984) views on art in cartography (see Chapter 1) offer a hint of the directions such analysis might take (although Keates does not couch his argument in semiotic terms).

How Signs Signify: Specificity or Levels of Meaning

While an icon–index–symbol typology of signs focuses on the sign-vehicle to referent link and a typology of discourse focuses on how and why signs are used, attention can also be given to the directness/explicitness of the link between sign-vehicle and interpretant. From a logical perspective–typological perspective, Morris (1946/1971) proposed eight categories of sign (or sign-vehicle) distinguished primarily on the basis of consistency or specificity of meaning.[9] Hervey (1982) discounts the possibility of the first category (sign-vehicles that are not part of a sign family), but goes on the summarize the remaining seven. These are defined as:

Singular sign: The interpretant permits only one referent (for a map, an example would be "capital of the United States in 1990").

General sign: The interpretant permits any number of individual referents (a map example might be "river" but could also be "Columbus"—a name-sign for which several referents exist within the United States).

Interpersonal sign: Several interpreters share the same signification (a good map example might be signs dealing with geologic structure for which those trained in geology share a common understanding of the map sign-vehicles, their referents, and their interpretants).

Comsign: Has the same signification for the producing organism and the interpreter (the goal of most communication-model-oriented cartographic research was to develop, discover, or teach comsigns).

Vague sign: Does not allow a determination of whether a particular entity is or is not a referent of the sign (on a map, "forest" without the necessary spatial qualifiers that determine the smallest area that will qualify and the required density of trees is a vague sign).

Unambiguous sign-vehicle: Has one interpretant (this is the goal of most map symbology, but it is only met in those situations where we limit the definition of interpretant to an explicit meaning specified in a legend).

Ambiguous sign-vehicle: Has several interpretants (as discussed in Chapter 7, sign-vehicles on maps are probably ambiguous in this sense more often than cartographers have cared to admit).

Issues of vague signs and ambiguous sign vehicles are also taken up by Guiraud (1975) in relation to what he terms the "codification of signs." According to Guiraud, all signification (the relation between sign-vehicle and interpretant) is codified, or defined by a "code" or convention between individuals for whom the sign serves a communication function.[10] That code may be explicit or implicit. Guiraud's concept of convention, then, is one of degrees. Codification is viewed as "an agreement among the users of a sign [who] recognize the relation between the signifier [sign-vehicle] and the signified [interpretant] and respect it in practice. Such agreement may be more or less inclusive and more or less precise" (Guiraud, 1975, p. 25). Signs, then, can be *monosemic* (i.e., unambiguous) and precise, or *polysemic* (i.e., ambiguous) and imprecise. Similarly, signs can be *explicit* versus *implicit*, *conscious* versus *unconscious*, and *denotational* versus *connotational*.

For natural language, polysemic codes are the rule. With a polysemic code, a single sign-vehicle has multiple referents (e.g., "table" can represent an object to dine on, something that can be done to a decision at a board meeting, or an organized listing of numbers). Scientific languages and signaling systems (e.g., naval signal flags), along with other "logical" codes, are cited by Guiraud (1975) as the only monosemic sign systems. Bertin (1981) contends that cartography is a monosemic system of codes, but many arguments to the contrary can be made. This issue of the extent to which cartography is monosemic or polysemic will be considered in Chapter 7.

According to a number of semioticians, particularly those following in the Saussurean tradition (e.g., Barthes), all codes are polysemic in the sense of having two (or more) "levels" of meaning. The first is the primary, conscious, explicit meaning that can be defined as a sign's "denotation." To this can be added a secondary, implicit and (perhaps) unconscious meaning, the sign's "connotation."[11] Guiraud (1975, p. 28) provides a relatively clear example of the distinction in relation to the sign function of military uniforms: "A uniform denotes rank and function; it connotes the prestige and authority attached to rank and function." He goes on to suggest that "scientific codes, being essentially monosemic, eliminate possibilities of stylistic and connotative variation which abound in poetic codes."

This elimination of possibilities for multiple connotations was clearly the goal of a communication-model-oriented cartography that viewed maps as "scientific," and therefore objective and free from valuative connotations. As will be discussed fully in Chapter 7, Harley (1989) contends that cartographers have for too long presented their maps as scientific and free from multiple connotative meanings. Both in semiotics and cartography there has been a growing realization that the separation between science and art is not as clear-cut as science would like to believe and that most signs, scientific or otherwise, carry connotative meaning.

Hjelmslev's theory of connotation serves as a basis for several current semiotic approaches to the denotation-connotation distinction (Nöth, 1990). The key feature of this theory was Hjelmslev's linguistically motivated expression–content dyadic sign model. Signs, in this model (as in Saussure's), were considered to be a relation between an expression and a content. What Hjelmslev added was that signs themselves could serve as either the expression or the content of other signs. He used the label "metalanguage" for signs as content and the label "connotation" for signs as expression. Barthes (1967), building on Hjelmslev, formalized this idea as a graphic model (Figure 5.6).

For Barthes, signs denote via convention (generally accepted relational rules), but connote via *signification*. Signification is, for Barthes (1967), "*a property of objects that do not declare openly their possession of signification*." In summarizing Barthes's perspective on connotation, Hervey (1982, p. 136) suggests that Barthes' " 'connotation' is appropriately used in cases where a sign acquires a 'higher' level of signification, functioning thereby as a 'secondary' sign that hints at a partially concealed, but all the more conspicuous, not to say insidious, message." For Barthes, there are no "innocent" facts.

In relation to visual images (in advertising), Barthes (1977, p. 37) distinguishes between literal and symbolic (cultural) messages based on knowing what things are versus what they stand for. The "literal image is *denoted* and the symbolic image *connoted*." Barthes contends that the system of connotation "takes over" that of denotation and suggests that we often use language to prevent or limit this "taking-over." Linguistic messages (text on the image) are said to "fix" a "floating chain of signifieds,"

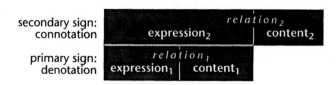

FIGURE 5.6. A depiction of Barthes's model of connotation as a semantic extension of a denotative sign. *Derived from Barthes (1967, p. 90)*.

or to "anchor" an otherwise polysemic sign. As an apparent complement to his view that connotations are often hidden in signs so that they exert their influence unconsciously, Barthes (1977, p. 198) suggests that text in images can "remote control [the viewer] toward a meaning chosen in advance." It can prevent connotations from being achieved—precisely its goal on many maps where text is added to define, reduce ambiguity, and generally try to ensure a monosemic system of signifying.

Traditional semantics treats connotative meaning as a secondary meaning that a sign may have in addition to its primary denotative meaning (Nöth, 1990). This idea (apparent in much of Barthes's work cited above) opens up a variety of possibilities for kinds of inference that might be made to achieve this secondary meaning level. Recent deconstructionist ideas suggest that it may actually be impossible to determine which is the denotation and which the connotation, which is the primary meaning and which must be inferred from it. Eco, as early as 1968, seemed to imply as much with his typology of connotations (originally published in Italian, with a German translation; Eco, 1972; cited in Nöth, 1990, p. 102):

1. Connotation of definitional meaning (e.g., Venus = Morning Star).
2. Connotations of the constituent elements (e.g., Lat. *luna* connotes "feminine").
3. Ideological connotation.
4. Emotional connotation.
5. Connotations derived from hyponomy (*tulip* connotes "flower"), hyperonomy (*flower* may connote "tulip") or antonymy (*husband* connotes "wife").
6. Connotations by intersemiotic translation (e.g., a word sign connoting a picture sign).
7. Connotations of rhetorical figures (e.g., metaphors).
8. Rhetorical–stylistic connotations.
9. Global axiological connotations (referring to values).

Regardless of the precision, strength, or multiplicity of levels of the conventional relation or signification between sign-vehicle and interpretant(s), the relation can be one of two types: motivated or unmotivated (i.e., arbitrary). Motivation here is used in roughly the same sense as in Chapter 4; a motivated relationship is one in which there is cognitive economy in recognizing similarity on some criteria. According to Guiraud (1975), motivation is a natural relation that can be either analogical (related to substance) or homological (related to form).[12] Motivated signs are equated to icons as defined above. The important distinctions between Guiraud and most other authors, however, are that motivation or iconicity should be considered a continuum (rather than the discrete categories proposed by Peirce, Sebeok, and others) and that this continuum is a concept that applies to both denotation and connotation. For denotation, Guiraud's concept of a sign motivation continuum is similar to that suggested by Knowlton's variable semiotic triangle cited above.

Typology of Comprehension (or Miscomprehension)

If we are interested in "how maps work," we must consider how signs, at all levels, work. One aspect of this question is whether signs are *comprehended*.

Prieto's theory of semiotic acts (as described in Hervey, 1982) devotes considerable attention to the success and failure of sign "comprehension." The fundamental principle of his theory is that for a sign to function, a person comprehending it must recognize that the perceptible sign-vehicle belongs to a particular class and infer from it that some other indicated entity (the interpretant) belongs to a specific class.[13] Both the sign-vehicle and the interpretant exist in a separate "Universe of Discourse" (defined as the possible sign-vehicle and interpretants). The sign-vehicle (which Prieto calls the "indicator" or "signal" occupies a Universe of Discourse termed a "sematic field" consisting of all the alternatives with which it significantly contrasts. On a U.S. National Park Service map, for example, a pictorial point symbol will have a relatively limited sematic class consisting of the 74 possible symbols in the complete set. The interpretant's Universe of Discourse is termed the "noetic field." For the same Park Service map, the noetic field could vary tremendously in size depending upon how familiar a person was with the features represented on the maps (From some small number of possible interpretants to some indefinitely large number). Comprehension involves comparing these two fields (or Universes of Discourse).

In relation to this general framework, Hervey (1982) provides a concise statement of how Prieto evaluates comprehension.

> Taking cognisance of the fact that the sender intends to convey a message, and perceiving the signal [sign-vehicle] as identifying a particular Sematic field to which it belongs, lead to a state of uncertainty in the receptor. This uncertainty is given specificity by the fact that, on recognizing the appropriate Sematic field, the receptor is, by

automatic association, made aware also of the corresponding Noetic field....

Within the Noetic field which the receptor of a signal [sign-vehicle] identifies..., the receptor's uncertainty has the precise form of an indecision as to which of a (perhaps indefinite) number of mutually exclusive classes in the Noetic field he could fix on as the class to which the sender's message belongs. Comprehension, therefore, can be seen as the dispelling (in part or totally) of the particular uncertainty in question, ideally by identifying the "narrowest" Noetic class that corresponds to the signal in question. (p. 67)

It should be clear from the above statement that several potential levels of comprehension present themselves, based on both the kind and the level of success in matching sematic and noetic classes. Complete success requires not only correct, but total comprehension. Prieto establishes a typology of sorts to delineate a set of comprehension possibilities. They are as follows:

1. *Complete success*: The interpreter has narrowed the noetic field down to a single class corresponding to the class of the sign-vehicle, and the choice is an exact correspondence (e.g., on a five-class choropleth map of mean income, recognizing the third darkest gray tone as the middle category and successfully matching this to a concept associated with middle income defined by a specified income range.
2. *Partial failure*: Identification of an appropriate superordinate class in the noetic field but failure to be able to narrow the choices down to a single class, thus retaining a level of uncertainty in the sign-vehicle—interpretant match (e.g., on the same choropleth map, recognizing the third darkest category as representing income levels, but being uncertain about which of the three central categories it represents).
3. *Total failure*: The interpreter has narrowed the noetic field down to a single class corresponding to the class of the sign-vehicle, and the choice is wrong (e.g., incorrectly matching the third darkest category with the second highest income level).
4. *Failure due to situational factors*: Specifically situations in which the originator of the sign-vehicle is either not as precise as the situation allows the percipient to be, or in which the originator is quite specific, but there are more interpretants than anticipated (e.g., an interpretation of the middle and next highest category as "middle income" in the context of a news story about "middle income" Americans).

Cartographically, the latter case might just as easily be considered a hypersuccess as a failure because the map percipient might be as likely to infer something useful as to infer the wrong thing.

THE NATURE OF SIGN SYSTEMS

In the context of cartography, a study of signs independent of how they interact with one another (the study of sign systems) would be little more than an intellectual curiosity. The key to productively applying a semiotic approach to map representation is to use that approach to consider how the individual-cognitive aspects of map representation discussed in Part I link to the public function and lexical process to be considered in Chapters 6 and 7. While understanding how signs denote and connote is an important piece in the puzzle, the puzzle is not complete until we consider how signs relate to one another. Issues of "the nature of signs" considered thus far closely parallel those of mental categories discussed in Chapter 4. Those of sign systems (discussed in this section) closely parallel those of map schemata. If maps are to work, mental categories and categories indicated by sign-vehicles need to correspond in some logical way and map schemata must link to the sign systems created by cartographers.

Morris seems to have had the most impact (at least in North America and certainly within cartography) on thinking about this question of sign interrelationships. His *three dimensions of semiosis, syntactics, semantics,* and *pragmatics,* provide the needed framework for addressing this question. A number of cartographers have tried to adapt these concepts to understanding map representation (Board, 1973; Morrison, 1974; Keates, 1982; Wolodtschenko and Pravda, 1993). Before I consider their efforts (in Chapter 6) a synopsis of Morris's dimensions of semiosis and some other attempts to address issues of semiotic systems is required.

Dimensions of Semiosis

For Morris (1938), semantics studies how sign-vehicles and their referents are related and pragmatics deals with sign-vehicle–interpretant relations. Thus each of these focuses on individual signs (as we did above). The third proposed relationship, syntactics, is probably the most important for cartography, but has also been the most controversial. According to Morris, syntactics is the relation between a given sign-vehicle and other sign-vehicles. There is a critical distinction here (that many cartographers have missed) between Morris's "syntactics" and the linguistic subcategory of "syntax." While syntax puts emphasis on word order and parsing (i.e., on a linear sequence), syntactics is much broader in scope. Syntactics allows for consideration of any kind of among-sign relationships.[14] Morris (1938, p. 16) makes this point explicitly in his statement that there are " syntactical problems in the fields of perceptual signs, aesthetic signs, the practical use of signs, and general linguistics." He provides an intriguing graphic depiction of his conception of the three dimensions of semiosis applied to the three attributes of a sign (Figure 5.7). Most of what Morris considered to be semantics and pragmatics has been alluded to above, in considering "the nature of signs." Although I will return to these dimensions of semiosis in the next chapters, here I will focus on syntactics.

At least three kinds of sign relationships seem to fall under Morris's umbrella of syntactics (Posner, 1985, in French; cited in Nöth, 1990, p. 51). These include: (1) "the consideration of signs and sign combinations so far as they are subject to syntactical rules" (Morris, 1938, p. 14), (2) "the way in which signs of various classes are combined to form compound signs" (Morris, 1946/1971, p. 367), and (3) "the formal relations of signs to one another" (Morris, 1938, p. 6). Cartographically these perspectives on sign relationships emphasize, respectively, issues such as: (1) development of logic for map legends, (2) rules about combining nominal with ordinal sign-vehicles, and (3) principles for matching graphic variables to differential versus ordinal data.

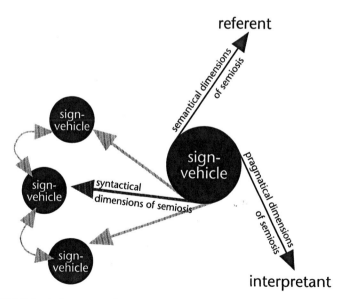

FIGURE 5.7. A depiction of a sign as an entity linking its three components via the relations of syntactics, semantics, and pragmatics. *Derived from Morris (1939/1971, p. 417)*.

Robinson and Petchenik (1976) presented a convincing argument that maps have no "syntax." Implicit in their argument was an assumption that linguistic syntax was equivalent to syntactics. Most subsequent cartographers (particularly within North America) accepted the argument against a map syntax and took for granted the equivalence of syntax and syntactics. As a result, North American cartography has largely ignored the concept of syntactics.

Robinson and Petchenik's discounting of syntax as a viable cartographic issue was part of a broader argument against a linguistic approach to cartography. They (rightly) pointed out that individual maps have no predetermined reading sequence, and therefore no "word" order comparable to that considered under the linguistic concept of "syntax." In addition, they asserted that maps have no equivalent to "words" and are not "discursive." To demonstrate the weakness of the mapping-as-language analogy, they pointed to two possible aspects of mapping that might be equated with syntax, both of which were intended to demonstrate that the possible link between mapping and language is tenuous at best. First, they suggested that syntax (for a static map) can only be defined in relation to spatial structure (either horizontally across geographic space or vertically in terms of visual perceptual levels). They went on to concede that there is something weakly analogous to linguistic "syntax" in the structuring of visual levels. One could argue, in fact, that if carefully designed, visual levels could be used to lead a percipient through a series of stages from global patterns to local details. They also conceded that animated maps can have a kind of syntactical structure related to temporal order (but contended that this structure "has nothing to do with the map per se") (Robinson and Petchenik, 1976, p. 56).

As should be clear by now, the point that Robinson and Petchenik (1976) missed is that while most maps do not have syntax in the narrow sense of structured reading order, they do (or should) have a carefully structured syntactics in terms of the interrelationships among signs they are composed of. Most potential applications of syntactics to map representation relate to Morris's broad, nonlinguistic approach. Since 1976, when Robinson and Petchenik developed their argument against map syntax, however, technological changes have resulted in practical tools for the design of animated maps. It is therefore now considerably more important to question their contention that the temporal syntax of animated maps has nothing to do with the map. It is my contention that the temporal syntax of animated maps has everything to do with the map!

When maps play out over time, as a map movie, it may prove useful to borrow some ideas from film analysis to address the many new issues that arise. From a semiotic perspective, the most interesting possibility of film analysis with map animation applications is Metz's (1968/1974) *filmic syntax*. Metz developed a model (or typology) of "syntagmatic types" to characterize the temporal–visual manipulation possible in film, but not in other visual media. As Korac (1988) points out, temporal manipulation results in a range of possible syntactic relations from those that mimic real time–space relations (and result in a motivated or iconic sign system) to those that have arbitrary relations with real-world time–space relations (thus resulting in an arbitrary or symbolic sign system). Metz's model is hierarchically organized with a major division into relatively simplistic films consisting of a single coherent sequence (single-shot units) and those more complex films in which there are multiple units (to which the filmmaker exerts varying kinds and degrees of manipulation) (Figure 5.8). The categories depicted can be defined as follows:

> *Sequence shot*: The most "motivated" temporal–visual organization in which filmic and chronological time are *identical* and visual plane has continuity (i.e., full duration from single point of view).
> *Scene*: A highly motivated temporal–visual organization in which filmic and chronological time are identical, but visual continuity is interrupted (i.e., full duration, but from several different viewpoints as typical in a TV portrayal of a sporting event).
> *Ordinary sequence*: Contains both visual and temporal discontinuity,

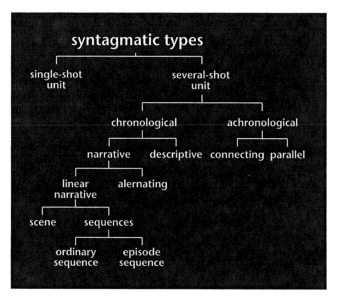

FIGURE 5.8. Metz's semiotic typology of syntagmatic types for film. *Derived from Metz (1968/1974, p. 146) and Korac (1988, Fig. 1, p. 78).*

having changes in duration, gaps in time, and changes in place, but with temporal order maintained.

Episode sequence: Like above, but more extreme. Filmic time is compressed to an extreme degree (e.g., several years in a few minutes created through a sequence of potentially time-compressed shots interposed with large jumps in time).

Alternating syntagm: Events actually taking place at the same time are presented alternately in the film (a variation is alternating flashback and flashforward, producing a psychological rather than a physical sense of simultaneity).

Descriptive syntagm: The use of a time sequence to describe an area (rather than presenting a temporal sequence).

Connecting syntagm: A linear sequence of spatially and temporally different shots depicting objects or events that belong to the same category or class, thereby implying links of ideas due to juxtaposition in film time (e.g., horrors of war signified through a sequence showing bombs dropping, destroyed houses, dead or wounded, etc., regardless of which war is shown). Korac (1988, p. 80) labeled this syntagm "the stuttering filmic equivalent of categorization." This technique uses temporal proximity (a Gestalt property) to suggest grouping due to family resemblance (one aspect of the prototype theory of categorization).

Parallel syntagm: An alternating sequence of shots that depict changes in place but not time, typically places that belong to opposite categories where the temporal juxtaposition of the scenes puts emphasis on contrasts.

The Metz typology has several similarities to the approach to dynamic mapping that some colleagues and I developed (DiBiase et al., 1992). In particular, identification of both temporal and nontemporal uses of time sequence and the attention given to manipulation of time as a sign corresponds to our approach to treating time in a manner analogous to the visual variables of map symbolization (size, shape, orientation, etc.).[15]

Systemology

Morris's tripartite typology of semiosis (into semantics, pragmatics, and syntactics) is not the only cartographically relevant attempt to formalize the study of semiotic systems. Hervey (1982) describes a complementary classification, termed *systemology*, that he defines as a "deductive classification of semiotic systems." Systemology is derived from principles of *ax-*

iomatic semiotics, a variation on *functional semiotics* attributable to Mulder and Hervey (1980). As described in Hervey (1982), both axiomatic and functional semiotics are subfields of (or perspectives on) semiotics that limit their attention to semiological systems and typologies of indices (signs), specifically signs that have an intention to communicate. Differences between the two branches of semiotics are primarily in terms of limits on scope of inquiry, and need not concern us here.

Hervey initially describes why the medium of communication in which signs are used (e.g., sight, sound, etc.) and the pragmatic use of the system (e.g., for the sighted vs. the blind) are not semiotically relevant. He then goes on to contend that *semiotic economy* provides the only important dimension on which to classify semiotic systems.

Semiotic Economy

"Semiotic economy" is defined in relation to two levels of entity in a semiotic system. A distinction is made between the "figurae" and the "signa." This distinction can be interpreted as that between the sign-vehicle and the sign of which it is a part, or perhaps the mark that becomes the sign-vehicle (prior to its becoming part of a sign) and the sign (or sign-vehicle–interpretant relationship). Hervey (1982) uses a simple example to clarify the terms. The example involves the letters H, W, and C as they often appear on single-knob bathroom shower controls. The letters by themselves are figurae. On the faucet, if the proper sign-vehicle–interpretant connection is made, the result is the signa of: H = hot water, W = warm water, and C = cold water.

According to Hervey's systemology, figurae occupy the *cenological* (from the Greek indicating "empty" of meaning) level of signification. Signa, on the other hand, occupy the *plerological* (or "full" of meaning) level. Semiotic economy is calculated by determining the ratio of figurae in the cenological level to the signa of the plerological level. For the water faucet example, the semiotic economy of the system is 1:1 because there are exactly the same number of potential figurae as there are signa. Such a system is considered "simple" in contrast to "complex" systems (i.e., those having a one-to-many relationship). A good example of the later is Morse code in which two figurae in varied combinations (a dot and a dash) represent all letters and numbers. Written language, of course, has much greater economy with (in English) only 26 letters able to produce all possible words of the language (native Hawaiian achieves this with only a dozen letters).

Cartographically, if we consider subcomponents of maps separately (e.g., just those symbols forming a system for representing point locations) we can find examples of both simple and complex semiotic systems. The U.S. National Park Service system of pictorial symbols for depicting feature locations is a simple 1:1 system in which each symbol is a different shape. Semiotic economy is achieved when pairs of graphic variables (e.g., shape and color hue) are used in combinations where one graphic variable acts as a qualifier of the others. One example is the map of Motor Vehicle Manufacturers in the United States for 1986 found in the National Geographic Society's *Historical Atlas of the United States, Centennial Edition*. On this map, seven point symbol shapes are used to represent categories of manufacturers (the four major U.S. firms at the time plus other American, European, and Japanese manufacturers) and three color hues are used to represent type of facility (assembly, parts, headquarters/R&D). The ten figurae can result in 7 × 3 signa, or 21 disparate signs, a semiotic economy of 1:2.1. For maps as a whole, like written language, it becomes clear that we have a system with tremendous semiotic economy. Whether one accepts Bertin's (1967/1983) contention that there are seven graphic variables, or the expanded set of eleven identified in the next chapter, we have a system with a *figurae:signa* ratio of *few:indefinite*. Individual maps, of course, vary in the way they take advantage of the potential economy and we know little if anything at this point about the relationship between semiotic economy of maps and the cognitive aspects of map representation discussed in Part I (i.e., does semiotic economy help or hinder a user's ability to identify visual categories and/or apply appropriate schemata?).

Simultaneity versus Articulation

In addition to the simple–complex dichotomy of semiotic economy, Hervey (1982, p. 193) proposes a second dichotomy, "that between the formation of *simultaneous bundles* and the formation of *articulated constructions*." The two concepts are presented in relation to linear semiotic systems. With simultaneous bundles, for example, the order of combination is irrelevant (e.g., on encountering a pair of highway signs while approaching an intersection, one indicating "Stop" and the other "No left turn," the order in which the signs are encountered has no effect on their meaning). In contrast, when dealing with Morse code with its two units of expression, the order in which dots and dashes (and pauses between them) are encountered provides the basis for meaning. Morse code is one of Hervey's (1982) examples of a prototypic articulated system. Hervey considers articulated systems to be more powerful than those whose sign-vehicle combinations result in simultaneous bundles. Articulated systems can be seen as making use of human pattern recognition abilities (see

Part I). Although articulation is a concept primarily associated with order of presentation or encounter with signs, it has also been applied to nonlinear constructions such as traffic "signs."

For dynamic maps, the concept of articulated versus simultaneous combinations becomes particularly relevant. As demonstrated in DiBiase et al. (1992) and described in the next chapter, if we treat time as a cartographic variable (instead of just something to map), the meaning of a particular set of dynamic signs will be determined, in part, by the order of presentation. A set of cartographic primitives including temporal order as an operator (or figurae, in Hervey's terms) would thus be considered an articulated semiotic system, at least in those cases for which order is explicitly used as a sign-vehicle.

Combinatorial Relations

Hervey (1982) suggests that semiotic systems can be articulated at either the cenological or the plerological level. This combination of two dichotomies results in four possible types of subsystem (Figure 5.9). The highway-sign example above represents a simultaneous–plerological subsystem because combinations are among bundles of "signa" (or signs). Morse code is an articulated–cenological system because in it, figurae (devoid of independent meaning) are combined. Hervey cites arabic numbers as an example of an articulated–plerological subsystem in which the signa (0, 1, 2, . . . , 9) have independent meaning and their combinations are dependent upon how they are arranged (e.g., 123 is not the same as 321). Using a diamond, circle, and triangle on a map to represent hotels, restaurants, and theaters would be an example of a simultaneous–cenological system. The figurae or sign-vehicles have no predetermined meaning and whether you see • ♦ ▽ next to a point on the map or ♦ ▽ •, the meaning is the same (although consistent ordering will probably facilitate more purely perceptual tasks such as visual search). Hervey goes on to contend that some systems exhibit all four subsystems

	componential subsystem	articulation
cenological subsystem	cenematics	cenotactics
plerological subsystem	plerematics	plerotactics

FIGURE 5.9. Hervey's semiological subsystems. *After Hervey (1982, Fig. 7.5, p. 197). Adapted by permission of Routledge, Chapman & Hall.*

at various levels of analysis. Human languages are shown to feature the interlocking of all four, and it seems likely that all could be identified in some dynamic maps.

APPLICATION OF THE SEMIOTIC APPROACH TO MAP REPRESENTATION

This chapter has provided an abbreviated synopsis of selected issues in the field of semiotics. It is my belief that semiotics has tremendous potential as a tool for systematizing our approach to maps as representations and for developing logical systems of, and transformations among, representations. In addition, a semiotic perspective offers a structured way to consider the interaction of the explicit and implicit meanings with which maps are imbued. The remaining two chapters of Part II will make extensive use of this introduction to semiotics as it relates to functional and lexical aspects of map representation. Chapter 6 addresses the structure of cartographic representation as a set of hierarchically interlocking sign systems in which attention can be directed to a range of issues from how individual symbols represent to how entire maps serve as a sign for a particular worldview. Although it is not possible to completely separate Morris's three dimensions of semiosis (semantics, syntactics, and pragmatics), the chapter emphasizes the first two of these, the semantics and syntactics of map representation. Chapter 7 continues from this base to emphasize the multiple levels and kinds of meaning in map sign relationships and the processes by which these multiple meanings arise, thus, the pragmatics of maps.

NOTES

1. The spelling was changed from Peirce's original "semeiotic" and the use of the plural form was officially adopted at this time.

2. For those wishing to pursue a semiotic approach to cartography further, Nöth's (1990) tabulations of terminology should prove useful in comparing ideas by different authors using different terminology. He provides a table listing terms used by 15 scholars who have adopted the dyadic model of "sign" and another table of terms from 10 authors who have adopted the triadic model.

3. Axelson and Jones (1987) and Wood and Fels (1986) point to the ways in which this assumed real-world referent for topographic and other large-scale reference maps can hide and distort other kinds of signifying relationships.

4. Peirce frequently used the term "sign" in both a broad sense of the overall relationship and in a narrower sense corresponding to "sign-vehicle" as defined above, and at times talks about the interpretant as a "sign" in the mind of

the interpreter. Quoting from Peirce, therefore, presents interpretation difficulties. To minimize the confusion, I have inserted in brackets [ ] the appropriate interpretation for "sign" in the succeeding passages taken from his work.

5. Nöth's interpretation is based on a 1977 edition of *Zeichen: Einführung in einen Bergriff und seine Geschichte* (Frankfurt: Suhrkamp), which is a translation of a 1973 publication.

6. Peirce also cites a photograph as a kind of image hypoicon. This dual categorization is evidence that Peirce did not consider his typology a mutually exclusive categorization of sign-vehicles, but a categorization of attributes that sign relations could have, individually or in conjunction. It seems, then, that Peirce approached categorization from what later would be identified as a prototypic, rather than a classical, perspective. Although he did not link his approach to category theory (which was developed several decades after Peirce's death), Peirce was quite clear about the difficulties involved in classifying signs (Hartshorne and Weiss, 1931).

7. Saussure's use of symbol fits in the latter category and may be one of several reasons for the lack of interaction between the two traditions.

8. As Guiraud (1975) points out, this typology is derived from Jakobson (1960).

9. He also proposed two categories of relations among signs and one dealing with interpreters of signs (interpreter-family as a group of interpreters for whom a particular sign has the same interpretant). The categories relating signs will be taken up below.

10. Following from Saussure, signs do not exist for Guiraud unless there is an intention to communicate.

11. If we accept Harley's (1989) or Wood and Fels's (1986) critiques of cartography, these secondary connotations are an intimate part of mapping, are one reason for the importance of maps in society, and are anything but unconscious for the cartographers. These issues will be taken up in detail in Chapter 7.

12. Guiraurd's homological motivation seems to be roughly equivalent to Peirce's diagram hypoicon, and his analogical seems equivalent to Peirce's image hypoicon or a combination of image and metaphor hypoicons.

13. Prieto's work is thus far available only in French. The discussion presented here is based on a synopsis of Prieto's theory presented by Hervey (1982). It remains unclear from Hervey's presentation whether Prieto considers the "indicated entity" an interpretant or a referent. Since Prieto seems to have developed his ideas within a Saussurean context, however, it is probably safe to assume a dyadic sign model with the "indicated entity" being an interpretant rather than a referent. See Martinet (1990) for more on Prieto.

14. Throughout this book, the adjective "syntactic" relates to the broader concept of syntactics rather than to syntax.

15. Our approach to dynamic variables for animated maps will be considered in Chapter 6 and the similarities and differences between Metz's typology and our conceptual approach to map animation will be considered in Part III.

CHAPTER SIX

A Functional Approach to Map Representation
THE SEMANTICS AND SYNTACTICS OF MAP SIGNS

This chapter deals with functional relations within and among map signs and sign systems and, thus, with the structure of semiotic relationships. Initial emphasis is on categorizing "stand-for relationships" inherent in mapping (emphasizing map sign semantics). This is followed by consideration of sign system specification (map sign syntactics). In both cases, it becomes clear that the relevant issues can be addressed at multiple levels from multiple perspectives, and (as contended throughout this book) such a multifaceted approach is essential to understanding any particular aspect of map representation.

As noted in Chapter 5, the nature of signs can be considered from several points of view. Possible category systems for signs in general can be grouped on the basis of which element of the sign relationship (sign-vehicle, interpretant, or referent) is treated as the mediator between the other two. Each perspective is addressed as it applies to understanding how map signs work. The link of sign-vehicle to referent through the mediator of an interpretant is given particular emphasis. Several sign typologies have been proposed for categorizing kinds of map sign-vehicle to referent links. After reviewing a few key typologies, I argue for a continuum of map sign relationships based on the Ogden–Richards semiotic triangle.

Discussing the nature of map signs leads naturally (as it did in the previous chapter) to consideration of map sign systems. Among the issues here is the question of what the units of construction are for the sign-vehicle component of the sign relationship and how connections among these units correspond to connections among interpretants and/or referents. I begin this section with examination of Bertin's "semiology" of graphic variables and several extensions. Most authors who have considered map symbolization at the level of fundamental graphic variables have offered a set of rules for matching the graphic variables to referents. These rules specify syntactical relationships among graphic variables that can be matched to syntactical relationships among referents (or interpretants). Attempts to specify a map sign syntactics have provided the underlying basis for various attempts to standardize or systematize map symbolization (efforts stimulated by attempts to internationalize map symbology and by attempts to develop knowledge-based systems for map symbol selection). As with the range of perspectives on the nature of map signs detailed above, we cannot limit ourselves to considering map sign systems in isolation. We must also consider multiple interlocking systems if we are to really understand how maps structure knowledge.

THE NATURE OF MAP SIGNS–
MAP SEMANTICS

Cartographic research in general, and cartographic application of semiotics specifically, has paid little attention to the distinction between dyadic and triadic sign models. As a result, the difference between what a map sign means and what it represents has become blurred. This blurring, in turn, has lead to charges that cartographers foster a view of maps as objective mirrors on reality rather than as representations of selected aspects of reality (or of aspects of an intermediate representation of reality) (see Harley, 1989, for such charges stated in nonsemiotic terms). In the discussion below, I will explicitly follow a triadic model of sign relations, thus recognizing a clear distinction between interpretant and referent. This distinction results in the recognition that sign-vehicle–referent, interpretant–referent, and sign-vehicle–interpretant relationships are not only distinct, but that all are within the realm of cartographic inquiry.

A triadic model of map signs (and maps as signs) considers the relations identified above (between the three pairs of sign components) from three perspectives. These perspectives are based on three interpretations of the semiotic triangle, each of which takes one sign component to be the mediator between the other two. Each perspective on sign relationships puts emphasis on particular cartographic issues.

Sign-Vehicle as Mediator

Considering map signs from the perspective of the sign-vehicle emphasizes their role as the link between thing and meaning (Figure 6.1). Attention is directed to the aspect or aspects of a referent that the sign-vehicle represents and the category of meaning or intention into which the interpretant might be placed (e.g., appraising vs. prescribing). At a "functional" level of analysis, map marks act as sign-vehicles that mediate between the sign's referent and an associated meaning of that referent in a particular context.

The context for interpretation is provided by a map schema (as detailed in Chapter 4) which is prompted by visual processing of the sign-vehicles and suggests the *aspect* of the referent for which an interpretant is needed. At a "cognitive" level of analysis, the map mark (the functional sign-vehicle) actually becomes the referent, with a node of the map schema acting as a virtual sign-vehicle that mediates between the physical map marks and the categories or mental constructs they bring to mind. Sign systems are therefore integrally related to knowledge schemata at multiple levels of representation.

Drawing from Morris's modes of signifying and dimensions of use, some additions to Peirce's typology of signs proposed by Sebeok (a Morris student), and Guiraud's functions of communication, we can distinguish two major aspects of referent–interpretant relations for which a map sign-vehicle can be mediator. These aspects (or sign functions) can be categorized as those that *apprise* (apprisive sign aspects) versus those that *stimulate some reaction* (stimulative sign aspects) (Figure 6.2). It should be kept in mind that sign aspects are not meant to be mutually exclusive (either within or between categories), but instead are likely to occur in varying degrees in any sign.

Apprisive map signs are those that purport to provide attribute

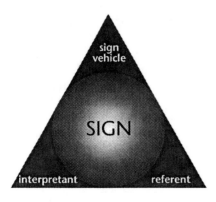

FIGURE 6.1. Sign relation with sign-vehicle as mediator.

Sign Aspects	
apprise	stimulate
designate appraise indicate label relate	prescribe emotive connote poetic/ aesthetic

FIGURE 6.2. Categories of map sign aspects.

and/or location information about objects or concepts. In relation to Morris's dimensions of sign use, the apprisive aspect of a sign-vehicle (as referent–interpretant mediator) can be further broken into informative (purporting to be value-free) and valuative. In general, signs that can be considered apprisive in aspect include those that designate (including Guiraud's referential and meaning/information functions), appraise, indicate (Morris's identificative, Peirce's indexical, and Guiraud's attention), label (Sebeok's name and Guiraud's metalinguistic), and relate (Morris's formative) (see Chapter 5 for discussion of typologies by Guiraud, Peirce, Sebeok, and Morris).

In contrast to signs with a primarily apprisive aspect, stimulative signs are those for which a behavior is the primary aspect. Morris's incitive and systemic dimensions of use can be considered subcategories of stimulative sign aspect. Signs characterized as emotive, prescriptive (Sebeok's signals and Guiraud's injunctive), connotative (Sebeok's symptoms), and poetic/aesthetic can be considered to emphasize the stimulative aspect.

Designative map signs specify explicit properties or attributes. They refer to the property and to its meaning in the context of the overall map. The designative aspect of map sign relations is by far the one most studied by cartographers. Circle sizes or gray tones that designate specific quantities or a quantitative category, color hues to designate vegetation types, point symbol shapes to designate points of interest or location of services on travel maps, and so on—all clearly exhibit a designative aspect (Figure 6.3). For each there is an "objective" property or attribute of a referent that is linked by the sign-vehicle to an interpretation or meaning (the interpretant). A given circle size at a particular location on a population map, for example, might represent (i.e., designate) "500,000 people" which, in the context of the United States might be interpreted as a "medium-sized city." Much of the research stimulated by the communication approach to cartography seems to have been directed toward identifying the optimal sign-vehicles for specific referent–interpretant "designations."

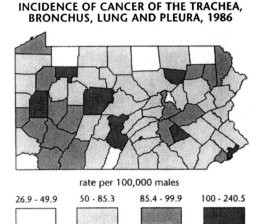

FIGURE 6.3. Sign-vehicle with a dominant designative aspect. In this case, color values designate the four quantitative ranges for aggregate county data.

Although not as dominant on maps as designative functions, map signs having an appraisive aspect are relatively common. This aspect of map signs emphasizes a judgment about a place (rather than "objective" attributes of that place). Typical explicit examples are: "scenic" route, point of "interest," "primary" road, "major" cities, land "suitable" for development, regions of "high" environmental risk, and the like (Figure 6.4). The sign-vehicle in these cases links a referent that is a judgment (rather than an attribute) with a realm of discourse in which the interpretation of a particular judgment can be made (e.g., for the vacation traveler, a scenic route is a positive appraisal, while for a business traveler it probably means delay and frustration). In contrast to the designative aspect of a sign, the appraisive aspect of map signs is generally considered to be more subjective (and thus open to wider interpretation). The designative aspect of map signs, however, is probably less objective than it is generally taken to be (a topic taken up in Chapter 7).

Sign aspects of indicating, labeling, and relating are similar in a map-representation context in that they provide context information within which other sign aspects can be interpreted. In the case of signs that indicate, the sign-vehicle might "indicate" that a position has the coordinates of 44°N and 95°W (the referent) which is matched to an understanding of latitude–longitude specification that provides the context to interpret this specification. The overall graticule system thus indicated provides the geographic context (the spatial structure) for interpreting various geographic relations (adjacency, proximity, density, regular vs. ir-

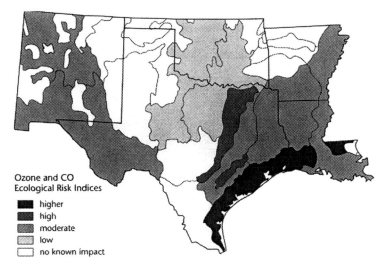

FIGURE 6.4. Sign-vehicle with a dominant appraisive aspect. This map depicts an appraisal of ecological risk due to ozone and CO for EPA Region 6. *After Map 10 in U.S. Environmental Protection Agency Region 6 Comparative Risk Project, Ecological Report (1990).*

regular distribution, etc.). Similarly, signs that label can define the attribute context depicted (e.g., "soils") or the specific category within this context (e.g., histosols—soils high in organic matter). Map signs that label can also specify that a member of a category (e.g., river) exists at a specified location or that this member is one specific entity (e.g., the "New" River). Map signs that relate provide the graphic equivalent of concepts such as "union" or "intersection." They can specify attribute or spatial links among groups of map signs. A relative aspect may be the result of co-occurrence of two or more designative or appraisive signs (Figure 6.5) or may be the dominant aspect of a sign (see Figure 5.5, in which spatial relation is directly signified by the hook-like sign-vehicle).

In contrast to signs that apprise, those that are intended to stimulate a response can only be evaluated from a behavioral perspective. Instead of linking a "meaning" with a thing or concept, sign-vehicles for these behavior-inducing signs link a meaning to an action or feeling. A sign with a prescriptive aspect makes this link explicit (e.g., an arrow next to a street on a city map that has a prescribed movement in space as its referent and "street can only be traveled by vehicle in the specified direction" as the interpretant). In general, emotive, connotative, and poetic sign aspects are implicit. They are therefore particularly dependent upon the context in which the map representation is encountered and the socio-

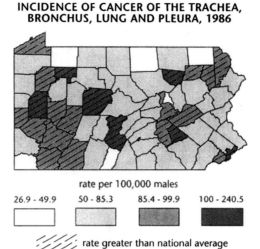

FIGURE 6.5. Attribute sign aspects can relate by application of some visually similar element (e.g., line orientation) to multiple map symbols (in this case indicating shared membership in a higher level category).

cultural context to which the percipient belongs. Their discussion, therefore, will be left to the next chapter (in which lexical–pragmatic issues of map representation will be taken up).

Referent as Mediator

A view of map signs from sign-vehicle through referent to intepretant considers the link of two abstract representations (one physical and usually visual and the other mental and perhaps imagistic) by virtue of their correspondence with the thing they both refer to (Figure 6.6). This perspective stresses the fact that maps and map marks are just one of many possible representations of their referent and that the representation they provide is generally to be understood and judged (at least in part) from the perspective of some other representation (e.g., language or propositional knowledge structures). In addition, a referent-as-mediator perspective draws attention to the importance of considering various categories of referent. Five modes of referent categorization relevant to map semantics will be noted here.

Perhaps the most important categorization of map referents is that between geographic and nongeographic information. There are two ways of looking at this issue. Keates (1982) proposed a distinction between "lo-

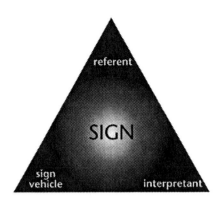

FIGURE 6.6. Sign with referent as mediator.

cational" and "substantive" information, a distinction between where an object is (in geographic coordinate terms) and characteristics of that object (including spatial characteristics such as size and shape). Lyutty (cited in Schlichtmann, 1985) distinguished between "plan" and "plan-free" information. This dichotomy separates spatial questions (i.e., questions about where, what size, what shape, etc.) from attribute questions (i.e., questions about what kind, what color, how much, for how long, etc.).

Schlichtmann (1985, 1991) has adopted Lyutty's plan/plan-free approach and elaborated it considerably. This approach seems preferable, particularly in the context of digital databases in which there is often a physical separation and different organizational structure for plan and plan-free information. Among the conceptual additions that Schlichtmann (1991) makes is a subcategorization of plan information into external and internal traits. External traits are mathematically imposed characterizations that include dimensionality, location, size, shape, and orientation. Internal traits (for which two further subcategories—cover form and boundary form—are cited) deal with spatial variability within, across, and between objects (e.g., boundary distinctness, the discrete–continuous character of an area object's contents, range of occurrence of dispersed objects, etc.).

In the context of digital data, Nyerges (1991a, 1991b) has also pointed out that sign relations have spatial and nonspatial components (which he labels "space" and "theme"). Critical differences from Schlichtmann (1985) are apparent. Nyerges adds a third aspect of sign relations, time. Also, any particular geographic sign vehicle is considered an "entity with an attached bundle of aspects," while a referent is considered "a phenomenon with its bundle of properties" (Nyerges, 1991b, p. 68). In both cases, the bundle includes theme, space, and time. In the context of geographic databases, Nyerges contends that these three sign aspects represent an "inseparable bundle." Although one aspect might dominate in a particular situation, Nyerges argues that a geographic entity cannot be adequately defined (i.e., signified) without all three.

Traditional approaches to cartography have (understandably) put emphasis on the spatial aspect of map signs and maps as signs. As a result, one important categorization of referents has been in terms of spatial dimensionality (an aspect of Schlichtmann's external plan traits). Spatial dimensionality of potential map sign referents is generally considered to include a distinction among point, line, area, and volume features (a distinction first noted by Wright, 1944, and made common practice by Robinson, 1953, in his first text). The distinction is not as clearcut as it might first seem because the determination of whether a referent should be considered an area (vs. a point or a line) is dependent upon scale of analysis. Cities, for example, are points at a global scale but areas at a local scale.

As mapping expands into a dynamic realm (i.e., animated maps) considerations of spatial dimensionality need to be expanded to space–time dimensionality. The ways in which time is matched to space will define the kinds of possible referents in space–time. Treating time as points, for example, results in space–time slices having zero time thickness (comparable to the zero dimensions of a point in space). This representation strategy is typical for depicting change in attributes of locations (Figure 6.7). If time is dealt with as intervals, slices in space–time can then have thickness resulting in the potential for four-dimensional space–time slabs (if a three-dimensional spatial referent is considered over some time interval). It also creates the interesting possibility of three-dimensional space–time referents generated from linking a two-dimensional spatial object (e.g., hurricane eye position) with a time interval (Figure 6.8). This latter view has the effect of treating time as an attribute of space, just as Jenks's (1967) data model concept treats population density as the Z-axis for a three-dimensional population surface. A question deserving more attention in this regard is, In what ways does the sign specifying some aspect of this three-dimensional referent differ from one specifying a referent that is three-dimensional in space with zero time thickness?[1]

A third major categorization of referents involves their distributional form. This distinction relates to what Schlichtmann (1991) identified as the internal traits of the plan aspect of referents. Hsu (1979) distinguished between phenomenon distributed continuously across space and those that are discrete (Figure 6.9). Discrete phenomena were further categorized as concentrated or dispersed. Hsu proposed that a natural link exists between discrete and positional phenomena and between continuous and area/volume phenomena. Although she proposed a "sequential"

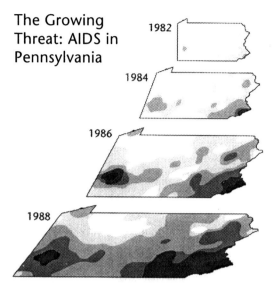

FIGURE 6.7. Time "layers" resulting from a "time-as-a-point" perspective on temporal cartographic objects.

distributional form as the match with linear phenomena, it seems that linear phenomena can also be characterized as either discrete or continuous (e.g., towers for high tension electric lines occur in a discrete linear arrangement across the landscape while the electric lines themselves are continuous). As will be discussed in more detail below, David DiBiase and I have argued that distributional form cannot be completely characterized on the basis of a discrete–continuous dichotomy (MacEachren and DiBiase, 1991). We suggest, instead, that a two-dimensional phenomenon space defined by a discrete–continuous and an abrupt–smooth axis provides a more complete description (at least in the case of areal phenomena).[2]

Complementing the distinctions among referents based on spatial dimensionality and distributional form is another based on level of measurement (a plan-free aspect of referents). Most cartographic texts have emphasized a division into nominal, ordinal, interval, and ratio levels of measurement. As I have pointed out elsewhere (MacEachren, 1994), these four categories are usually grouped into two higher level categories of categorical information (including nominal and ordinal) and numerical information (including interval and ratio level). Functionally this distinction between nonmetrical information and metrical makes sense, but cognitively, ordinal categories are, perhaps, more naturally grouped with the numerical categories of interval and ratio-level information. Al-

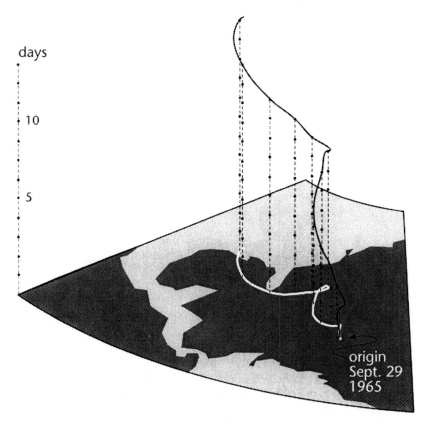

FIGURE 6.8. A three-dimensional space–time cube in which time is treated as the Z-axis.

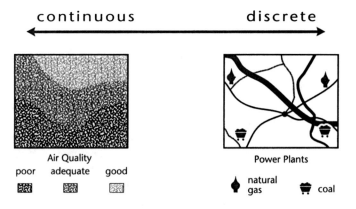

FIGURE 6.9. Depictions of discrete and continuous phenomena.

though both nominal and ordinal information can be mentally organized through use of a container schema, ordinal information shares the use of the up–down image schema with interval and ratio-level mental organization. For example, the up–down image schema applies with equal felicity to high-, medium-, and low-income areas as to temperature in degrees Celsius or pressure in millibars.

Another important distinction among referents was proposed by John Ganter in the context of cartographic visualization.[3] Ganter argued that a key issue in understanding the role of map representation as prompts to mental visualization was the recognition of a difference between phenomenon-representations (which he termed P-reps) and concept-representations (which he termed C-reps). A P-rep is a representation of an aspect of the physical world, of an entity (or judgment about it) that could be experienced (a mountain, rainfall, vegetation, atmospheric pressure, scenic route, etc.). Rather than being derived directly from reality, as is a P-rep, a C-rep is a derived from conceptions about reality (or about any alternative realities that a human can conceptualize). The referent for a C-rep may be a hypotheses about the state of an actual phenomenon (e.g., predictions about the vote totals in an upcoming presidential election) or it may be an abstraction that does not exist outside of a particular theoretical framework (e.g., a linear trend surface representing fiscal flows among federal banks in the United States).[4]

The importance of this distinction for map representation is that it makes explicit the fact that not all maps (or map signs) are intended to represent the physical world. For C-reps, the referent is not a thing but an idea. It therefore is no more concrete than the interpretant that it links the sign-vehicle with. The distinction between P-rep and C-rep could be interpreted as a reversal of interpretant and referent. In the case of a P-rep, a sign-vehicle stands for a phenomenon in the world (the referent) while a mental concept (the interpretant) is used to link the referent and the sign-vehicle. With a C-rep, the concept becomes the object of reference for or entity denoted by the sign relationship (i.e., the referent). The world, however, provides the interpretant—the exemplars (and sources of metaphors) through which the concept-as-referent is given meaning.

The P-rep–C-rep reversal of position between interpretant and referent is particularly apparent in map applications associated with modeling and simulation. For example, maps often serve as sign-vehicles that represent the theoretical concepts behind global climate models while our experience with the real situation of past climates provides the mechanism through which we interpret what is depicted on the map.

One issue that cartography needs to address, then, is whether it is reasonable to treat C-reps and P-reps as if they were the same kind of referent, as we devise sign-vehicles for maps. This issue is particularly important in the context of scientific uses of maps for which signs dealing with nonphysical entities are probably the norm. As will be considered in Part III, the advent of computer tools associated with virtual reality threatens to blur the distinction between P-reps and C-reps even further, with uncertain consequences concerning the function of maps as signifying representations.

Interpretant as Mediator

Our third perspective, a view from sign-vehicle through interpretant to referent, puts emphasis on the role of map signs as a means of shared understanding between cartographer and percipient (Figure 6.10). The sign-vehicle (i.e., map mark or entire map) represents a referent by virtue of an agreed code that specifies an interpretation (i.e., interpretant) for some aspect of that referent. A key issue here for map signs seems to be the nature of the agreed code linking a sign–vehicle with a referent.

Several typologies have been proposed for map signs. Keates (1982), drawing directly from the semiotic approach of Peirce, Morris, and their proponents, presents the basic icon–indexical sign–symbol typology in the context of mapping. He points out that "conventional sign" is a term long used in cartography to refer to standard map "symbols." He contends, however, that most map symbols are not symbols in the strict semiotic sense of having only a conventional (i.e., arbitrary) link to their referent. The example of a black square on a topographic map is used to illustrate that what at first glance might appear to be a completely abstract assignment of sign–vehicle to referent has some element of iconicity. Keates (1982, p. 67, fig. 26) provides examples of sign-vehicles that he

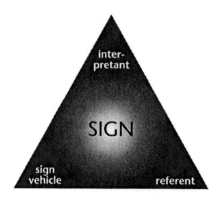

FIGURE 6.10. Sign relations with interpretant as mediator.

considers to have iconic versus conventional links with their referents (Figure 6.11).

The idea of degrees of iconicity for map sign-vehicles has been expressed more explicitly by Robinson et al., (1984). In relation to mapping "place" data with point sign-vehicles (i.e., point map symbols), they present three categories: *pictorial, associative,* and *geometric* (Figure 6.12). The three categories are described as "occupying successive positions on a continuum of symbolization ranging from the analogical or mimetic at one end to the purely arbitrary at the other" (p. 288). Although Robinson et al. do not provide strict criteria for identifying pictorial sign–vehicles, the illustrations they provide suggest that pictorial sign-vehicles correspond to Peirce's image hypoicons in that they are similar in appearance to their referent. As pointed out, however, pictorial sign-vehicles can vary considerably in how stylized they are. To be successful, "pictorial symbols should communicate without the necessity for a legend" (Robinson et al., 1984, p. 286). As with their pictorial category, Robinson et al. (1984, p. 287) do not explicitly define associative sign-vehicles other than to say that "symbols in this class employ a combination of geometric and pictorial characteristics to produce easily identifiable symbols." They also suggest that associative sign-vehicles may be "quite diagrammatic compared to pictorial symbols" (p. 287). The latter statement implies a commonality with Peirce's diagrammatic hypoicon in which the sign-vehicle represents the referent through some analogous relation of parts. The example of a box with a cross on top to represent a church fits this description, while the example of crossed pickaxes to represent a mine seems to be a case of metonymy (in which a part stands for the whole). In either case, associative sign-vehicles (in general) seem to

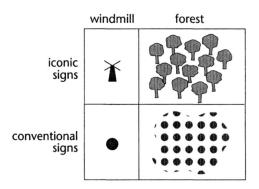

FIGURE 6.11. Examples of iconic versus conventional sign relations. *Derived from Keates (1982, Fig. 26, p. 67).*

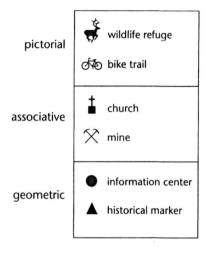

FIGURE 6.12. Examples of pictorial, associative, and geometric nominal point symbols. *Derived from Robinson et al. (1984).*

be less iconic than are pictorial sign-vehicles. The third form of point sign-vehicle identified by Robinson et al. (1984), geometric, is presented as "purely arbitrary" in its relation to a referent. As Keates (1982) points out, however, few map sign-vehicles are "purely" arbitrary.

The textbook categorization of point sign-vehicles described above seems to be grounded in Robinson and Petchenik's (1976) more detailed analysis of "map mark"–referent relationships. Here, it is proposed that the "concept of representation and meaning in mapping requires that we distinguish clearly between those marks that are visually completely arbitrary, and those that retain some graphic characteristic that can be visually or conceptually related to the referent" (p. 61). They suggest a continuum from mimetic to arbitrary map marks (as an alternative to discrete mutually exclusive categories such as icon and symbol). Examples are provided of the application of such a continuum to positional map marks, place labels, and lettering placement (Figure 6.13). They also seem to be the first to suggest that evaluating sign relations in terms of their iconicity is applicable not only to individual map marks (sign-vehicles), but to entire maps.[5]

One of the critical issues brought up by Robinson and Petchenik (1976) in their discussion of sign iconicity is that a sign can be iconic in various ways, some of which are more obvious than others. For example, they note that while a cartogram is mimetic (iconic) in matching geographic size to population size (a metaphorical hypoicon in Peirce's terms), it is not mimetic in the expected sense of space standing for space.[6]

For graphics in general, Knowlton (1966) (a decade earlier) proposed a continuum of sign relationships that is in many ways similar to

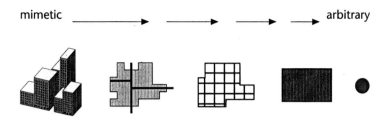

FIGURE 6.13. Example of the mimetic to arbitrary continuum of map marks for a city. After Robinson and Petchenik (1976, Fig. 3.1, p. 62). Copyright 1976 by the University of Chicago Press. Adapted by permission.

Robinson and Petchenik's (1976) mimetic–arbitrary continuum (and provides a way to formalize their continuum). His continuum of sign function is derived directly from the Ogden–Richards (1923) semiotic triangle, and therefore matches well with the semiotic discussion above. This *arbitrariness* continuum extends from arbitrary sign processes at one extreme to direct perceptual (i.e., nonsign) processes at the other. Knowlton, as discussed in Chapter 5, depicts the sign relations as changing from no overlap of sign-vehicle and referent at the arbitrary extreme to virtually complete overlap at the perceptual extreme (at which point the sign-vehicle and the referent become confused and the sign relationship no longer exists) (see Figure 5.4).

Recently, Ucar (1993) suggested an abstractness continuum similar to those of Robinson and Petchenik (1976) and Knowlton (1966). He described a category of signs (somewhat confusingly labeled "iconic signs") that grades from high abstraction (i.e., not iconic) to high iconization. Ucar presents this "iconic" sign continuum as one of two components of a higher level sign category labeled "artificial" signs (Figure 6.14). The other component of his artificial sign category is labeled "symbols," and seems to be reminiscent of Krampen's (1965) "emblem" (a highly conventional sign such as a cross to represent Christianity). Artificial signs are then matched at a yet higher level by a category of signs that represent "dimensions of the plane." The distinction here seems to be the same as that made by Bertin (1983) with his X–Y dimensions of the plane matched with the "retinal variables" and Lyutyy's plan and plan-free sign aspects (cited in Schlichtmann, 1991).

An important feature of Knowlton's (1966) presentation of sign relations is depicted via the circles corresponding to sign-vehicle and referent positions in his diagrams (see Figure 5.4). This notation is not a simple visual convenience that allows us to see the extent of overlap. The circles signify that the relation is between sign-vehicle category and referent category (not between an individual sign-vehicle and a particular in-

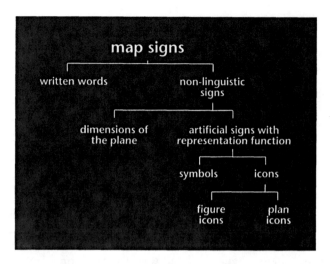

FIGURE 6.14. Ucar's typology of map signs. *Derived from Ucar (1993, Fig. 3, p. 774)*.

stance of referent). The implication is that any sign-vehicle having sufficient similarity to a prototype (perhaps shown in the map legend) stands for a kind of referent (e.g., any blue line on the map stands for a river, as does any incidence of "River" in blue).

Nyerges (1991a) has explicitly emphasized the links between a semiotic approach to geographic representation and theories of categorization. His interest is in sign "meaning" in the context of geographic abstraction that underlies a spatial database model. An adaptation of the Ogden–Richards semiotic triangle incorporating the concept of bundled space–theme–time sign aspects is presented (Figure 6.15). Nyerges treats sign meaning as a conceptual category and suggests that the nature of meaning for an individual (or in a digital database) will depend on the model for defining concepts (categories). Three category models (taken from Smith and Medin, 1981) are presented: classical, prototype, and probabilistic. The first two are discussed in detail in Chapter 4. The probabilistic alternative treats concepts as defined "by a collection of features . . . everything that has a preponderance of those features is an instance of that concept" (Nyerges, 1991a, p. 1488). Nyerges points out that it is convenient to understand types of geographic entities in terms of prototypes but easier, in a database context, to define them in terms of classical definitions.

A prototype-based interpretation of Knowlton's iconicity continuum has interesting implications for the Robinson et al. (1984) discrete pictorial–associative–geometric classification of qualitative point symbols (described above). As Knowlton (1966) pointed out, a useful taxonomy

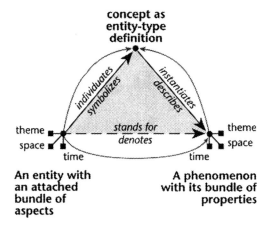

FIGURE 6.15. Nyerges's meaning triangle. The inner triangle is based on Ogden and Richards (1923) and the outer triangle on Brachman (1985). *After Nyerges (1991a, Fig. 1, p. 1487). Adapted by permission of the author and Pion Limited, London.*

of signs should be based not on attributes of the sign-vehicles, but on how signs are linked to their referents. He contends that "iconicity of a sign must be determined with reference to the *critical* attributes that are common to sign vehicle and an exemplar of the sign's referent category" (p. 165). It seems clear that the Robinson et al. categories do not neatly "occupy successive positions on a continuum" (p. 288) because the categories are not classical categories. Within each, a range of abstractness is evident and selected individual sign-vehicles are often difficult to classify. The categories probably have prototypes that fit the specified abstractness order, but the ranges of the category extensions overlap (Figure 6.16).

If iconicity depends upon the mediating interpretant (as it seems it must), then the iconicity of particular sign relations depends on human mental-categorization processes. As a result, iconicity is likely to have some definite origin and etymology as well as to differ across cultures, over time, and from individual to individual. Following this logic, Krampen (1965) suggested a continuum of "symbolism" defined on the basis of extent of collective agreement within a society (i.e., convention). He proposes the "quasi-symbol" as a sign relation that is ambiguous (e.g., the possibility of either slow or slimy being associated with a pictorial sign-vehicle representing a snail), a "true symbol" as a sign relation that is well established (e.g., the Republican elephant), and an "emblem" as a "symbol" that is universally understood within a culture to the point where the sign-vehicle assumes the meaning of its referent (e.g., cross = Christianity). Krampen's three categories seem to match well with the most

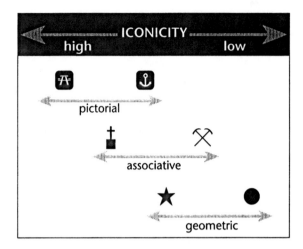

FIGURE 6.16. The relative abstractness of pictorial, associative, and geometric nominal point symbols.

iconic three of Knowlton's four categories of arbitrariness and suggest one mechanism by which signs might shift along this continuum over time.

The linking of increasingly conventional signs with decreasing arbitrariness disagrees with some semioticians' definitions of conventionality (which they equate with arbitrariness). Eco (1985b), however, has argued that "'conventional' does not necessarily mean 'arbitrary,' and 'motivation' does not exclude 'cultural agreement'" (p. 181). A convention (such as points on a compass) can rely on a motivation "which requires the expression [sign-vehicle] to embody the same form as the content [referent] *in some respect or capacity*" (p. 181).[7]

An issue that remains unanswered by Knowlton's variable semiotic triangle approach, and by other approaches to sign iconicity as a continuum, is how to judge the position of a particular sign relation on the continuum at a given time. Even when we focus on categories of sign-vehicle to referent relations (rather than on individual signs), it is not immediately clear which kinds of relations are more or less iconic. Is, for example, an image hypoicon (a sign-vehicle that shares similarity in appearance with its referent) always less arbitrary (more iconic) than a diagrammatic hypoicon (a sign-vehicle that shares similar structure with its referent)? Where does metaphorical or metonymic correspondence fit in? A related issue is Krampen's (1965) distinction between direct and indirect reference. An example of a direct sign might be a pictograph of a picnic table that stands for a place to eat a picnic lunch. In contrast, an example of an indirect sign is a drawing that stands for a domed building which in turn stands for a state capital on a map. It is logical to assume indirect reference to be less iconic than direct reference (when both have the same referent). What is far less clear, however, is how to compare an indirect sign (For which the first relation is highly iconic and the second completely arbitrary) with a direct sign for which the single relation is at the midrange of iconicity. If we follow Knowlton's suggestion that judging iconicity will depend upon identifying the critical common attributes between sign-vehicle and referent, we must also pay heed to his caution that "criticality of attributes is partly dependent on the interpreter of the vehicle" (Knowlton, 1966, p. 165). Again, we see the unique perspective that the interpretant as mediator provides; it makes clear that sign relations are not necessarily the same for everyone.

Another aspect of map sign iconicity is brought up by Knowlton (1966) in a discussion reminiscent of the C-rep versus P-rep distinction made in relation to the referent-as-mediator perspective. Knowlton asked whether it is "useful to speak of iconicity when a referent has no tangible existence" (p. 179). He contends that it is, but that iconicity can only be defined in terms of "isomorphism" between sign-vehicle and idea. Iconicity in terms of isomorphism in such instances (i.e., for C-reps) "must be an isomorphism between the sign vehicle and *one particular way*, of several possible ways, that ideas might be logically (hierarchically) arranged in *thought*" (Knowlton, 1966, p. 179). Knowlton talks of a "referent" with no tangible existence. His subsequent reference to iconicity as isomorphism between sign-vehicle and "ideas" suggests that when the referent has no tangible existence, the interpretant and referent become one.

In spite of some uncertainty about how to judge iconicity of specific signs, or whether C-reps can be iconic, an advantage of considering sign relations as a continuum rather than as discrete categories is that the same continuum seems applicable to both local and global levels of representation (i.e., individual map marks and entire visual displays). John Ganter and I (Ganter and MacEachren, 1989) adapted Knowlton's conceptual model to deal explicitly with displays in geographic visualization. This adaptation is presented in modified form here (Figure 6.17).[8] It should be immediately apparent that the continuum of arbitrariness (or "abstractness," to use a more common cartographic term) matches precisely with one axis of my model of maps as radial categories (see figure 4.3). Viewed from the semiotic perspective of a sign relation having degrees of iconicity, then, it is quite clear why there are no distinct boundaries between map and image or between map and diagram (graphic). That empirical evidence exists to support the radial nature of map as a category provides support for Knowlton's (and Robinson and Petchenik's, 1976) contention that sign relations for information graphics do not fall in the discrete categories that Peirce suggested with his icon, index, and symbol, but are characterized by degrees of iconicity.

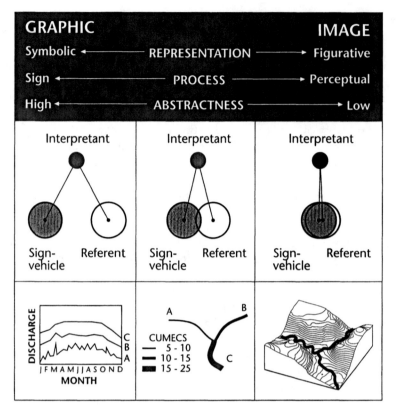

FIGURE 6.17. Application of Knowlton's (1966) "degree of arbitrariness of signs" concept to georeferenced signs.

Attention to iconicity and directness of sign relations is of interest beyond a role in defining the bounds of "map" as a radial fuzzy category. Comparison of different map forms on these dimensions can provide insight into the kinds of map schemata required to interpret the map form as a sign. The image-to-diagram (or graphic) continuum suggests differences in kind of stand-for relationships. The need for conceptual models (interpretation schemata) increases as we move from image toward graphic. We can expect, then, that different general schemata will be applied to different positions on the image–graphic continuum, leading to different expectations about what can be obtained. These differences in expectation, in turn, suggest that different meanings (at various levels) can be embedded in and extracted from the graphic variables used to reveal the display.

An interesting example of the implications of iconicity and directness of the display as sign is encountered when we consider map forms used to depict enumerated quantities. In 1967, Jenks proposed what he termed the *data model* concept for statistical mapping. This concept was based on the idea that a phenomena measurable at an interval or ratio level could be characterized (or signified) by a graphic model combining spatial and attribute information in a three-dimensional depiction. These graphic depictions were based on underlying mathematical conceptualizations. Jenks (1967) initially proposed a graphic data model to explain the information characteristic of choropleth maps, and then went on (Jenks and Caspall, 1971) to suggest several related data models that can be thought to underlie various forms of isarithmic map (Figure 6.18). From a data model perspective, choropleth and isopleth maps are indirect

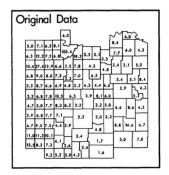

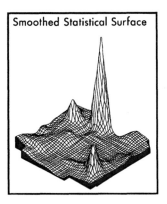

FIGURE 6.18. Jenks's data models for choropleth and isopleth maps. *Reproduced from Jenks (1963, Figs. 2, 4, and 7, pp. 16, 17, and 19). Reprinted by permission of Blackwell Publishers and the Association of American Geographers.*

signs that represent their data models which in turn represent the data (which in turn represent the phenomenon). Each of the data models proposed by Jenks is iconic in its use of height to represent magnitude (a metaphorical hypoicon) and in the way spatial variation in the model surface is used to represent assumed abrupt versus smooth spatial variation of the depicted phenomenon (a diagrammatic hypoicon).

Each of Jenks's three-dimensional data models can be transformed into a standard two-dimensional map depiction (e.g., as a choropleth or shaded isopleth map) (Figure 6.19). This transformation, in relation to the above model of iconicity, moves the representation toward the arbitrary end of the continuum. A more advanced map schema is therefore required to understand the kinds of similarity that exist between the two-dimensional map depictions and their referents. For example, rather than up = more, the less direct image schema extension of darker = more must be applied for both choropleth and shaded isopleth maps. In addition, on the choropleth map, data will typically be classified, resulting in relatively similar referents being assigned the same sign-vehicle, thus implying that they do not differ at all. The basic surface texture is signified in a similar way on the choropleth data model and the choropleth map, so therefore there is little change in iconicity in relation to this aspect of spatial variability (i.e., drastic changes along boundaries will still appear drastic). For the isopleth map, an isoline interval must be selected and this interval will determine which surface features are emphasized and which are suppressed. The isopleth map also requires the percipient to understand the more abstract (or arbitrary) concept that isolines are lines of equal value from which a three-dimensional surface can be visualized.

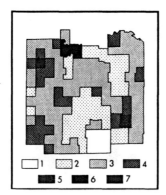 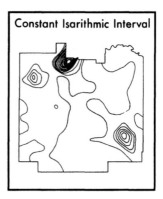

FIGURE 6.19. The 2-D equivalents of the data models in the previous figure. *Reproduced from Jenks (1963, Figs. 5 and 6, pp. 18 and 19). Reprinted by permission of Blackwell Publishers and the Association of American Geographers.*

Rather than a surface used to represent a surface, the isoline map (without shading) uses lines to represent a surface—this results in a decrease in iconicity.

The above discussion of data models emphasizes differences in directness and iconicity of sign relations for different kinds of map. Even on a single map, the abstractness of sign relations may not be consistent within or among levels of representation. This poses interesting questions about the complexity of map schemata needed to deal with various map types. Maps are unusual in their juxtaposition and combination of highly motivated (mimetic) with highly conventional (abstract) signs, both within and between levels of analysis. As Schlichtmann (1979) points out, most maps use space in a directly motivated way to stand for space.[9] Maps, however, can at the same time contain "symbols" whose individual spatial characteristics (size, shape, orientation) have no obvious correspondence to spatial or other attributes of the referent represented.

That multiple degrees of iconicity can exist within the same map suggests that multiple map (and map sign-vehicle) schemata are required to interpret these maps. It is likely that these multiple schemata will (of necessity) make use of the same image schemata in different ways (through the use of different metaphorical extensions). An example is the bivariate pollutant map shown in Figure 3.39. On this map, an up–down schema is metaphorically extended to the geographic space depicted by the map as a whole, resulting in up = north and down = south. The symbols on the map, however, also use an up–down image schema, but one extended to a different domain of up = more and down = less. Few experienced adult map readers are likely to have trouble with this dual use of a standard image schemata. This is just the kind of inconsistency, however, that can be expected to confuse children who are just learning to grapple with stand-for relations and developing the ability to use metaphorical extensions of image schemata to alternative domains.[10]

To deal with the potential combinations of iconic and noniconic elements in information graphics ("pictures"), Knowlton (1966) proposed a multivariate taxonomy of pictures. This taxonomy is based on the contention that visual–iconic representation has three parts: the *elements*, their *pattern of arrangement*, and their *order of connection*. The kind of picture is determined by the various possible combinations of iconicity on each part. In an apparent attempt to make tractable a three-variable taxonomy in which the variables are continuous functions, Knowlton proposes three discrete categories of reference: iconic, analogical, and arbitrary relationships between sign-vehicle and referent. The iconic–arbitrary distinction matches the two ends of the continuum described above, while the analogical category seems to be equivalent to Krampen's "indirect" reference. The possible combinations of three picture parts and three kinds of relation for each part yields 27 different states. By grouping those states for which "elements" have a common kind of sign-vehicle–referent relationship, Knowlton identifies what he considers to be three basic picture types: *realistic, analogical,* and *logical* pictures (Figure 6.20).

Knowlton's realistic pictures share the feature of having iconic elements. They are illustrations that would be categorized as "images" in the context of the radial map category (e.g., a bird's-eye photo) *and* as having direct reference to what is pictured. An analogical picture, in contrast, has an analogical relation between sign-vehicles and referents of elements. It uses an image to refer to something else (e.g., a 3-D surface representation that looks like a modeled terrain surface to represent a population density "surface"). The reference, therefore, is indirect. A logical picture is one in which elements of the picture are represented in an arbitrary fashion ("while pattern and/or connection are isomorphic with the state of affairs represented"; Knowlton, 1966, p. 178).

Knowlton considered both electrical schematics and highway maps to be logical pictures. This categorization of highway maps ignores the fact that "elements" in maps are often not depicted at the same level of iconicity. While sign-vehicles for cities might be arbitrary (e.g., a circle or a star), those for places of interest may be iconic (e.g., a fish to represent a fish hatchery, an airplane to represent an airport, lines on the map to represent linear features such as roads, etc.). To be useful for maps, Knowl-

	e	p	c	e	p	c	e	p	c
realistic pictures	I	I	I	I	A	I	I	X	I
	I	I	A	I	A	A	I	X	A
	I	I	X	I	A	X	I	X	X
analogical pictures	A	I	I	A	A	I	A	X	I
	A	I	A	A	A	A	A	X	A
	A	I	X	A	A	X	A	X	X
logical pictures	X	I	I	X	A	I	X	X	I
	X	I	A	X	A	A	X	X	A
	X	I	X	X	A	X	X	X	X

FIGURE 6.20. Knowlton's typology of visual–iconic signs. In the figure, "e" = elements, "p" = pattern, "c" = order, "A" = analogical, "I" = iconic, and "X" = arbitrary. *After Knowlton (1966, Table 1, p. 175). Adapted by permission of the Association for Educational Communications and Technology, Copyright 1966, Washington, DC.*

ton's typology would have to be extended to allow for separate assessment of different categories of map element within the same map.

THE NATURE OF MAP SIGN SYSTEMS–
MAP SYNTACTICS: LOGICAL INTERRELATIONSHIPS

Cartographers have worked to formalize symbol–referent relationships by developing typologies of symbol categories and rules for matching these symbol categories to categories of geographic features (map syntactics). These typologies are devised on the basis of evidence concerning human vision and visual cognition, experience and intuition concerning what works and what does not, and personal preferences of individual cartographers.

An important point to consider is that while not all maps are intended to serve a communication function (beyond the private realm), all map sign-vehicles (and maps as sign-vehicles) are intended to match with categories of mapped information in a clearly defined way. All map goals depend upon percipients being able to distinguish categories that are similar to one another from those that are different. While this general goal does not imply any particular emphasis on specific kinds of similarity, it seems clear that a common understanding between mapmaker and map percipient concerning symbol syntactics is required.

There are several levels at which map sign typologies and related map syntactics have been proposed. This section begins with the most fundamental: the level of graphic variables. Graphic variables are the categories of sign-vehicle features that can be individually manipulated by the cartographer or graphic designer and include size, shape, color hue, and the like. After introducing Bertin's original set of graphic variables and a range of extensions subsequently offered by several authors, a mapping syntactics based on their logical application is suggested. Extensions of these graphic variable–syntactics ideas into tactile, dynamic, and audio realms is then considered.

The analysis at a graphic variable level is followed with consideration of map symbol (sign-vehicle) sets and "rules" for matching symbol differences to differences in the entities they signify. Issues of map critique, standardization, and expert systems are surveyed. At the next level of analysis, interrelations among sign systems embedded in the same map are addressed. The section concludes with presentation of a typology of map forms that treats the map itself as a sign-vehicle whose spatial characteristics should match those of the phenomenon (or concept) type depicted (the map's referent).

Visual Variables and Syntactic Rules

There can probably be no complete agreement within cartography (or information graphics as a whole) concerning what fundamental variables we have to work with. To some extent, the delineation of these variables is subject to choice of analysis scale. In addition, the potential variables are dependent upon the technology available for producing sign-vehicles and entire displays. Since Bertin is usually credited with being the first to recognize the value of examining map signs at the level of visual variables, I will begin with his typology, then discuss more recent modifications and extensions of it. Once a set of graphic variables is delineated, the issue becomes how they relate to one another and how these relationships might be matched to those of referents. From a functional perspective, there are several basic syntactic rules that underlie "appropriate" matches of visual variable distinctions to referent distinctions. Looked at from a cognitive point of view, these rules correspond to the necessary map schemata that a percipient should apply in order to discover the intended sign relations.

Static Visual Maps

Bertin (1967/1983) appears to be the first cartographic author to formally propose a set of fundamental visual variables that serve as the building blocks for all map sign-vehicles. He proposed seven fundamental *visual variables*. They include: geographic position in the plane (X–Y location), size, (color) value, texture,[11] color (hue), orientation, and shape (Figure 6.21). I will call those variables other than geographic position "graphic" variables. These were termed "retinal" variables by Bertin, due to an assumption that humans have automatic, preconceptual, reactions to these variables at the level of retinal processing—or the level of the visual array.

For each visual variable, Bertin proposed rules for their appropriate use. These rules can be considered a sign-vehicle syntactics that defines how variations within each visual variable can be matched to corresponding variations among referents. Bertin categorized each visual variable in terms of whether or not it is appropriate for depicting quantitative (i.e., numerical) information (he did not distinguish between interval and ratio levels), ordered information, or nominal information.[12] The standard interpretation of Bertin's contentions concerning application of his seven visual variables is presented in Figure 6.22.

Although Bertin's typology-syntactics has had a tremendous impact on cartographic thinking, few authors have accepted his pronouncements

A Functional Approach to Map Representation 271

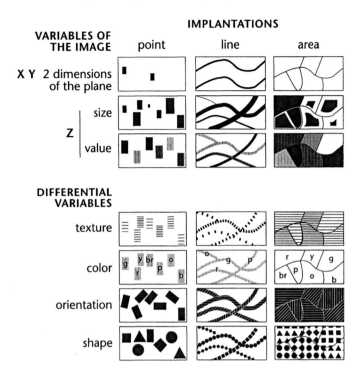

FIGURE 6.21. Bertin's graphic variables. *After Bertin (1977/1981, p. 187)*. Adapted from Graphics and Graphic Information Processing *by permission of Walter de Gruyter & Co.*

as given. Bertin's contentions about the appropriateness of different graphic constructions are presented as fact. He propounds a single formula for matching graphic constructions to particular information types. Cartographers and information designers have found his model useful, but flawed. It simply does not always work and (in relation to Bertin's *Semiology of Graphics* [1967/1983]) one critic went so far as to tell me, "I do not believe that there is a single good illustration in the entire book."[13] While few cartographers are this critical, most would agree that Bertin's system represents a theory about how maps and graphics should be designed and it is a theory that has been subjected to only limited empirical verification.

One possible interpretation of Bertin's theory, if we treat it as such, is that his contentions are contentions about schemata that typical map and graph readers bring to the process and that these schemata may be typical even for novices because they are founded upon embodied image schema related to spatial relations. Following arguments presented in Chapter 4, however, we would expect that even if Bertin's contentions

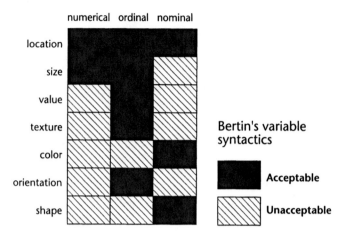

FIGURE 6.22. Bertin's graphic variable syntactics relating level of measurement to graphic variables. *Derived from Bertin (1967/1983, Fig. 2, p. 69, and the associated discussion in Section C, The Retinal Variables, pp. 60–97).*

represent a prototypic map schema, there will be considerable variability in schemata across viewers. In addition, a variety of individual factors will influence the effectiveness with which any particular schema (even if it matches Bertin's) will be brought to bear on a particular information extraction task.

In addition to critiques of the dogmatic nature of Bertin's syntactics (discussed in more detail below), a number of authors have argued that his typology is incomplete. Among the first to suggest additions was Morrison (1974). He added arrangement and the third dimension of color, saturation, to the list. While Bertin mentioned saturation, he chose to merge it with hue into a single category called "color." With the advent of computer-specified map colors, treating color saturation as a separate visual variable has become particularly important because most graphic design software used for computer-assisted map production today allows for separate control of all three attributes of color (value, hue, and saturation).

That the specification of a single set of fundamental visual variables can be problematic is also evident in Caivano's (1990) presentation of "texture" as a semiotic system. He treats texture not simply as the spacing of pattern elements (as Morrison does), or as equivalent to Bertin's grain, but as a tripartite variable analogous to color. Although Caivano's system is largely limited to area sign-vehicles, the distinctions he makes are worth considering. Three dimensions of texture are proposed: directionality (the ratio of length to width of texture units), size (of the texture units), and density (ratio of texture units to the background) (Figure 6.23). As he points out, these texture dimensions can be manipulated individually or together, resulting in a system of possible variations (Figure 6.24). Bertin's view of grain as a variable is a composite of two Caivano components: size and density of elements.

Caivano's texture semiotics illustrates the fact that identification of fundamental versus composite visual variables is a complex issue. He proposes texture as a composite variable rather than an individual one (and suggests the analogy to color with its three components). As an alternative, I propose that we consider "pattern" as the higher level visual variable consisting of units that have shape, size, orientation, texture (in Bertin's sense of grain), and arrangement. Without the variable of arrangement suggested by Morrison, neither Bertin's nor Caivano's system allows us to distinguish between the patterns depicted in Figure 6.25, although they are clearly quite different. The use of pattern arrangement proved to be important in the design of shading patterns produced on relatively low resolution computer output devices (Slocum and McMaster, 1986; Lavin, 1986).

With the addition of color saturation and pattern arrangement to the set, Morrison proposed a typology of nine visual variables that he matched to two categories of information in a system of map sign syntactics. His syntactic divisions assigned each visual variable to either nominal- or ordinal-interval-ratio applications. Morrison (1974, p. 123)

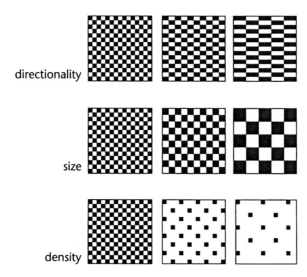

FIGURE 6.23. Caivano's texture dimensions. *After Caivano (1990, Figs. 4–6, pp. 244–245). Adapted from* Semiotica *by permission of Walter de Gruyter & Co.*

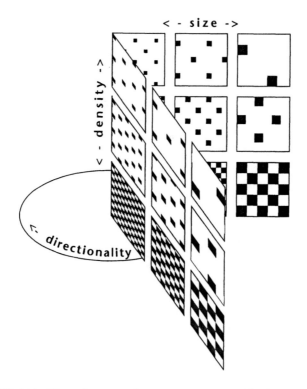

FIGURE 6.24. Three-dimensional texture space proposed by Caivano. *Derived from Caivano (1990, Fig. 9, p. 246).*

dropped Bertin's syntactical distinction between ordinal and numerical because of the opinion that "marks on a map can visually connote [denote] only two categories or scales of measurement: nominal (qualitative), or ordinal or higher scales (quantitative)." He does allow, however, that "when imbued with meaning via a legend," all four measurement scales are possible.[14] Perhaps more significant than the reduction of Bertin's three syntactical categories to two is a hint at the possibility of degrees of appropriateness for matches between variables and levels of measurement. Matches between graphic variables and the two levels of measurement are depicted as "useable," "possible," and "impossible" (Figure 6.26).

To make explicit the difference between the four levels of measurement available for referents and the two proposed for sign-vehicles, Morrison (1984) subsequently proposed the terms "differential" and "ordered" (the latter term borrowed from Bertin) for the two categories. Differential applications, then, are those in which referents differ in kind, and or-

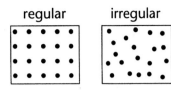

FIGURE 6.25. Two patterns that differ only in arrangement of component elements.

dered are those in which they differ in *at least* rank order. Two interesting features of this revised presentation are that Morrison seems to have dropped the partially applicable (possible) category from his syntactics and that the syntactics is described as "a schema for analyzing symbolization in cartographic representation" (Morrison, 1984, p. 46). In the context of map critique, the visual variable syntactics becomes the knowledge structure used to assess existing maps. As described below, Morrison demonstrated the usefulness of this approach in relation to symbolization in national atlases.

In the context of visualization of uncertainty, I have argued that a further addition to Bertin's original seven variables is needed: focus (MacEachren, 1992b).[15] In that paper, I proposed four kinds of focus, two dealing with sharpness of detail (edge and fill crispness), one dealing with resolution of base information, and one dealing with transparency of an intervening layer (called "fog"). After further study of these ideas, I now propose an alternative term, *clarity*, and a tripartite approach (analogous to the three aspects of color or Caivano's three aspects of texture). The

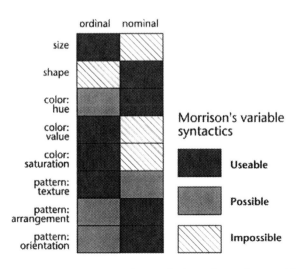

FIGURE 6.26. Morrison's syntactic relations for his eight graphic variables. He does not specify syntactics for the location variable. *Derived from Morrison (1974, Table 3, p. 124).*

three visual variables (or subdivisions of clarity) are crispness, resolution, and transparency.[16] I now conceive of *crispness* as a variable that allows adjustment of the visible detail of a map via selective spatial filtering of edges, fill, or both (Figure 6.27). *Resolution* deals with spatial precision change. In a digital environment, this translates to a grid size at which raster data will be displayed or vector data will be plotted (Figure 6.28). Resolution is the equivalent of texture applied to the geography rather than to the attributes of that geography. To match the terminology of rendering tools offered by recent advances in computer graphic technology, the term *transparency* has replaced my original "fog." I have seen transparency used quite effectively as part of a four-variable map dealing with global environmental hazards (Treinish, 1993). When applied to the depiction of uncertainty, the term "transparency" represents a "fog" that obscures the map theme in proportion to the uncertainty in data about the theme (Figure 6.29).

From my perspective, Morrison was correct in his initial loosening of Bertin's syntactical rules to allow for degrees of appropriateness in matching visual variables with phenomena characteristics, but perhaps incorrect in eliminating the distinction between numerical and ordered phenomena. I, therefore, propose the visual variable syntactics depicted in Figure 6.30 as a tool for selecting and evaluating appropriateness of specific graphic variables in relation to phenomena that can be characterized as numerical, ordered, and nominal.

Static Tactile Maps

Although most maps are designed to represent visually, there has been a growing concern with the spatial information needs of the visually impaired (see, e.g., Weidel, 1983; Tatham and Dodds, 1988; Tatham, 1991; Coulson, 1991; and Andrews et al., 1991). As one step toward meeting these needs, attention has been given to development of a tactual "language for mapping," a set of fundamental tactual symbols.

At least one of the tactual symbol design efforts has started with a semiotic basis extending from Bertin's graphic "semiology." Vasconcellos

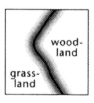

FIGURE 6.27. Examples of boundaries (left) and area fills (right) with low "crispness."

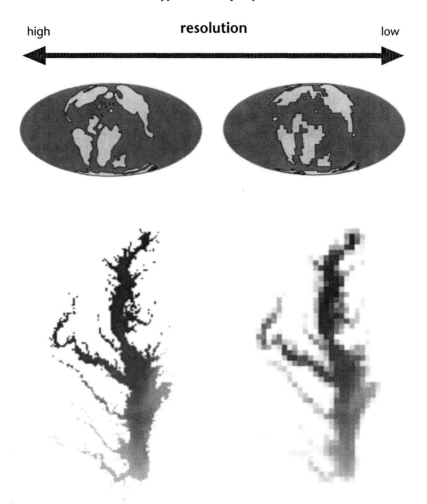

FIGURE 6.28. Examples of resolution change in vector and raster environments. Use of resolution may be most effective (at least in depicting uncertainty) if applied to sign-vehicles representing geographic base information (i.e., boundary or coast lines) on which the thematic sign-vehicles are plotted.

(1991) proposed a set of seven tactual variables for mapping comparable to Bertin's original visual variable set. Position in the plane, size, value, texture, form, and orientation all have direct tactual complements (Figure 6.31). Although Bertin's variable of "color" (hue + saturation) does not exist in a nonvisual world, Vasconcellos identified a new variable to take its place: elevation. All tactile maps are three-dimensional, at least to the extent that sign-vehicles are raised from the surface of the map. If they were not, there would be no way for the map user to sense them.

FIGURE 6.29. Examination of some risk data through the fog of uncertainty. *After MacEachren (1992b, Fig. 6, p. 15). Adapted by permission of* Cartographic Perspectives.

The amount that the sign-vehicles are raised is variable (to some extent), resulting in a new variable that might have also been considered a subcategory of "size" (in Z rather than in X–Y).

Vasconcellos (1992, p. 207) points out that "it is important to analyze the nature of the data (quantitative, ordered and differential)," but does not discuss how her tactual variable set should be matched with these levels of measurement. The implication is that, because of the logical correspondence between her tactual variables and Bertin's visual variables, the syntactic rules for use will be the same. This interpretation suggests that elevation, being a variant on size, should be applied to quantitative differences. A question that remains to be addressed is whether any tactual variables can impart information at a numerical level, or whether ordinal distinctions are the best that can be expected.

Dynamic Visual Maps

As part of a developing research effort directed to scientific visualization of geographic data, I taught a seminar on map animation during the spring of 1991. One of the principle outcomes of this seminar was an extension of Bertin's visual variable taxonomy/syntactics into the realm of dynamic maps. Although Bertin explicitly stated that his principles concerning visual variables were not applicable to dynamic display, our initial explorations suggest otherwise (DiBiase et al., 1991b, 1992). Bertin (1967/1983, p. 42) argued that "although movement introduces only one additional variable, it is an overwhelming one; it so dominates perception that it severely limits attention which can be given to the meaning of the other variables." Bertin's contention is supported in part by the view, presented in Chapter 3, that time and space are indispensable variables.

A Functional Approach to Map Representation

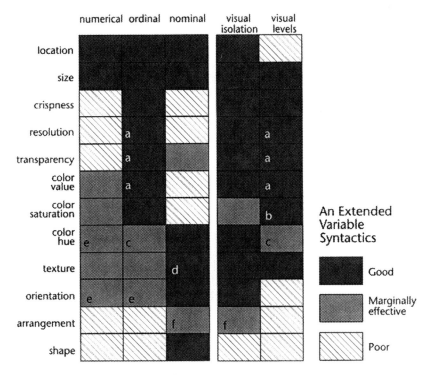

FIGURE 6.30. A visual variable syntactics representing the expanded variable typology presented here. The two columns on the right represent Bertin's concepts of "associativity" and "selectivity," as discussed in Chapter 2. Letters in boxes refer to qualifications on the good–poor assessments, as follows: (a) The clarity variables of crispness, transparency, and resolution can be used for no more than two or three categories. These variables are untested, but assumed to be most useful for representation of uncertainty. They may prove to be most practical in an interactive setting in which an analyst is able to toggle them on and off when needed. (b) Purer, more saturated colors appear to be in the foreground, while dull, desaturated colors fade into the background. (c) Hues must be carefully selected for an order of hierarchy to be apparent (e.g., the part-spectral sequence from yellow through orange to red). Hues interact with one another in sometimes unpredictable ways, so it is often difficult to determine which hues will dominate others. (d) Pattern texture is good for only two, or perhaps three, identifiable categories. (e) Orientation provides limited ability to communicate numerical or ordered information—glyphs based on a clock face or geologic strike and dip symbols are successful examples. (f) Pattern arrangement is best as a redundant variable to make a visual difference between cetegories more obvious. Derived from MacEachren (1994a, Fig. 2.28, p. 33); see Chapter 2 of that publication for more detailed discussion of most of the proposed visual variables.

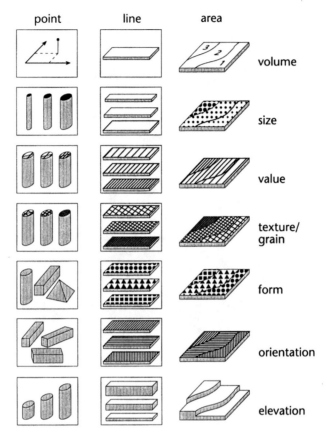

FIGURE 6.31. Tactile map variables matched to Bertin's original seven graphic variables. *After Vasconcellos (1991, Fig. 1, p. 208). Adapted by permission of the International Cartographic Association.*

Based on this idea, we would expect movement to be a powerful map "variable" (because it combines the two indispensable variables of time and space). As can be seen on static maps, however, an indispensable variable (space) does not completely overpower other variables (e.g., Rogers and Groop, 1981, presented results showing that map percipients could sort out three separate patterns on a trivariate dot map when the dots representing the three variables differed in color hue). What is probably true, however, is that on a dynamic map things that change attract more attention than things that do not and things that move probably attract more attention than things that change in place. The fact that the use of time gives the map designer a powerful new graphic tool is a reason to explore that tool in detail rather than to exclude it from consideration.

In our examination of animated maps, we concluded that Bertin's statement that time "introduces only one additional variable" is wrong—if we treat variables in the same way as we have done for the static visual variables described above. Initial consideration and experimentation led our animation group to propose three *dynamic variables* for mapping. These are *duration, rate of change*, and *order* (DiBiase et al., 1992). Subsequent contemplation of the kinds of manipulation available to the designer of a dynamic map (and viewing of many more examples of map animation) leads me to propose three additional dynamic variables, *display date, frequency*, and *synchronization*. The resulting six dynamic variables can be defined as follows:

1. *Display date*: The time at which some display change is initiated. The display "date" can be linked directly to a temporal location (chronological date). Display date is to chronological date as display coordinates are to geographic coordinates. Display date is specified in relation to a time span that is either linear or cyclical, with units that are ordinal, interval, or ratio (and generally specified in integer values corresponding to "frames"). On a dynamic map, the display date at which a feature is turned "on" can indicate creation of a phenomenon (e.g., on a historical settlement map a railroad might get turned on to correspond with the time of its completion) (Figure 6.32). For repeating cycles (e.g., months of the year), display date might be used to depict which month during each year had the highest number of tornadoes, births, cases of measles, or whatever (Figure 6.33). Display date can also be used to highlight particular places. In this case, display date represents something other than chronological date. An example would be on a "narrative" map delineating regions of the country in which endangered species exist. Species lo-

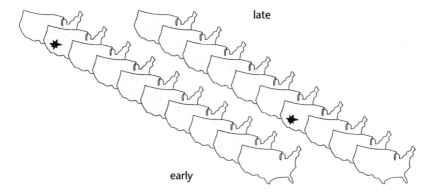

FIGURE 6.32. Display date specified in terms of frame number in an animation.

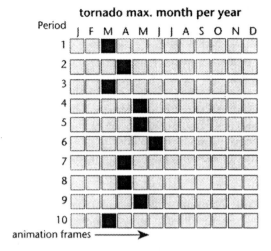

FIGURE 6.33. Display date applied to 10-year cycles of a 12-month period. Each box represents both a frame in an animation and a month in the cycle.

cations might be sequentially turned on while their plight is described via a set of graphs, voiceovers, or the like.

 2. *Duration*: The length of time between two identifiable states.

 a. Duration can be applied to individual *frames* of an animation. These frames are usually of fixed duration throughout a particular animation. In this case, short duration frames result in smooth change over time while long duration frames result in a choppy appearance reminiscent of a silent film.

 b. Duration can also be applied to what Szegö (1987) termed a *scene*, a sequence of frames with no change from frame to frame. In this case, the duration of the scene can be manipulated by controlling the number of frames (Figure 6.34).

 c. A coherent sequence of scenes can be referred to as an *episode*.[17] Episodes in a cycle can be thought of as phases. Episodes and phases obviously have duration, but from the perspective of dynamic variables, this duration is usually controlled indirectly by controlling the duration of frames.

 d. When episodes are phases in a repeating cycle, duration can also be applied to the *period* of the cycle, the interval between repetitions of a phase. The simplest application of controlling period duration is in the binary cycling of on–off used in "blinking." In this technique, the value and/or hue of a map mark is changed back and forth between two states to draw attention to a place. Monmonier (1992) has applied this technique to highlighting individual categories of a choropleth map to draw

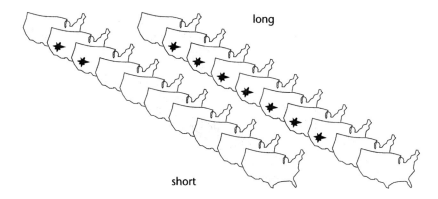

FIGURE 6.34. A depiction of scene duration.

attention to regional patterns (or the lack of them). Our Pennsylvania State University animation team has suggested a similar application as a figure–ground tool for enhancing the prominence of small point features (indicating earthquake epicenters) without the need for large symbols (DiBiase et al., 1992). In both cases, period duration was held constant for the cycle between phases. Changing the period can result in an appearance of more or less urgency. A related application of duration to period of a cycle is in the technique of color cycling (Figure 6.35). With

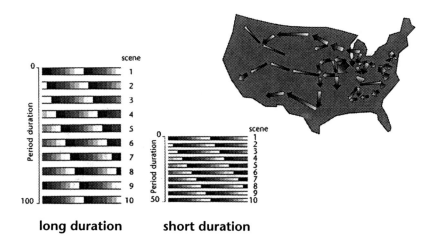

FIGURE 6.35. A still frame from a series that produces color cycling, an appearance of movement through the arrows depicting fiscal transfer (map derived from Tobler, 1981, Fig. 3, p. 421). The schematic depictions at the lower left illustrate the changes of value over time at relative positions within each arrow. The shorter-period example would produce an appearance of more rapid flow through the arrows.

color cycling, a range of color values or hues is "cycled" through a symbol. Color cycling results in apparent movement or flow through the symbol in a particular direction (with the apparent speed dependent upon the duration of each period). Duration can also be applied to period of a natural cycle (e.g., months of the year) as used in our Penn State analysis of global climate model predictions of temperature and precipitation. For a yearly cycle of monthly data, duration of the period can be divided evenly among the months (resulting in equal scene durations) or proportionally based on data values for each month (resulting in scene duration that indicates data magnitudes).

3. *Order*: The sequence of frames or scenes. Time is inherently ordered. When depicting change over time, order of presentation becomes a highly iconic sign-vehicle when matched with chronological order of the events depicted. Order, however, remains iconic when matched to nontemporal order (although probably metaphorically iconic rather than imagistically so). Slocum et al. (1990), Monmonier (1992), and DiBiase et al. (1992) all demonstrate successful use of presentation order as a variable matched to numerical order of some quantity other than chronological time (Figure 6.36).

4. *Rate of change*: The difference in magnitude of change/unit time for each of a sequence of frames or scenes (or m/d). Rate of change can be constant or variable. It will, of course, be zero if there is no change (because m will be zero). Assuming that neither m nor d is zero, either or both can be controlled to produce an increasing, constant, or decreasing rate of change (Figure 6.37). If, for example, change in position of some map sign decreases at a decreasing rate, the feature will appear to abruptly slow down and ease to a halt. If the decrease occurs at a constant rate,

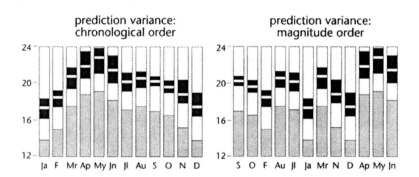

FIGURE 6.36. A comparison of chronological and attribute order of temperature graphs used in analyzing the uncertainty in global climate model produced predictions. *After MacEachren (1994b, Fig. 13.8, p. 125). Adapted by permission of John Wiley & Sons, Ltd., from* Visualization in GIS.

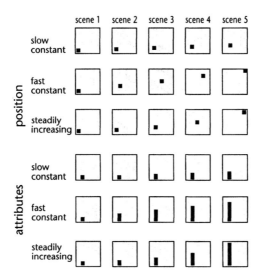

FIGURE 6.37. Rate of change applied to position information (for which changes in the geographic location variable are controlled) and applied to attribute information (for which changes in graphic variables are controlled). *After MacEachren (1994b, Fig. 13.2, p. 117). Adapted by permission of John Wiley & Sons, Ltd., from Visualization in GIS.*

the initial slowing will not be as abrupt, but the final halt may appear much more sudden.

5. *Frequency*: The number of identifiable states per unit time—temporal texture. In the same way that spatial texture is fundamentally a ratio of two sizes (the distance from center to center of symbol elements compared to a unit distance), temporal frequency is a ratio of two durations (the time from scene to scene, episode to episode, or phase to phase compared to a unit time). It is worth treating as a separate dynamic variable because (as with spatial texture) humans react to frequency as if it were an independent variable. Frequency is clearly a variable in the technique of color cycling cited above. The frequency with which colors change during the period of a cycle will determine the extent to which flow appears smooth (or appears at all) (Figure 6.38).

6. *Synchronization* (phase correspondence): The temporal correspondence of two or more time series. It can apply to matching chronological "dates" of two or more data sets precisely. If, for example, field samples are taken by two teams of researchers who both record spatial and temporal coordinates for their measurements, some postprocessing adjustments in both space and time may be needed to achieve perfect correspondence of the data sets. If peaks and troughs of the time series

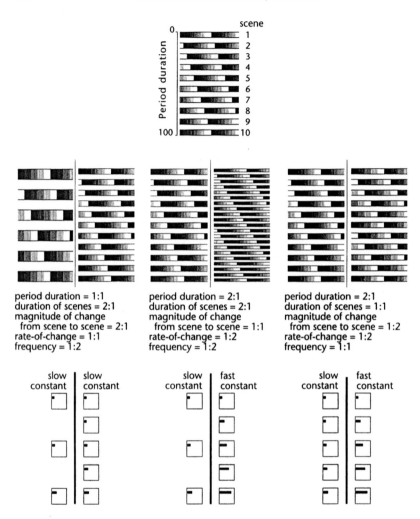

FIGURE 6.38. Frequency variation can be used to create jumpy or smooth cycling. The technique of color cycling discussed above is actually the result of manipulation and/or control of period duration, scene duration, magnitude and rate of change from scene to scene, and frequency. Examples of the interaction among these three variables are illustrated for the schematic color cycling scenes.

correspond, the series are said to be "in phase." If they do not correspond, they are termed "out of phase." Things that are in phase seem to be related. A set of point symbols blinking in phase, for example, will be seen as a related group or "figure" (due to the Gestalt principle of *common fate*). In the realm of exploratory visualization, manipulation of synchronization can also be a representational tool that allows otherwise hidden patterns to be noticed. If natural patterns are out of phase, for example, adjusting their synchronization at the display stage may uncover causal links that would not be apparent if chronological dates are closely adhered to (Figure 6.39). For example, time series of climate phases are likely to match those of vegetation phases in frequency and duration, but be out of phase (out of sync) because the climate event always precedes the vegetation change (e.g., rain precedes crop growth).

As for the typology of static visual variables, a syntactics of dynamic sign-vehicle–referent relations can be suggested. Display date is clearly a nominal variable. It can be used to show that a feature is or is not in a location at a particular point in time (dates in relation to one another are, of course, ordinal). Duration is measured on a ratio scale, so it seems clear that it (like size of geographic position) is suited to both ordered and numerical data. While duration by itself is ordinal or numerical, patterns produced by changing duration can suggest nominal distinctions (just as spatial patterns produced by combinations of size, shape, texture, and orientation in an area can be nominally different). Examples of nominal temporal pattern categories that might result are "fuzzy," "jittery," "pulsating," and so on. Order is clearly ordered but offers no way to signify numerical differences. Rate of change, even if constant, shows a numerically measurable difference from scene to scene. It, therefore, is suited to ordered and numerical depiction. Since we could measure, on a ratio scale, the degree to which two time series matched phases, synchronization is (theoretically) capable of depicting ratio level differences. In practice, however, it seems that synchronization (or lack of it) produces two nominal categories of "in phase" and "out of phase." The above suggested dynamic variable syntactics is summarized in Figure 6.40.

Dynamic Audio Maps

Once time is added to maps we have the possibility for sound as well as visual change. Building from our efforts to devise a typology and syntac-

FIGURE 6.39. An out-of-phase pair of graphs makes the temporal correspondence of two patterns difficult to notice. Shifting time so that the patterns are in phase, however, makes the correspondence impossible to miss.

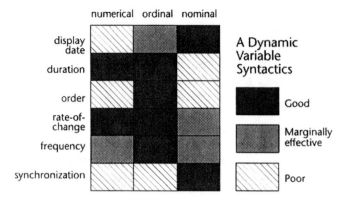

FIGURE 6.40. A dynamic variable syntactics. It must be made clear that this syntactics is a suggestion based on creation and observation of numerous animated maps and graphics, but no empirical testing. Like the various syntactics offered for static visual variables, if both cartographers and map percipients recognize an inherent logic, the syntactics will probably lead to successful maps. If the logic is apparent only to the cartographer, the map percipient is likely to apply an alternative map schema leading to quite different interpretation of the map sign-vehicles.

tics of dynamic variables John Krygier (a Ph.D. candidate at Pennsylvania State University at the time) explored a further extension—into the audio realm.[18] Krygier (1994) identifies an extensive range of "abstract" sound variables defined at a level similar to that of Bertin's graphic variables (Figure 6.41).

Each of the constant sound variables is defined below:

Location: The location of a sound in a two- or three-dimensional sound space.
Loudness: The magnitude of a sound.
Pitch: The highness or lowness (frequency) of a sound.
Register: The relative location of a pitch in a given range of pitches.
Timbre: The general prevailing quality or characteristic of a sound.
Duration: The length of time a sound is (or is not) heard.
Rate of change: The relation between the durations of sound and silence over time.
Order: The sequence of sounds over time.
Attack/decay: The time it takes a sound to reach its maximum/minimum.

With the exception of register, the first five variables are roughly analogous to static visual variables (location with location, loudness with size, pitch with value, and timbre with shape). Krygier suggests that regis-

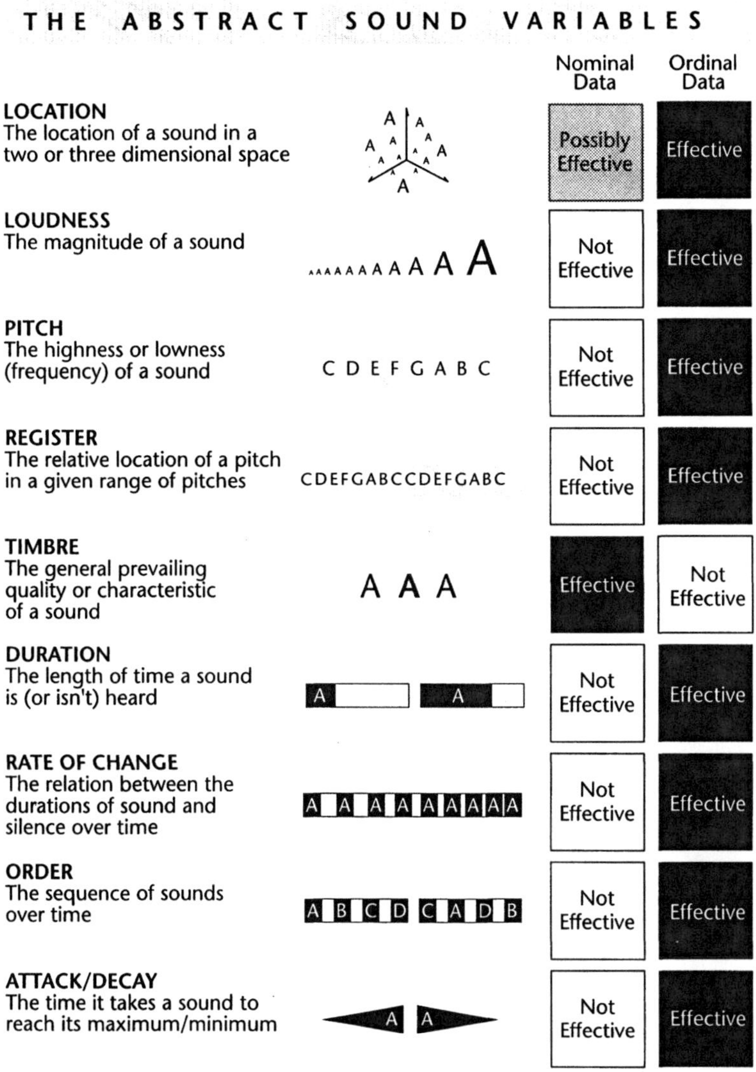

THE ABSTRACT SOUND VARIABLES

			Nominal Data	Ordinal Data
LOCATION	The location of a sound in a two or three dimensional space		Possibly Effective	Effective
LOUDNESS	The magnitude of a sound	ᴀᴀᴀᴀᴀAAA A A	Not Effective	Effective
PITCH	The highness or lowness (frequency) of a sound	C D E F G A B C	Not Effective	Effective
REGISTER	The relative location of a pitch in a given range of pitches	CDEFGABCCDEFGABC	Not Effective	Effective
TIMBRE	The general prevailing quality or characteristic of a sound	A A A	Effective	Not Effective
DURATION	The length of time a sound is (or isn't) heard	[A] [A]	Not Effective	Effective
RATE OF CHANGE	The relation between the durations of sound and silence over time	A A A A A A A A A	Not Effective	Effective
ORDER	The sequence of sounds over time	A B C D C A D B	Not Effective	Effective
ATTACK/DECAY	The time it takes a sound to reach its maximum/minimum	◄ A A ►	Not Effective	Effective

FIGURE 6.41. Krygier's sound variable typology and syntactics. It should be noted that the suggested syntactics of sound variables does not distinguish between ordinal and numerical information. *After Krygier (1994, Fig. 1, p. 153). Adapted by permission of Elsevier Science Publishing Company.*

ter may prove to be a particularly important variable because it has no obvious match among the static visual variables. Register together with pitch provides a nested hierarchy of sound variables that might be useful for signifying equivalent hierarchies of referents. Of the remaining variables, the first three (duration, rate of change, and order) are direct matches to dynamic visual variables detailed above. Krygier's "attack/decay" is analogous to frequency. Although Krygier did not propose them, it seems reasonable to suggest that there are also sonic equivalents to the dynamic variables of display date and synchronization.

Sign-Vehicle Sets

The visual, dynamic, and auditory variables identified above can be grouped in various combinations to produce individual map sign-vehicles and sign-vehicle sets. It is with the sets of sign-vehicles that syntactic relations become relevant. A number of authors have considered sign-vehicle sets from a semiotic perspective or from a perspective that can be evaluated in semiotic terms. At the highest syntactic level, all such systems share one "rule": *similar referents should be depicted by similar sign-vehicles and different referents by different sign-vehicles.* Thus, even in those cases where individual sign-vehicles might be judged completely arbitrary, their relations should be to some degree iconic. Following this "rule" is expected to allow map percipients to apply the Gestalt principle of similarity in order to visually group similar phenomena.

Schlichtmann (1979) argued that meanings (interpretants) should be expressed by marks (sign-vehicles) organized into hierarchies and "paradigms" (i.e., a set of features within the same basic category, such as "roads"). Each paradigm should be distinguished by keeping one visual variable constant (e.g., all cities on a population map might be depicted with circles whose size is then varied to correspond with variations in population). This idea of hierarchically arranged paradigms served as the basis of Ratajski's (1971) attempts at standardization of sign-vehicles for economic maps (discussed in Chapter 2). The choice of which graphic variable(s) to hold constant at each level of a hierarchy should be made on the basis of principles of perceptual organization together with syntactic principles related to appropriate matches between specific visual variables and levels of measurement.

Ratajski's (1971) syntactic approach to the standardization of sign-vehicles for economic maps has been largely overlooked in the cartographic literature, perhaps because few cartographers really have accepted the notion that standardization of sign-vehicles is a good idea. If we disregard the emphasis on standardization, however, and consider Ratajski's syntactic analysis, we find one of the earliest systematic applications of semiotics to the level of sign-vehicle sets—and an important complement to Bertin's analysis at the level of graphic variables. Among the important concepts presented was a view that matching a sign-vehicle set to a particular phenomenon (economic activity in this case) requires a classification not only of sign-vehicles, but of the phenomenon itself. This contention matches the one stated above that a "good" set of map sign-vehicles will be one in which logical relations within a referent set (Prieto's noetic field; see Chapter 5) are matched to corresponding logical relations within a sign-vehicle set (Prieto's sematic field).

Another contention made by Ratajski (1971), having implications for cartographic approaches to rules for sign-vehicle sets, is that syntactics devised by cartographers should draw on cultural agreement (conventions of the type Krampen, 1965, described as either symbolic or emblematic—that are almost automatic) and/or psychophysical processes (translated into our terminology here we might say principles of visual processing, perceptual organization, and fundamental applications of image schemata). In the first group, Ratajski includes some basic elements identified in Chapter 4 as part of the general map schemata used in most cultures (e.g., that north is up). As examples of the latter group, more fundamental image schemata such as line = linear, value range = linear order, and red = near are given.

A third principle of map sign syntactics propounded by Ratajski (1971) is that the simplest and most abstract sign-vehicles should be assigned to the highest levels of classification while the lowest levels should be assigned the most iconic sign-vehicles. This principle corresponds to issues of hierarchical categorization and basic-level categories examined in Chapter 4. That high-level categories are not likely to be basic level means that it would be hard to find iconic sign-vehicles to represent them. Basic-level categories are the highest level for which all category elements have similar overall appearance, so they are the highest level for which visually iconic sign-vehicles would be possible. One point of potential disagreement between my discussion of basic-level categories and Ratajski's view (that the lowest level should have the most iconic sign-vehicles) is that as we go lower than basic level, grouping criteria become less visually based. This suggests, therefore, that if we have a typical hierarchical category system (in which basic-level categories are in the middle), it is this middle level that should be distinguished by visually iconic sign-vehicles with different degrees of nonvisual iconicity on either end.

Where Ratajski's use of syntactic analysis was directed to developing a single standard set of sign-vehicles for a particular thematic map type, Morrison (1984) provided perhaps the most comprehensive semiotic critique of existing reference maps. As a preliminary step in developing a sign-vehicle set for a planned second edition of the U.S. National Atlas, Morrison analyzed six national reference atlases in relation to how they depicted point, line, and area information. Like Ratajski, Morrison considered both the logic and the perceptual characteristics of sign-vehicle sets in each atlas. Particular attention was given to whether ordered and differential sign-vehicles were matched to appropriate phenomenon types and whether differences of particular graphic variables exceeded least practical difference thresholds. The analysis criteria developed (and modified through application to the six sample atlases) were then applied to the first edition of the U.S. National Atlas. Several flaws in symbology used in this atlas were identified. As an example, for point symbols, Morrison suggested that the apparent qualitative difference between cities depicted with open or yellow circles and filled black circles was inappropriate, as was that between county seats and other administrative centers; that a qualitative difference should be used to distinguish abandoned from inhabited sites; and that the number of city sizes depicted exceeded LPDs at the map's scale (Figure 6.42). Based on this critique, Morrison proposed what he considered to be both a more logical and a more perceptible alternative for use in the new atlas (a project which never received funding) (Figure 6.43).

Both Morrison and Ratajski emphasized symbolization for particular kinds of phenomena. Approaching the map syntactics problem from the other direction (graphic variables) several authors (Eastman, 1986, 1987; Gibson, 1987; Rader, 1989; Mersey, 1990; Brewer, 1992) have focused on developing syntactic logic for color attributes. These efforts all attempt to identify structural relations among sign-vehicles that can be matched to structural relations among referents, and all take what we know about color perception into account.

Attempts at derivation of a color syntactics for mapping have approached the task by considering the variety of ways in which color might be categorized. The principle common denominator in these efforts has been acceptance of the trivariate model of color using hue, value, and saturation. Matching hue with nominal, value with ordered, and saturation with ordered have been generally accepted as logical associations. Most authors addressing the color syntactics issue, however, have questioned Bertin's contention that hue is completely inappropriate for quantitative data. They have also recognized the fact that not all phenomena mapped can be represented through the use of a single graphic variable. As a result, some applications may need combinations of hue and value or hue and saturation.

Gibson (1987), Eastman (1987), and Mersey (1990) all suggest that using a section of the hue range (in contrast to the full range of visible colors) can result in a logically ordered color set. Gibson contends that a hue range as large as 120° will be seen by most observers as intuitively or-

FIGURE 6.42. The point symbol set from the first edition of the U.S. National Atlas (the area fill for Detroit, the Memphis circle, and the state capital circle were printed in yellow). *After Morrison (1984, Fig. 35, p. 76). Adapted from* Cartographica *by permission of University of Toronto Press, Inc. Copyright 1994 by University of Toronto Press, Inc.*

dered. She suggests using red, blue, and yellow (the three subtractive primaries) as break points for three subdivisions of hue that can be matched to ordered data. Any of these hue sets is considered a particularly good match to proportional data where data represent a percentage of a total and the inverse is also meaningful (e.g., percent female—the inverse of which is percent male). At least from a cartographer's or printer's point of view, this sign-vehicle–referent link is iconic, because of knowledge about how the primary hues at either end of the scale will mix (e.g., blue + yellow = green).

Where Gibson looked to principles of color mixing at the display level for a syntactic logic of color assignment, Eastman looked to theories of neurological color mixing. He based his proposed color syntactics on opponent-process theory (OPT) with its four "unique" hues of red, yellow, green, and blue (Figure 6.44).[19] The predicted advantage of this

FIGURE 6.43. Morrison's suggested alternative sign-vehicle set for point features on the general reference maps of the U.S. National Atlas (the fill of the area mark for Chicago will remain in yellow). After Morrison (1984, Fig. 36, p. 77). *Adapted from* Cartographica *by permission of University of Toronto Press, Inc. Copyright 1994 by University of Toronto Press, Inc.*

OPT-based color syntactics (over a color-mixing model basis) is that it should not require learned knowledge of color mixing because it is linked to precognitive (neurophysiologically based) color processing mechanisms. Eastman proposed that proportional data of the kind considered by Gibson, which Eastman called *balance* schemes, could be logically linked to a color set that *spans* any of the four sides of the OPT color model. In contrast, *bipolar* data, in which values vary in two directions from a central point (e.g., positive and negative population change), are logically linked to a color set that *crosses* a unique color. Eastman went on to test his color syntactics experimentally and found that, while the logic matched performance in some cases, color value changes dominated the color logic in others. Some of the "span" schemes (e.g., yellow–orange–red), due to a clear value range, were more easily associated with a simple order (i.e., a linear-order schema) than with mixture (i.e., an overlap schema). What Eastman's study shows is that functional and cognitive approaches to representation can address the same cartographic questions but will not always arrive at the same answers. The implication is that the most powerful map syntactics will be one that is iconic (at least in terms

FIGURE 6.44. The OPT color space as diagramed by Eastman. *Derived from Eastman (1986, Fig. 3, p. 326).*

of an isomorphism of structure between the sematic and noetic field) and one for which this iconicity is derived from preconceptual image schemata (particularly those emanating from neurophysiological processes).

Perhaps the most sophisticated color syntactics to date (although she did not discuss it in semiotic terms) is Brewer's (1992) system of color selection for geographic data analysis and visualization.[20] This syntactics emphasizes thematic maps but is more complete in the kinds of thematic maps considered than either those by Gibson or Eastman. Issues of referent structure are balanced with those of sign-vehicle structure to produce a comprehensive typology-syntactics for both one-variable and two-variable mapping problems. One-variable problems are divided into four subcategories based on kinds of data represented: qualitative, binary, quantitative sequential, and quantitative diverging. These are then matched to hue and value (termed lightness) differences (Figure 6.45). Saturation is added as an alternative for a second quantitative category in two-variable problems and is suggested to be applicable for data quality representation. For these, three main categories are proposed with one having three subordinate categories (Figure 6.46). A complete syntactics matched to this two-variable typology was not provided, but the general hue-value rules presented for one-variable topics seem to be followed.

Multiple Linked Sign Systems

In discussing visual semiology, Saint-Martin (1989, p. 308) suggests that "it is apparent that the products of syntactic analysis of visual language can be represented by graphs of a certain order of complexity altogether different from verbal syntactic 'trees.'" Understanding visual syntactics,

data display emphasis	Perceptual characteristics	
	hue	lightness
qualitative	hue steps (not ordered)	similar (not ordered)
binary	neutrals, one, or one hue step	single step
sequential	neutrals, one, or hue transition	single sequence
diverging	two, one and neutrals, or two hue transition	two diverging sequences

FIGURE 6.45. Brewer's one-variable color syntactics. *Derived from Brewer (1992, table, p. 3).*

then, requires a nondigital spatial model that describes not only structural interrelations among visual variables but that determines "the spatial 'undulation' produced by the integration in a context of *simultaneity* of the many-leveled nets of signifiers." This perspective of visual semiotics suggests that we must consider not just individual sign relations or individual sematic fields for which a well-defined syntactics can be developed, but the interaction among map signs and sematic fields that a two- rather than a one-dimensional medium of presentation allows. Several cartographers have pursued a semiotic analysis at this level.

The syntactic analyses of color presented by Gibson and Eastman were directed to the thematic content of single-topic maps (particularly of the kind amenable to choropleth depiction). Rader (1989) presents a broader perspective in which he tries to devise a syntactic structure applicable to all components of multi- as well as single-topic maps and to reference as well as thematic maps. His syntactics also emphasizes color choice and follows the standard logic of hue differences for nominal data and value or saturation differences for ordered data. The principle innovation in Rader's syntactics is the inclusion of a distinction between global and local aspects of analysis (Figure 6.47). The local aspect deals with individual topics as considered above. The global aspect deals with creation of overall visual levels (see Chapter 3). Rader argues that although value and saturation are most appropriate for depicting individual referent hierarchies, hue can be useful in establishing a hierarchy of visual levels.

A Functional Approach to Map Representation

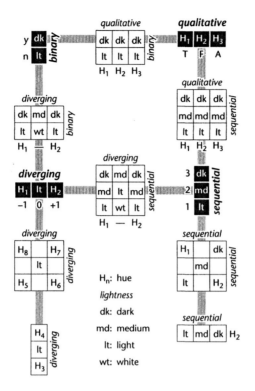

FIGURE 6.46. The graphic typology of one- and two-variable mapping problems in relation to differential and ordered phenomena. *After Brewer (1994, Fig. 7.2, p. 128). Adapted by permission of Elsevier Science Publishing Company.*

Rader offered a graphic model of his syntactics and described how it might be used to plan color assignments on both thematic maps and reference maps. In the thematic case, the graphic model provides a flexible way to make appropriate sign-vehicle–referent color assignments and to adjust the relative visual dominance of the theme in relation to context information (Figure 6.48). For the reference map, several examples are provided that demonstrate the potential for manipulating the relative prominence of different map features (Figure 6.49). In addition to pedagogic applications, Rader suggests that his approach is applicable to the development of interactive design environments and expert systems. In relation to interactive design, he points out that the "dynamism of both the conceptual and graphic structures is suited to exploring differences in both message and design parameters" (1989, p. 107). This suggests application of the ideas to interactive visualization, a topic to be considered in more detail in Part III.

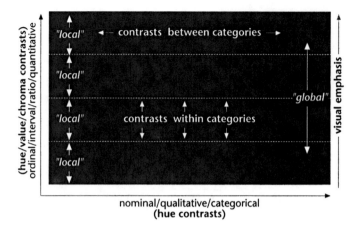

FIGURE 6.47. A color syntactics for mapping that provides for the local aspect of specific semantic fields and the global aspect of visual map levels. *Derived from Rader (1989, Fig. 20, p. 94)*.

Perhaps the most comprehensive approach to multiple linked sign systems is that offered by Wood and Fels (1986). They delineate five kinds of *intrasignificant codes*, or "codes that the map exploits." These are contrasted to *extrasignificant codes*, or codes "by virtue of which the map is exploited." The latter will be taken up when discussing the lexical level of representation in Chapter 7.

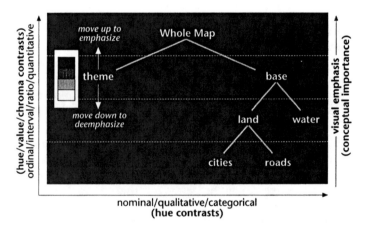

FIGURE 6.48. Application of Rader's color syntactics to a typical thematic map. *Derived from Rader (1989, Fig. 22, p. 98)*.

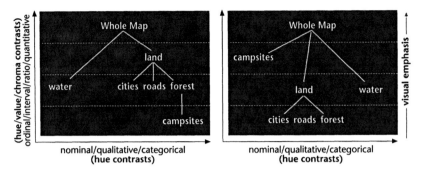

FIGURE 6.49. Two perspectives on the same map. *Derived from Rader (1989, Fig. 23, p. 99).*

In relation to the perspectives on map signs presented earlier in this chapter, Wood and Fels's typology of intrasignificant codes can be considered a typology based on the "sign-vehicle as mediator" point of view. Emphasis is on qualitatively different kinds of sign-vehicle exploited by maps, how they function as independent sign systems, and how each sign system relates to the others. Wood and Fels define the following five intrasignificant codes:

Iconic: Things and events "with whose relative location the map is enrapt."
Linguistic: The code of names (both places and themes).
Tectonic: Spatial or locational codes, "the code of finding, . . . the code of getting there."
Temporal: The code of time.
Presentational: The code of visual organization and arrangement.

The term "iconic" is used here in a rather different sense than defined above. Wood and Fels's (1986, p. 68) iconic code is any code "of the inventory, of the world's fragmentation." It is the code linking all categories of graphic sign-vehicles representing the map theme (regardless of their degree of iconicity) with their referent categories. Examples provided are codes for signs of streets, rates of cancer deaths, airways, Napoleon's losses in Russia, and so on. These are the codes that cartographers have given the most attention to.

"The linguistic code is the code of classification, of ownership: identifying, naming, assigning" (Wood and Fels, 1986, p. 68). Examples presented include reactive hydrocarbons, Orange County, age-adjusted rate by county. Although both iconic and linguistic codes involve categorization, linguistic codes (through the map legend) serve the role of "interpreter between the unique semiological system of individual map and the cultural universal system of language" (Wood and Fels, 1986, p. 78). This view of the role of linguistic and iconic codes in maps corresponds directly to Bunn's (1981) presentation of map-models as a dual-sign function (Figure 6.50).

The tectonic code is presented as including two components, *topological* relations and *scalar* (metric) specifications.[21] Wood and Fels point out that these codes have often not been treated as codes, perhaps because they have no visible sign-vehicle. They are, instead, made visible through association with iconic and linguistic codes. Another reason why tectonic codes may often be overlooked as codes is that their function is such an integral part of most people's general map schema that they are taken for granted. This "taking for granted" of tectonic codes is one reason why the controversy over the Peters projection (discussed in the next chapter) was in fact a controversy.

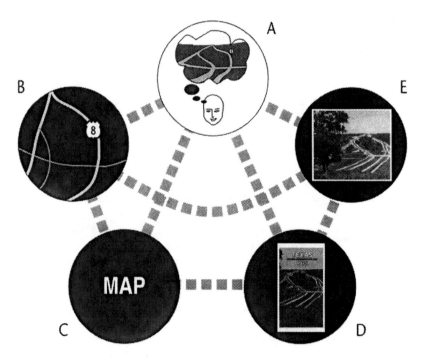

FIGURE 6.50. A dual semiotic triangle that depicts the link between graphic (iconic) sign-vehicle and referent together with the link between name (linguistic) sign-vehicle and referent, between name sign-vehicle and graphic sign-vehicle, and so on. *Derived from Bunn (1981, Figs. 7 and 8, p. 31).*

The temporal code is, in many ways, analogous to the tectonic. Like the tectonic, it is a code without visible presence on the map except as defined through iconic and linguistic codes. In addition, the temporal code has a nonmetric component, its *tense* (at least roughly matched, in a dynamic environment, by the variable of order), and a metric component, its *duration* (matched, in a dynamic environment, by the variable of duration). Duration is described in terms of the time "thickness" depicted by the map, and therefore can be thought of as analogous to the spatial extent of the tectonic code. The invisibility of the temporal code is, of course, only true for static maps.

Although Wood and Fels (1986) do not explicitly deal with dynamic space–time mapping environments, their discussion of temporal codes has a significant implication for such environments. Most cartographers working on space–time representation problems have treated time as a fourth dimension (e.g., Gersmehl, 1990; Szegö, 1987; Dorling, 1992b). This view has a long history in mathematics and in philosophy. Wood and Fels, however, seem to treat time as a complement to space, but not necessarily in the simplistic unidimensional way that the time-as-a-dimension perspective implies. The view that time is potentially more complicated than the one-dimensional view implies has been echoed (in the GIS context) by Peuquet (1994). She points out that although time shares properties with space, time also possesses a number of special properties that treating time as a fourth dimension cannot capture, yet which are essential for temporal–geographical analysis.

Wood and Fels's (1986) resistance of the time-as-a-fourth dimension perspective allows them to recognize some of the complexity of time in map representation. Their focus on static maps, however, causes them to ignore some aspects of temporal code complexity. Those temporal issues that are most difficult to deal with in a static environment are omitted or dealt with only superficially. These include the recognition and representation of temporal texture (how often?), rate of change (how fast or how much difference?), sequence (what order), and phase (in synch or out of synch?). The failure to address these issues directly can be explained, in part, by the fact that Wood and Fels's two categories of temporal code do not take into account the added graphic variables that become available in a dynamic environment (frequency, rate of change, order, and synchronization).

All of Wood and Fels's codes (the iconic, linguistic, tectonic, and temporal) are joined through the presentational code. The presentational code interacts with processes of perceptual organization and figure–ground to signify what is important and what is not, which phenomena are related and which are not, whether the map should be treated as reliable or whether it should not, and so on. An example of the latter is to use resolution as a graphic variable to signify limited precision of thematic information (see Figure 6.26—the global climate grid representation). Rader's (1989) inclusion of global-level syntactics in his color syntactics model is one example of an attempt at formalization of attention to the presentational code.

Maps as Signs

Just as we can consider how one system of codes (one sign system) can interact with another within a single map, we can also consider how maps as codes (or complex sign systems) can be treated as elements in a system of related maps as signs. One example of such a maps-as-signs analysis is my attempt, with the aid of David DiBiase, to devise a typology–syntactics of map types matched to phenomenon types in the context of quantitative enumerated phenomena (MacEachren and DiBiase, 1991). This typology–syntactics is an extension of Jenks's data model concept described above. One goal of this model is to draw attention to differences between how something is and how it is conceived to be. The fact that data are aggregated to enumeration units has often lead to the assumption that the most appropriate graphic display will emphasize those units. This assumption is due to confusion between the representation form of data and the nature of the phenomenon the data depict.

My approach to a typology–syntactics for quantitative spatial data is based on the contention that there should be an iconicity in spatial structure between a referent phenomena (or the aspects of the phenomena being emphasized in a particular context) and the map as sign-vehicle representing that phenomena. The first step, then, was to develop a typology of phenomena based on characterization of spatial attributes. After considering several alternatives, we settled on a typology that characterizes phenomena on two orthogonal dimensions: continuity and spatial (in)dependence. Continuity refers to the spatial completeness of the phenomenon: is it defined everywhere (e.g., population density) or does it occur at distinct separate locations (e.g., number of gas stations per county)? The second dimension, spatial (in)dependence, refers to the smoothness or abruptness of variation from place to place (Figure 6.51). If adjacent locations are independent, variation from place to place can be abrupt. Spatial variation will be smooth, however, if adjacent locations are dependent.

The typology of phenomena defined above is then linked to a corresponding typology of graphic data models (of the kind proposed by Jenks, 1963, 1967) (Figure 6.52). These data models correspond to a set of two-dimensional map forms (Figure 6.53). The relative position of these map

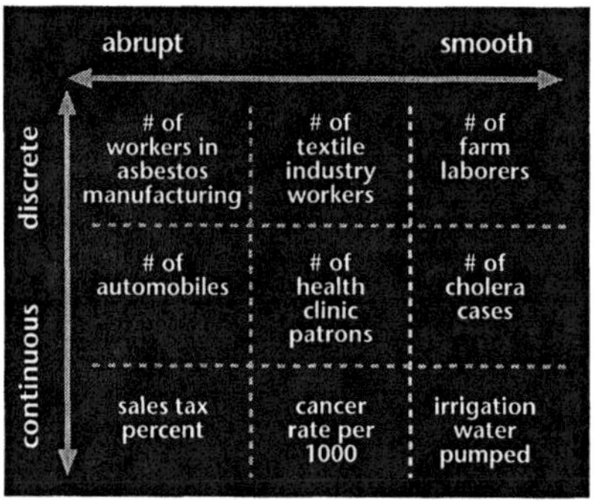

A Functional Approach to Map Representation

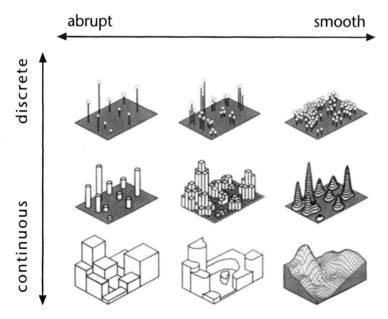

FIGURE 6.51. A typology of phenomena with examples. *Derived from MacEachren and DiBiase (1991, Fig. 2, p. 223).*

forms in the space defined by the spatial phenomena dimensions results in a map-level syntactics that describes how each map form is related to every other form in the typology and suggests the appropriate map schema for interpreting each map form. Recognizing that choropleth maps suggest a more abrupt spatial texture than isopleth maps or that graduated symbol maps suggest a spatially discrete distribution provides

FIGURE 6.52. A typology of data models. *Reproduced from MacEachren (1992b, Fig. 8, p. 16). Reprinted by permission of* Cartographic Perspectives.

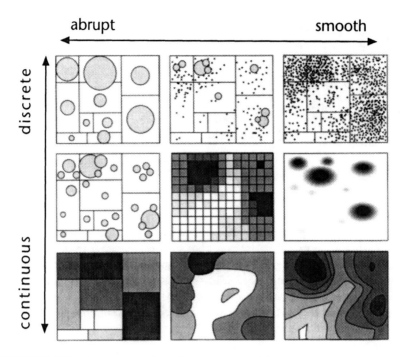

FIGURE 6.53. A syntactics of map forms. *Reproduced from MacEachren (1992b, Fig. 9, p. 16). Reprinted by permission of* Cartographic Perspectives.

the mapmaker with a logical basis for selecting the presentation form (setting a presentational code). Map form can be selected on the basis of the aspect of the phenomena that is of interest at the time, not on the basis of convenience (as is often done).

In addition to a role in comparing, and selecting among, symbol forms for quantitative maps, the idea of maps as signs should be considered in the context of narrative presentations involving multiple maps in a dynamic environment. Monmonier's (1992) concept of "Atlas Touring" is perhaps the best developed such environment to date. The "Atlas Touring" concept is founded on the ideas of graphic scripts that contain graphic phrases. The latter are coherent sequences of related images (maps, text, graphics, etc.) that present (or signify) a limited concept or idea. In this context, each graphic phase could be treated as a sign and the way graphic phrases relate over time could be examined from the film syntactic perspectives introduced in the previous chapter. Because Monmonier's graphic script approach has been developed initially as an exploratory visualization tool, detailed discussion will be picked up in Part III on geographic visualization.

Map Sign Comprehension

As suggested in the previous chapter, to comprehend a sign involves recognizing the sematic field (the set of relevant sign-vehicles) to which it belongs, matching that sematic field to a corresponding noetic field (the set of relevant referents), and matching the sign-vehicle class with a referent class. For maps, this entails recognizing the set of map marks that correspond in terms of some visual variable (e.g., they are all point symbols), identifying the variable that signifies differences among the sign-vehicles (e.g., shape), matching the set of map marks with a set of referents (e.g., industries) and linking specific variations in the relevant sign-vehicle variable with variations in the set of referents (e.g., different kinds of shape with different kinds of industry). For maps as a whole, the process involves identifying the relevant universe of map types (e.g., those that can depict enumerated quantities), identifying the nature of differences among those map types, matching the map type identified with the category of referent to which it applies, and linking specific variations in the sign-vehicle (map type) with corresponding variations in the referent (phenomena type).

To evaluate the success of map sign comprehension, we could apply Prieto's typology of comprehension (discussed in Chapter 5). This would allow us to distinguish among complete success, total failure, partial failure, and failure due to situational factors. While no cartographer has yet attempted to evaluate map comprehension in precisely these terms, Gerber's (1981) framework for exploring "competence and performance in cartographic language" approaches comprehension from a related perspective.

Cartographic performance, according to Gerber, has to do with understanding how to use cartographic signs to draw maps, while competence has to do with understanding cartographic sign systems. Interpreted in terms of Part I of this book, both can be linked to knowledge of or ability to generate appropriate map schemata to suit a particular map-sign context. One important distinction that Gerber (1981) makes is between *errors* (that the percipient does not have the knowledge to correct) and *mistakes* (which the percipient does have knowledge to correct). Errors can be perceptual (e.g., misidentification of colors due to simultaneous contrast) or conceptual (e.g., lack of an appropriate map schema to interpret a shaded isopleth map). Mistakes, on the other hand, seem restricted to conceptual problems such as accessing the wrong schema from memory or misapplication of a correct schema. Mistakes can be self-correcting (if the percipient takes the time to reconsider the initial evaluation of map signs). Correction of errors, however, requires additional knowledge.

Errors and mistakes can be identified at either or both of two levels that Gerber calls "identification" and "comprehension." In relation to our triadic model of signs, identification involves matches between sign-vehicle and referent. A percipient might, for example, succeed in matching a map sign-vehicle encountered on the map (a circle of particular size) with its intended referent (500 hectares). This identification, however, can take place without the percipient having any idea what "hectares" means. Comprehension requires the additional match of sign-vehicle to interpretant.

Gerber presents identification and comprehension as successive stages in sign processing. While comprehension clearly relies on higher level cognitive processing, it does not follow that comprehension requires identification. From a triadic sign–model point of view it is easy to see how the appropriate sign-vehicle to interpretant link might be made without successful identification. This is precisely the result of successful understanding of a legend prior to looking at the map. That a percipient has available the appropriate map schemata for linking a particular circle size with an interpretation such as 500 hectares does not guarantee that the percipient will make a correct identification when the sign-vehicle is encountered on the map. Perceptual errors of interpretation can occur in spite of error-free comprehension.

Gerber (1981) goes on to delineate a four-stage hierarchy of comprehension that can be viewed as the application of increasingly sophisticated schemata to interpretation of map signs. The stages of the hierarchy are defined as knowledge of the sign in context; knowledge of the sign out of context; understanding of the concept represented by the sign; and the ability to make inferences on the basis of the sign.[22] A knowledge of signs in context might be derived from a general map schema that allows a percipient who knows she is looking at a "map" rather than at a "diagram" to know that blue lines represent rivers. A more specific schema is probably required to interpret signs out of context (e.g., to understand any of the color syntactic models proposed above so that red can be equated with high disease incidence, orange with moderate incidence, and yellow with low). The third stage of comprehension that Gerber (1981, p. 107) identifies seems to deal with reaching a more complete interpretation of the sign-vehicle–referent link to the point of being "fully aware of the range of the acceptable attributes of the concept." Finally, the fourth stage involves an ability to make inferences based on the sign and its context. This involves the ability to link plan and plan-free aspects of individual signs and sign systems.

Gerber's stages of comprehension suggest the distinction between the functional approach to representation (the semantics and syntactics of signs) and the lexical approach of the next chapter (sign pragmatics).

DISCUSSION

The map and how it works has been approached in this chapter from a functional representation perspective—a perspective internal to cartography. The goal has been to identify a range of issues associated with individual sign relations on maps and with the logical interrelationships that are derived through interaction among map signs and sign systems.

Individual sign relations on maps are presented as tripartite, composed of sign-vehicles, their referents, and the interpretant that both share. Sign relations can be considered from the perspective of each sign component. In taking these separate perspectives, we can emphasize the aspect or functions of signs (e.g., to apprise, to stimulate to action, or to designate), the category of referent to which signs relate (e.g., location vs. attributes), or the nature of the meaning relationship established (particularly its level of abstraction). The various typologies presented should provide a base from which to study how cartography imposes semantic structure on maps, and thereby controls what characteristics of the world the map speaks to and what characteristics about which it is silent. At a more practical level, understanding the kinds of emphasis that are possible in map sign relations should facilitate successful design decisions.

At the level of the entire map, the study of syntactics (or logical interrelations among signs) is critical to making maps that work. This section of the chapter borrows from Bertin's approach to symbolization. Attention is directed to the fundamental variables with which the cartographer can build the appearance of similarity or difference across kinds of information. Where Bertin's emphasis was exclusively on static visual map displays, however, this chapter extends the approach into the realms of dynamic and nonvisual maps. In addition, attention is directed to development of integrated sign-vehicle sets and their combination as multiple linked sign systems.

The material presented in this chapter forms a base from which "rules" can be derived for map symbolization. Such rules are of particular interest in the context of the creation of expert systems that are designed to prevent mapping novices from making serious blunders likely to result in misleading maps. What has not been considered here is the map users' part in bringing meaning to maps—the lexical level of representation, largely external to cartography. This topic will be taken up in the next chapter.

NOTES

1. See Kraak (1988, 1989) for one perspective on these issues.
2. See Figures 6.51, 6.52, and 6.53 and related discussion below.

3. The following categorization of map representations was derived by Ganter and incorporated in an unpublished manuscript (Ganter and MacEachren, 1989).

4. The P-rep/C-rep dichotomy seems applicable to both plan and plan-free aspects of referents.

5. They do not actually use the terms "sign relations" or "iconicity," but talk of map mark–referent relationships.

6. What is "expected" is based on the prototypic "map" in our discussion of map as a radial category, for which the general map schema applies. This general schema, of course, includes the "rule" that space refers to space (see Figures 4.3 and 4.15).

7. This explanation of conventional signs corresponds directly with Giere's (1988) discussion of how scientific models signify aspects of the world. See Chapter 10 for details.

8. See MacEachren et al. (1992) for a slightly different perspective on the role of this mimetic (image) to arbitrary (graphic) continuum in the context of geographic visualization.

9. I would argue that this "direct" motivation is due to the fact that most peoples' general map schema contains this "map space represents real world space" assumption.

10. See Liben and Downs (1989) for consideration of the holistic (map) versus componential (sign-vehicle) understanding of representation by children.

11. Texture has been used in English translations of what might better be described as "grain." Bertin's interpretation of this variable is analogous to grain in film—the control of element size and spacing together, such that the percent inked area remains the same.

12. Bertin did not actually mention nominal (or qualitative) phenomena but his arguments have been taken to imply that those variables not suited to ordinal or numerical depiction are suited to distinguishing among nominal phenomena.

13. For obvious reasons, it is best to leave the quote anonymous.

14. The argument that size and geographic location (like value, saturation, and texture) can only represent ordinal information is interesting in relation to the fact that the most thoroughly studied research topic in cartography in 1974 was probably perceptual scaling of graduated circles. The underlying assumption of this research was that map readers could make numerical (interval or ratio) level judgments based on graduated symbol maps (if the symbols were scaled appropriately). The eventual result of this research effort seems to be that magnitude perception is too variable (both from individual to individual and within individual) for any scaling function to be adequate, thus apparently supporting Morrison's contention.

15. The idea for this graphic variable came up in a seminar I was teaching. I have been told by two participants in that seminar that focus was first proposed by David Woodward in a seminar at Wisconsin.

16. The reason for this change in terminology is that "focus" proved to be a confusing term in two ways. First, many students who encountered the term asso-

ciated it with only one aspect of what I had labeled focus—the sharpness of detail, or "crispness" of edges. Second, the term focus as a graphic variable is easily confused with the EDA technique of "focusing" (see Chapter 8 for more details on focusing). The switch from four components to three recognizes the inherent similarity of manipulating edge and fill crispness.

17. Mackaness (1993—abstract of Time-in-GIS workshop) used the term to refer to a generalized group of events. He seems to use "events" to mean the same thing as "scenes" here, thus my idea of episode may be the same as his.

18. Krygier's first "publication" of his sound variable typology was through an announcement on INGRAPHX, an electronic bulletin board (Krygier, 1991). That announcement attracted numerous requests for a copy of the paper. Based on feedback from that and other sources, Krygier developed a revised version of this audio syntactics upon which most of the comments in this section are based (Krygier, 1994).

19. See Chapter 3.

20. See Brewer (1994) for a more complete discussion of this color selection system. Brewer (1993) also produced a Supercard Macintosh tutorial that demonstrates her system.

21. This distinction differs from that of both Keates (locational/substantive) and Schlichtmann (plan/plan-free), but comes closest to Keates.

22. In Gerber's presentation, the first two stages seem to be linked with his concept of identification, while the latter two fit more clearly under his view of comprehension. All four, however, are presented as "components of the individual's comprehension."

CHAPTER SEVEN

A Lexical Approach to Map Representation
MAP PRAGMATICS

The lexical approach to how maps work is seen here as a complement (rather than an alternative) to the functional approach presented in the previous chapter. The functional approach emphasized the forms that map sign relations and systems can take. The goal was to suggest principles by which cartographers can build logical signs and sign systems for maps. Implicit in this limited goal (as it has been in most cartographic research on map symbolization) is an assumption that "proper" map signs will have a single semantic interpretation and that "proper" map sign systems will have a single (intuitively obvious) rule structure for understanding sign interrelationships. Most cartographers recognize that map representation is not this simple.

If our goal is to make effective maps, a functional approach to map representation offers a method for logical structuring of information. If, as in this book, we set for ourselves the broader goal of understanding how maps work, a functional approach alone (or even a functional plus cognitive approach) leaves us well short of that goal. To more fully understand how maps work we must consider how map representation as a semiotic system is embedded in a social milieu that provides a context within which sign meaning is established at multiple levels.

Past cartographic emphasis on cognitive and functional approaches to map representation implied (at least to recent cartographic critics) that cartographers in general do not recognize the potential for multiple meanings embedded in map representations. Whether or not these critics were right to suggest complicity among cartographers in fostering a map mythology that promotes maps as objective mirrors on the world, their critiques have directed attention to aspects of map representation often taken for granted. Recent work by Harley (1989), Wood (1992), and others has prompted both cartographers and the users of maps to expand their perspective on maps and map representation. Adding a lexical approach to cartographic research is, in my view, the principal result of this expanded perspective.

As argued in the present chapter, a lexical approach to cartographic representation accepts a potentially broad range of *legitimate meaning* for individual or groups of signs. It recognizes the multifaceted nature of map representation. The issue of correct or incorrect signs becomes secondary to that of exploring the various perspectives from which, or levels at which, map signs might be understood. While some map signs have single referents (and are in this sense singular signs), no map sign can have a single interpretant. As Harley (1989) has demonstrated, even the precise, unambiguous, apparently iconic representation of boundaries as definite lines on the map may be coopted (to support claims to territory).

An interesting example of the multiple kinds and levels of meaning to be found in even a narrow-purpose, seemingly unifunctional map is provided by Eco (1985b) in his discussion of sign typologies. Eco examines the *London Underground Map* as an example of multiple possible sign relationships embedded within the same map (or brought to the same map). He points out that the map is at once iconic (in its use of space to stand for space) and symbolic (in its use of straight unidimensional lines to stand for the fragmented route of the tracks). Its lines depict tracks and their intersection, but at the same time routes and their possible direction. The map as a whole can be a sign for the transportation system. At another level, it can also be a sign for a personal experience that involved use of that system. We might add that the map as a whole can serve as a sign for the social and economic structuring of the urban environment, a structuring imposed through control of movement and connections between places.

Although in Chapters 5 and 6, we discussed signs separately from sign systems, the sign's interpretant (its meaning) can only be defined within a system. Determining meaning of a map sign can be as much a process of establishing what the sign is not as it is a process of specifying what it is. Sign meaning, then, is a function of how that sign differs from other possible signs within a particular universe of discourse.

Meaning and map representation can be addressed from two perspectives, meanings *in* maps and meanings *of* maps. Meanings *in* maps are the primary or denotative meanings either specified precisely in a map legend or assumed to be part of the normal reader's general map schema (e.g., blue = water, or continuously varied symbols = continuously varied phenomena). These meanings relate to what the map is (or at least claims to be) about. Meanings *of* maps are the secondary or connotative meanings brought to map signs (or maps as signs) as a consequence of a denotative interpretant becoming a referent in its own right. That the lines in the map depict political boundaries which, literally, mean "borders between recognized political entities," can be taken to mean (at a secondary level of analysis) that some entity (political, corporate, etc.) recognizes both the existence of the other political entities depicted and their claims to territory contained within the lines. For the most part, denotative meaning of a map sign must be grasped before the sign can serve as the basis for secondary or connotative meaning. This distinction, between meanings *in* maps and meanings *of* maps provides the central organizing theme of this chapter.[1]

MEANING *IN* MAPS

When we look at how various sign systems inherent in maps provide explicit meaning to specific sign relations, several key issues present themselves. Perhaps the most important is to recognize that maps are powerful tools because they provide a means to meld three fundamentally different categories of meaning. These categories are meanings about space, about time, and about attributes in space–time.[2] Beyond this basic taxonomy of meaning are questions about the specificity of sign relations, directness of reference and associated literality of interpretants, differences in the concreteness of referents given meaning through signs, and the etymology of specific signs or groups of sign relations. Each of these issues is considered in turn below.

Space, Time, and Attribute Denotation

Wood and Fels (1986) argue that maps "exploit" five kinds of codes. Three of these, the "tectonic," the "temporal," and the "iconic" codes are directly associated with my space, time, attribute taxonomy of denotative meaning in maps. Maps are about things at particular places and times. The interpretants of map signs (or maps as signs) include not only interpretation of "what" the sign means but of "where" and "when" the meaning holds. On maps, Wood and Fels's remaining two systems of codes, the "linguistic" and the "presentational," are used primarily to make denotative meaning in space, time, and attributes possible and to limit the map percipient's options when determining the intended denotation for particular space, time, attribute map signs. Linguistic codes primarily link space, time, and attribute meaning to knowledge in the form of propositional representations. Presentational codes draw on visual cognition to link space, time, and attribute meaning with analogical representations (Figure 7.1).

Denoting Spatial Position

Maps use space to denote space, but map space is always a transformation and manipulation of world space. Interpretation of "where," then, depends on an understanding of the tectonic codes discussed by Wood and Fels (1986). I contend that this understanding is grounded in general and specific map schemata that provide the map percipient with a basic framework to which a particular unique map can be matched. In relation to space, the map denotes relative proximity of entities along with their size, relative direction, shape, and so on. Whether the spatial meaning derived by the percipient from individual or groups of sign-vehicles is consistent from percipient to percipient (i.e., the signs are interpersonal)

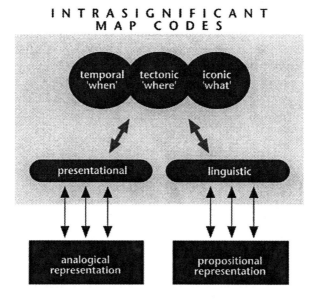

FIGURE 7.1. An interpretation of Wood and Fels's (1986) "intrasignificant" codes as they relate to image and propositional representations. Direct denotation of where, when, or what links through presentational and linguistic codes to the broader context of knowledge (in the form of propositional or analogical representations).

or from cartographer to percipient (i.e., the signs are comsigns) will depend on a common understanding of the tectonic codes and a common schema for interpreting relations among them. To avoid taking the map in Figure 7.2 to "mean" that Moscow is west of St. Petersburg requires an understanding that meridians are not parallel with the map border on the map projection used, nor are they parallel with one another.

The point of the above example is that, in spite of the apparently intuitive link between map space and world space, not all spatial denotations mean what they might at first seem to mean in relation to space. In addition, some maps "mean" more than others in relation to space because not all attributes of world space are denoted in specific maps.[3] Both maps depicted in Figure 7.3 can be interpreted to "mean" that warmer than normal temperatures were located primarily in the northern hemisphere in 1990. They differ, however, in their signification of globe area that was warmer than normal. The map at the right denotes (correctly) that the area of the globe experiencing warmer than normal temperatures is roughly balanced by areas of the globe experiencing cooler than normal temperatures. The map at the left, on the other hand, does not explicitly "denote" size of areas at all—but (incorrectly) "connotes" size and the impression that more of the globe had warmer than normal temperatures than had cooler than normal temperatures. Lack of denotation does not, therefore, prevent a map percipient from (incorrectly) assuming that the size on the map is a sign-vehicle representing size in the world. In relation to Prieto's typology of comprehension (cited in Harvey, 1982, and discussed in Chapter 6), the latter example can be considered a "failure due

FIGURE 7.2. Understanding east–west position requires the viewer to correctly interpret the denotative code provided by the graticule. If this code is ignored or misinterpreted, Moscow is likely to be judged west of St. Petersburg.

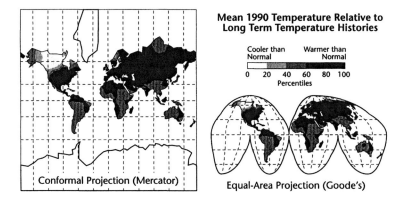

FIGURE 7.3. Mean 1990 temperatures relative to long-term temperature. The map at the left uses a Mercator projection (retaining angular relations, but distorting area wildly), while the map at the right uses Goode's projection (retaining correct area, while interrupting the oceans and presenting each pole as a multiple set of points.

to situational factors." If a person's general map schema includes the assumption that relative size on the map corresponds to relative size in the world, she is likely to (mis)interpret the sign-vehicles to mean something that they do not.

Denoting Temporal Position

Just as map meaning is linked to a place, it is also always relevant to a particular time (even when the map signs remain ambiguous about time). Although space can be portrayed as continuous through mapping of continuous world space to continuous map space, maps depict time as discrete.[4] Maps, then, denote time through one of several category structures by which the continuity of time is transformed into discrete mappable units.

Temporal codes can denote time as points or intervals on a linear continuum (For which the implied interpretant of map signs is limited to an instant or to a slice of time). Alternatively, time can be treated as a repeating cycle for which the denotation is a point on this cycle (e.g., noon or 6:00 p.m.), an interval on the cycle (e.g., an hourly mean), or an aggregate of points or intervals repeated across some time span (e.g., mean daily instant or interval for 1985). In relation to time series and cycles at various scales, there are several other aspects of time to be coded in, or derived from, both "real" and "virtual" maps.

- *Existence of an entity* (answering the question: *if* the entity exists at the specified time?): Temporal existence, as is spatial existence, is context-dependent. Entities only exist as cartographic objects (appear as sign-vehicles) if they have sufficient size in space and time at the resolution of the display (or they are of sufficient importance that spatial and/or temporal exaggeration operators are applied to increase their effective size in relation to that resolution).
- *Temporal location* (answering the question: *when* the entity exists?): Temporal location can be treated as a point or interval on a continuous function (e.g., 1:31 p.m. or May 1993), or an aggregate of points or intervals across repeated cycles (e.g., mean *noon* temperature or mean *June* precipitation).
- *Time interval* (answering the question: *how long* is the time span from beginning to end of the entity?): Time interval can apply to length along a temporal continuum or cycle or to the period between repeats of a recurring entity.
- *Temporal texture* (answering the question: *how often* does an entity occur?): It is a function of the time interval between entities in relation to some unit time (e.g., five times per hour, once in 100 years, etc.).
- *Rate of change* (answering the question: *how fast* is an entity changing or *how much difference* is there from entity to entity over time?): This aspect can be operationalized as being the ratio of magnitude difference at two locations in time and the time interval between them.
- *Sequence* (answering the question: in *what order* do entities appear?): Sequence can be applied to both temporal points (with time treated as a line) and intervals (with time treated as a cycle).
- *Synchronization* (answering the question: do entities occur *together*?): This aspect of time applies to correspondence of two or more sequences or cycles.

While geographic databases often code temporal features explicitly, temporal meaning is often implicit rather than explicit in maps (except with maps depicting time as an attribute or dynamic maps that use time as a sign-vehicle). When time is neither an attribute to be denoted or a sign-vehicle in its own right, the aspects of existence and temporal location predominate. For such maps, linguistic codes usually specify details of the temporal code (with varying levels of precision). For example, the only temporal information on a standard state highway map is the year of distribution and (possibly) the publication date (with temporal precision generally less than a month). Interpreted literally, this temporal sign can be taken to mean (probably incorrectly) that the conditions depicted will be true throughout the distribution year or were true (and confirmed) during the month of publication.

For maps in which time is treated as an attribute to be signified, the syntactic rules for static or dynamic sign-vehicles (detailed in the previous chapter) apply to the selection of specific sign-vehicles. For example, a quantitative variable such as size or duration should represent time intervals, and ordered variables such as value or order should represent chronological order. Cartographers, when limited to static maps, have devised a wide variety of creative methods for signifying different aspects of time. Monmonier (1990) provides the most complete account of these efforts and offers of typology of map types matched to aspects of time being depicted.

Denoting Attributes of Position in Space–Time

While time and place are often taken for granted in maps and not much thought is given to how they are denoted, the "what" of map sign denotation has been the major concern of those pursuing the cognitive approach to cartographic research. Attribute sign meaning on maps is often dependent upon linguistic signs, used in a legend or directly on the map, to provide a necessary link with prior knowledge. The more iconic (less arbitrary) the sign, however, the less need there is for linguistic signs as mediators. Cognitive–perceptual research conducted within the communication paradigm has paid particular attention to measuring the success of (nonarbitrary) map sign-vehicles in prompting an intended attribute interpretant.

Particular efforts have been directed to devising iconic (i.e., intuitively meaningful) sets of point sign-vehicles and/or evaluating existing sets on consistency of meaning. Gerber et al. (1990), for example, set out to devise a strategy for designing iconic positional sign-vehicles for tourism and recreation maps. They developed a set of guidelines for such sign-vehicles using a combination of cognitive and functional approaches to feature representation. Johnson's (1983) experiment described in Chapter 3, rather than generating new sign-vehicles, evaluated an existing set developed by the U.S. National Park Service. He evaluated this sign-vehicle set not only in terms of frequency of "correct" identification (in relation to the Park Service's assigned meaning), but in terms of the number and kind of alternative meanings that map readers assigned. He found the sign-vehicles to be generally successful but identified several problematic cases (Figure 7.4). Among these, one group of sign-vehicles were frequently misinterpreted due to visual confusion with other sign-

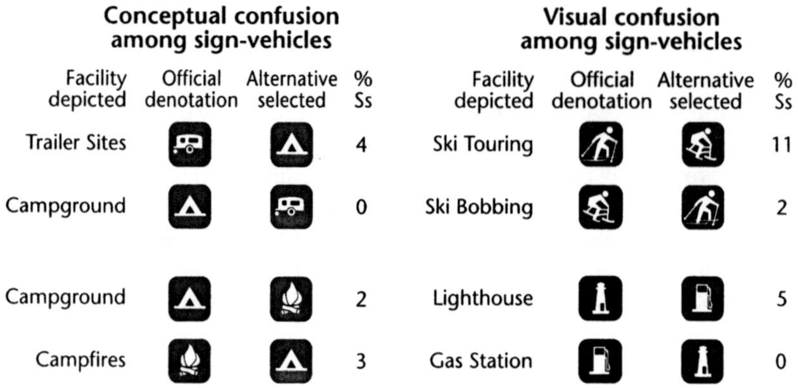

Conceptual confusion among sign-vehicles				Visual confusion among sign-vehicles			
Facility depicted	Official denotation	Alternative selected	% Ss	Facility depicted	Official denotation	Alternative selected	% Ss
Trailer Sites	🚐	⛺	4	Ski Touring	🎿	🎿	11
Campground	⛺	🚐	0	Ski Bobbing	🎿	🎿	2
Campground	⛺	🔥	2	Lighthouse	🗼	⛽	5
Campfires	🔥	⛺	3	Gas Station	⛽	🗼	0

FIGURE 7.4. Examples of conceptual and visual confusion found with National Park Service point sign-vehicles. *Derived from Johnson (1983, Tables IV–VI, pp. 113 and 115–117).*

vehicles (e.g., the gas pump sign-vehicle was often mistakenly taken to mean "lighthouse" rather than "gas station"). Another group prompted alternative interpretants to be associated with the sign-vehicle to referent link (e.g., the sign-vehicle intended to represent "fish hatchery" was taken to mean "fishing allowed").

For quantitative depiction on maps, meaning is often defined on the basis of some strict functional relationship between the graphic variable sets and the data sets being represented (e.g., size of graduated circles is made proportional to magnitude of data; therefore a circle twice as big as another is interpreted to mean that the entity to which it is linked has twice the magnitude of some attribute). For quantities, denotative meaning is primarily related to ordinal and (sometimes) to ratio judgments. Signs are intended to mean relatively "more or less" or to denote a ratio level comparison (e.g., "twice as much"). One ideal that has been put forward for judging sign-vehicle appropriateness is that the best quantitative sign-vehicles will be those that have the same interpretant (meaning in relation to order and/or ratio-level comparison) for cartographer and map percipient (i.e., are comsigns) without the need for mediating linguistic signs. Bertin (1967/1983), of course, contends that the visual variables of size and location intuitively denote both order and numerical differences, and that value and texture denote order but not numerical differences.

Experiments such as Cuff's (1973) on color logic for temperature maps have explored the (image schemata-based) notion that ordered color values "mean" rank ordered data to map readers and that dark values "mean" more while light values "mean" less. Similarly, McGranaghan (1989) has demonstrated that changing the surrounding background of a map displayed on a computer monitor can change the apparent meaning of color value ranges applied to choropleth maps. For some readers, dark always means more, while for others high contrast with the background means more while low contrast means less. Cuff also explored the issue of whether color hue could denote particular ends of the temperature scale in the absence of linguistic signs to specify the relationship. In 1973, when he conducted the experiment, Cuff found that (contrary to his expectations) red was not consistently taken to mean hot, nor blue to mean cold.

Patton and Crawford (1978) have also investigated the use of color hue to denote order. Specifically, they considered the use of full-spectral hue sets to denote elevation categories. They were particularly interested in alternative meanings that children using school atlases (in which the full spectral color set is quite common) might attach to different hues. Their subjects were students ranging from sixth grade to college. Each was asked to study a layer tint elevation map and respond to a series of questions. The map contained no recognizable linguistic codes except for an elevation legend labeled in meters. For any nonelevation questions, therefore, subjects were forced to rely on their own map schemata, the context within which questions were asked, and/or any other knowledge that seemed relevant. When asked to compare locations on the basis of elevation, subjects were reasonably accurate. Most subjects interpreted the spectral order of colors to mean elevation order. For questions related to land cover, the map did not denote anything. In spite of this, however, a preponderance of subjects interpreted green to mean "lush–vegetated" and brown to mean "dry–unvegetated." Patton and Crawford's specific conclusion, that children are likely to misinterpret hues on elevation maps to mean something about land cover (while at the same time correctly interpreting the elevations), may not be valid. Their experimental method probably led subjects to select an inappropriate denotative meaning by prompting them to think that the map denoted something other than elevation. The context provided by the questions prompted a schema in which color was linked with vegetation types. There is no evidence that children would select multiple meanings for the same map sign-vehicle if not prompted to think that the map must have vegetation information. In spite of this flaw, Patton and Crawford's evidence does support a view that certain color hues are consistently associated with specific categories. When the context suggests that land cover is denoted, green means "lush–vegetated" and brown means "dry–unvegetated."

As Wood and Fels (1986) point out, map signs often rely on context to establish denotative meaning. Blue "means" water in the context of a reference or highway map, even though most water is not blue and blue is used to depict things other than water on the same map. Another context issue is that provided by the position of the map as a whole on the image–graphic continuum. An unexplored, but important question, is whether the denotative meaning of a map changes due to its apparent position on this continuum.

Studies by Cuff (1973), Patton and Crawford (1978), and others make it clear that maps do not intuitively mean the same things to *all* readers in *all* contexts. For cartographers and readers to share interpretants for specific signs requires a common understanding of how sign-set variation is matched to data-set variation and/or an explicit definition for a sign relationship (usually provided via a legend). While legends typically employ linguistic codes to provide a link to propositional knowledge representations, presentational codes can also be used to link with analogical knowledge representations and image schemata. Ordering legend boxes for a choropleth map from low data values at the bottom to high data values at the top, for example, can help suggest the appropriate meaning for individual sign-vehicles (Figure 7.5). Hiller and DeLucia's (1982) "natural" legend concept provides another example in which presentational codes of a legend are used to make it clear that layer tint elevation symbols depict not only rank order but continuous variation (see Figure 4.18).

In addition to the definitional meaning of individual signs, visual hierarchies of map feature categories can be interpreted as an explicit sign for the relative significance of particular attributes in relation to others. Presentational codes are again used to draw upon both analogical knowledge representations and image schemata to denote relative significance of feature categories. Visual prominence is intended to mean more important. Rader (1989), applying a functional semiotic approach to map de-

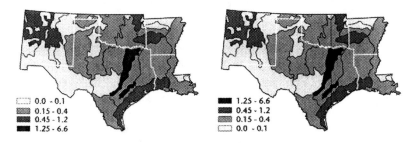

FIGURE 7.5. Map of ecological risk index values as derived by the U.S. Environmental Protection Agency's Region 6. The map depictions shown here are identical, but the legend on the right is arranged to match the natural schema of up = more. See MacEachren (1994a) for more detailed discussion of these data and their map representations.

sign, has formalized this application of presentational codes. His system allows map designers to logically manipulate these hierarchies, and as a result, to manipulate map meaning.

Specificity of Signs

Singular versus General Signs

As defined in Chapter 6, singular signs are those for which the interpretant has only one referent while general signs permit multiple referents. "U.S. capital" is an example of a singular name-sign, but only if our context is restricted to a particular time period during which the capital has been stationary. In contrast, "State capital" has multiple referents, each equally viable. "State capital," therefore, is a general sign.

Discrete categories of specific and general map signs impose the same awkward unrealistic limits that other strict categories often do. In practice, map signs will have increasing specificity (be less general) as the number of variables used in the sign-vehicle increases, and as linguistic codes provide links to detailed knowledge. For the sign to be a comsign (to have a common interpretant for mapmaker and map reader), both must agree on the level of sign specificity. If a sign-vehicle occurs in a legend, it can usually be assumed that its interpretant defines a set of possible referents, and that therefore the sign it is associated with is a general sign.

While the degree of generality (or specificity) will vary, virtually all map signs are general signs in the sense of permitting multiple referents. Blue lines of a particular width that are not smooth curves or straight lines (as with graticule) can have many instances of river as their referents. A river sign is usually not specific until the graphic sign-vehicle is joined to a linguistic sign-vehicle such as "Hudson River." Even in the latter case, a graphic sign-vehicle combined with a linguistic (or name) sign is often not entirely specific. While limiting the range of possible referents considerably, the addition of "Hudson" to "River," or to the blue line of particular width, may still permit several different referents within the United States. Like natural language, then, map signs sometimes are intended to be "singular" signs but depend on context (usually spatial context in the case of maps and temporal context in the case of language) to achieve that singularity. The map reader (if successful) realizes which Hudson River the cartographer was signifying due to the proximity of the linguistic and the graphic sign-vehicles and their location in relation to other features. This realization is dependent upon a general map schema that has as a central construct the idea that proximate graphic sign-vehicles and names on maps are likely to be linked and also includes an assumption that things are named on the map at a relative location corresponding to their map position.[5]

Issues of sign specificity seem to be directly related to the issues of category hierarchies discussed in Chapter 4. The most general signs will be linked to entities at the highest level in a category hierarchy. These high-level categories represented by general signs will (or should) usually make use of sign-vehicles distinguished on a single graphic variable. On a vegetation map, for example, vegetation type at Level I of the Anderson land-use classification system can be represented by a single graphic variable, like a color hue, assigned to a continuous and bounded map area. In order to become more specific, and thus depict lower levels of the hierarchy, another graphic variable (e.g., color saturation, pattern orientation, pattern shape, etc.) will be required.

As the above example suggests, sign specificity is directly linked to issues of map generalization. As the need to generalize increases (due to scale reduction or map purpose), the tendency will be to depict increasingly higher levels in feature-type hierarchies and, as a result, to use increasingly general signs (Figure 7.6). Carefully defining the relationships between intended level of generalization and sign specificity, then, would seem to be a critical issue in implementing automated generalization strategies.

In a geographic database context, Nyerges (1991b) directly addressed the issues of hierarchically organized levels of meaning as they re-

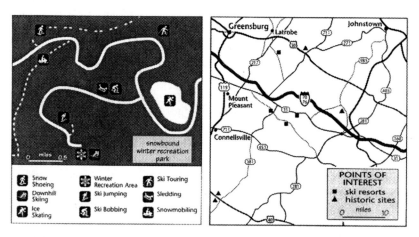

FIGURE 7.6. This tendency toward more general signs as map scale increases is evident when we compare a winter recreation park map and a regional map used to locate such parks and other points of interest.

late to levels of generalization. He contends that "a generalization as a database abstraction is expressed as a concept-type hierarchy that represents levels of concept specificity" (p. 71). Abstraction involves grouping "entities" into "entity types," a meaning concept that "defines" a collection of phenomena in the world as a phenomenon class and is "described by" or "symbolized by" an "entity class" in the database. The entity class describes only "aspects" of the phenomenon class (i.e., referent). The primary aspects identified by Nyerges (as noted above) are space, time, and theme.

Unambiguous versus Ambiguous Signs

As defined in Chapter 5, unambiguous signs are those for which each sign-vehicle has only one interpretant. Ambiguous signs, in contrast, allow multiple interpretants. A black star inside an open circle is the sign-vehicle for a "singular" sign with the interpretant of "state capital—Lansing" on the Michigan official highway map. On the Arizona official highway map, however, the same sign-vehicle is used in the "general" sign having "welcome center" as its interpretant. Similarly, a red square is used for a U.S. historic site on the Georgia highway map and for state hospitals and institutions on the Ohio highway map. Sign-vehicles, unlike real numbers or musical notes, may be unambiguous for an individual map or map series, but are always ambiguous for maps in general.

The graphic variables available to cartographers have no predetermined conventional meaning that can be relied upon to yield an unambiguous sign (although they do seem to have generally understood links to particular kinds of image schema). In contrast to iconic signs in which many meanings may come to mind simply due to experience with things in the world, arbitrary signs are linked to their referent primarily via convention (i.e., an argreement among users of the sign). As a result, the meaning of arbitrary signs must be explicitly assigned and the percipient is virtually forced to rely upon linguistic codes in the legend to derive meaning—thus rendering the sign unambiguous in the context of that particular map.

Ambiguity of map signs within individual maps seems to be an inverse function of their arbitrariness (as defined in Chapter 6). As signs are less arbitrary, or more iconic, the likelihood for multiple meanings escalates. This may counter initial expectations, if these expectations are due to an assumption (typically made by both cartographers and map percipients) that the meaning of iconic signs is obvious. Following such logic, cartographers often omit legend definitions for iconic signs and, if a legend is included, percipients often ignore it. Iconic sign-vehicles generally suggest multiple aspects of the referent that they stand for and the intended mediating interpretant can be any one or combination of these aspects. When a sign-vehicle has visual or structural similarities to its referent, it can also be easy to metaphorically transfer the apparent meaning to another context. The sign can then have both literal and metaphorical meanings. Familant and Detweiler (1993) make this point in relation to graphic "icons" used in graphical user interfaces. Their best example is a sign-vehicle with a rectangular body, ribs on the body, and a lid. This sign-vehicle can have the interpretant "trashcan" which links it to Familant and Detweiler's *sign referent* category or the interpretant "a place to throw things away" which links it to their *denotative referent*.[6]

Monosemic versus Polysemic Sign Systems

As noted in Chapter 5, the distinction between unambiguous and ambiguous signs is linked to that between monosemic and polysemic sign systems. Bertin (1967/1983) contends that the closer we are to an "image" versus a "graphic" (the more iconic vs. arbitrary), the more possible interpretations there will be (the more ambiguous individual signs will be). For Bertin, graphic signs are unambiguous while photographs are highly ambiguous signs. He argues that this is always true, by definition, because maps and other graphics are monosemic sign systems while photographs are polysemic. Monosemic systems have one-to-one links between sign-vehicle categories and referent categories. Polysemic systems, in contrast, have one-to-many (or even many-to-many) links. According to Bertin (1967/1983, p. 2), "a system is monosemic when the meaning of each sign is known **prior to** observation of the collection of signs." In contrast, "a system is polysemic when the meaning of individual signs **follows** and is deduced from consideration of the collection of signs"(p. 2).

Bertin's (1967/1983) contention that cartography is a monosemic system of reference is arguably true within individual maps or narrowly defined map series (e.g., U.S. Geological Survey topographic maps within a set scale range over a specified time period). Even for an individual map, however, the contention is only viable if we assume the signs are "comsigns," thus having the same sign-vehicle–interpretant–referent relationship for both cartographer and map percipient and that each sign has only one "intended" interpretant. There are innumerable possibilities for unintended interpretants (or intention of multiple interpretants) to be inferred from particular sign relations. The likelihood of cartographer and percipient using different codes is highest when a map percipient notes a similarity between a sign-vehicle and some possible interpretant that the cartographer had not anticipated (e.g., that a "Udults" soil depicted with a red area fill is always red). Even when there is no mismatch between mapmaker and map user in the interpretant attached to a particular sign relation, there can be multiple intended interpretants. On a temperature map, for example, the sign-vehicle "red" may occur between linear sign-vehicles that are interpreted as lines of equal value. The explicit link of red to a temperature range of 80–90°F can at the same time have the interpretants "hot," "higher than 70–80°," "average summer temperature," and so on (and these multiple interpretants may even be explicitly stated in the legend).

An assumption that cartographic codes are (or can sometimes be) monosemic is also dependent upon limiting consideration to "explicit," "denotative," and "conscious" codes. As will be considered below, all map signs (and entire maps) have the potential to carry both primary and secondary meanings. Multiple interpretants at varying levels of analysis, therefore, are always present. By arguing that maps are (or should be) monosemic, Bertin could be accused of trying to make claims for cartographic objectivity that are not supportable.

Beyond individual maps or limited map series, cartography must be considered polysemic. This is the case even for maps in a particular narrow category, such as U.S. state highway maps. Some cartographic codes are polysemic across all maps due to a lack of agreement among cartographers on standardized sign-vehicles. Other cartographic codes are polysemic in the same way that natural language is: the intended interpretant for a sign-vehicle varies from situation to situation (i.e., is dependent upon context). For example, even though the sign-vehicle is not depicted in the legend of the Georgia highway map, I can "see" that the blue lines on the map must be rivers. This ability to see what the cartographer intended me to see is based on application of an appropriate schema in which the sign relations are embedded in a context. Although some similar blue lines in the world map on my wall also appear to refer to rivers, others must refer to latitude and longitude. The scale of the map, the regular arrangement of lines, their presence on both land and oceans, and so on, provide the necessary cues that allow me to make this assumption in the absence of legend definitions. Maps establish context (and thus provide cues to appropriate schema) within three linked systems: space, time, and object (attribute). These three systems are linked through linguistic and presentational systems (what Wood and Fels, 1986, termed the linguistic and the presentational codes of maps).

Directness of Reference: Literality of Interpretants

The literal meaning of a sign is not always the one cartographers intend a map percipient to assume. While many meanings might be brought to the map through the social context within which the map is made and used, these secondary meanings will be considered in the second half of the chapter. Here, what I am referring to are those primary, intended denotations that involve an indirect reference, what Krampen (1965) calls an "indirect symbol." Indirect reference describes the situation in which one sign is appropriated as the sign-vehicle of another sign. Krampen gives the example of "elephant" or a picture of an elephant directly representing a very large four-legged mammal with large ears and a trunk which in turn indirectly represents the Republican party. The meaning of any sign, therefore, can be considered literally—where elephant means the animal—or metaphorically—where elephant means the animal which means the political party. A map user must rely on context provided within the map itself, context in which the map is used, and/or upon linguistic codes that explicitly define the appropriate level of meaning in order to determine whether the sign-vehicle on a map represents a zoo or a Republican political party headquarters.

Familant and Detweiler (1993) discuss the role of indirect reference in the context of graphic "icons" used in the graphical user interfaces (or GUIs) of many interactive computer programs. They term the literal reference a "sign referent" and the reference intended in the GUI the "denotative referent." They describe the sign relations, then, as consisting of two parts, the sign-vehicle to sign referent and the sign referent to denotative referent (Figure 7.7). The first relationship can make use of signs at any position on the arbitrariness continuum described in the previous chapter.[7] The sign created is then treated as the sign-vehicle of the denotative referent. Five logical relationships are identified by which a sign referent's feature set can be matched to the denotative referent's feature set (Figure 7.8). For signs used in GUIs, Familant and Detweiler contend

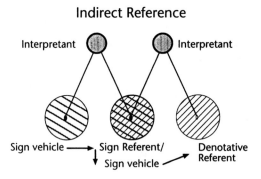

FIGURE 7.7. Graphic depiction of the concept of indirect reference. *Derived from Familant and Detweiler (1993, Fig. 5, p. 713).*

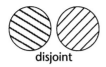 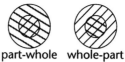
disjoint   part-part   part-whole   whole-part   identical

FIGURE 7.8. The five logical relationships possible between sign and denotative referent's feature sets. *After Familant and Detweiler (1993, Fig. 6, p. 715). Adapted by permission of Academic Press Limited.*

that the part–part and part–whole relationships are most common. For maps an example of indirect reference using the part–part relationship is exemplified by the standard church symbol used on U.S. Geological Survey topographic maps in which a part, the "cross," is used to stand for "Christian church," which in turn stands for the wider set of "places of worship." A part–whole relationship, on the other hand, is exemplified by the airplane sign-vehicle which stands for a jet which in turn stands for a commercial airport (Figure 7.9).

Concreteness of Signs: Concept versus Phenomenon Representations

As indicated in the previous chapter, a distinction can be made between signs in which the referent is a phenomenon in the world (or at least a category of phenomena that has physical existence in the world) and a phenomenon that is purely conceptual. Although we could easily become overwhelmed with philosophical arguments about whether there is a reality outside of our conception of it, most people recognize a distinction between tangible physical objects, like tables or rivers, and intangible concepts, like wealth or mean annual rainfall. While "concept" and "phenomenon" are clearly fuzzy categories, there seems to be an unambiguous difference between their prototypes. The cartographic–lexical question that should be asked is whether there is a difference in kind of meaning associated with signs for phenomena versus signs for concepts. If

FIGURE 7.9. A typical airport sign-vehicle as an example of a part–whole sign relationship.

there is a difference, how important is it to map understanding and what if anything should mapmakers or map users do to take it into account?

One intriguing issue that should be addressed is the interaction between choice of sign arbitrariness and recognition of sign concreteness. Highly iconic depictions of geographic base information on maps may assist a map percipient in selecting the appropriate interpretant for a phenomenon depiction. For example, terrain shading as an underlay to contour lines is likely to help a viewer go beyond the simple interpretant of "line of equal value" (as mediator between a line as sign-vehicle referring to ground locations that form an unbroken path sharing a common vertical separation from sea level). The terrain shading might help prompt extension of the simple interpretant above to incorporate an up–down image schema. The expanded interpretant may then be "line of constant value for which adjacent examples will be either lower or higher elevation and the spacing of which will reflect slope."

An even more direct implication of highly iconic (or realistic) base information on maps is that it will give authority to the apparently iconic use of "presence" and "position" on the map to represent presence and position in space and time. If the viewer sees a detailed rendering of terrain with a narrow valley containing a stream and a road running along the stream, the meaning of the sign thus created is certain to include both the location of that road relative to the stream (if the road is to the right of the stream on the map, there will be little doubt that when encountered in the world, the real road will be to the right of the real stream) and the assumption that the road really exists.

What, however, are the implications of realism for our understanding of signs for concepts or C-reps? Since a concept does not exist, except in the mind, an iconic sign-vehicle cannot look similar to its referent (at least not in the ways possible when a phenomenon is the referent). If it does look similar to something (e.g., roads), however, does the sign's interpretant take on some aspects of the "something" resembled? Does, for example, mapping the 100-year mean annual rainfall in shades of green "mean" consistently adequate rainfall in dark green areas, which in turn "means" continual lushness (both of which may be at odds with a reality of highly variable precipitation from year to year)? If isohyets used to depict the mean annual rainfall values are plotted on a composite image base derived from remotely sensed data collected during a high rainfall year, does this base add to the impression of consistent rainfall and/or suggest the supplemental meaning of "generally heavy rain"? These issues have not been addressed in the cartographic or the scientific visualization literature, but seem critical to our understanding of the implications of visualization tools in science. This topic will be revisited in more detail in the final chapter.

Etymology and Cultural Specificity of Meaning

The meaning attached to map signs differs across cultures and can change over time. In a monosemic system of unambiguous signs, sign meaning can be fixed through an extended period of time. On topographic maps, for example, the meaning (at least the explicit meaning) of individual signs can remain constant through many decades. On thematic maps, in contrast, although there have been a few attempts to standardize sign-vehicles (e.g., Ratajski, 1971), the links between sign-vehicle and referent and the meaning attached to these links is set individually for each map. Even in these cases, as Bertin (1967/1983) has pointed out, a set of expectations exist that establish some level of consistency in meaning (e.g., larger = more). As argued in Part I, these expectations are based on general and specific map schemata with roots in kinesthetic image-schemata.

Whether at the general level of graphic variables as sign-vehicles or the more specific level of graphic variable combinations (e.g., a black square with a cross on top to represent a church on a topographic map), common meanings associated with sign-vehicles can gradually evolve, making their use in a particular sign relation more or less arbitrary over time. An important example of variation in assumed sign-vehicle meaning over time is the use of color to represent quantity on maps.

Frutiger (1989) considers the evolution of a variety of sign-vehicles that are today considered conventional or arbitrary. One example he provides demonstrates how a highly iconic sign gradually transformed to a letter of our current Latin alphabet (Figure 7.10). A similar example is given for the gradual schematization of the Chinese character for horse (Figure 7.11).

Frutiger (1989) also makes a case for some level of common symbolic interpretation among humans that leads to similarities among sign-vehicles used by different groups. As examples, he cites the depiction of weapons such as arrows with arrows, mountains as triangles, and water as wavy lines. Frutiger admits that attempts to identify broad commonalities in sign systems across times and cultures have not met with much success. Wood's (1978) analysis of hill signs with the hypothesized parallel be-

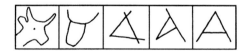

FIGURE 7.10. The transformation of an early pictorial sign for a bull ("aleph") through a series of stages culminating in our present letter "A." *Reproduced from Frutiger (1989, p. 114). Reprinted by permission of Weiss Verlag.*

FIGURE 7.11. The transformation from an archaic to a modern Chinese character for horse. *Reproduced from Frutiger (1989, p. 114). Reprinted by permission of Weiss Verlag.*

tween Mixtec-Aztec early signs and those of Roman to late Middle Ages European cartographers, is among the few attempts to demonstrate such commonalities for map sign-vehicles (and these commonalities are speculative at best).

The limited ability of either anthropologists or historical cartographers to demonstrate consistent commonalities across times and cultures in sign-vehicle to referent links suggests that Peirce (Innis, 1985) was right in his view that there are no true icons, only iconic signs (sign-vehicles that have some commonality with their referent). For any referent, there are many possible sign-vehicles that have some kind of commonality (in appearance, in structure, or through metaphoric transfer). The relation selected by a specific group at a particular time is probably quite arbitrary or the result of chance. What may be common is a developmental sequence in which initial sign-vehicles share a visual commonality with their referents and over time the sign vehicles (possibly for convenience of reproduction or other practical reasons) become gradually more arbitrary—or at least less dependent upon visual analogy to the referent. Cartographically, Wood's (1977) historical–developmental assessment of hill signs seems to fit this model. Wood goes on to present an argument that children and cartography have both gone through a similar developmental sequence in the way terrain features are represented. In both cases, Wood identifies a sequence from visually iconic elevations (egocentric views of an observer at ground level), through visually iconic oblique views (but ones that are less iconic than elevations because they represent a viewpoint that may not have been experienced), to diagrammatically (or metaphorically) iconic plan views using the abstract concept of contour lines.

MEANING OF MAPS

> The aim of criticism is to liberate the text and to restore its semantic richness by reconstituting the codes and the modes of signification which subtend it.
> —GUIRAUD (1975, p. 81)

Cartographers devise their typologies (and individual maps) within a sociocultural context. This context puts constraints on the potential categories that are considered appropriate to represent and on how the geography to be represented on maps is divided into categories. Perhaps because they work within the sociocultural context of the maps they create, rather than being able to critically assess the context from a distance, cartographers seem to take the features to be mapped as a given. They also tend to ignore broader issues of how the map as a whole functions as a symbol. Brian Harley (1989), Denis Wood (1992), and a growing number of writers (e.g., Crampton, 1992b; Edney, 1993; Rundstrom, 1993) have begun to redress this oversight. I maintain, as I did in the introductory chapter, that cartographic critique *by itself* does not lead to better maps (although it may lead to more informed map use). Integrated into an overarching theory of map representation, however, critical analysis of what map representations can or might mean provides a much needed balance to efforts toward making increasingly "better" maps. If we are to make maps work better, it is critical that we also address the question of "Better at what?," as well as the ethical implications of the answer.

Connotative Meaning of Map Signs

The main difference between denotative and connotative meaning of map signs (at least in Barthes's sense of the terms) is that between knowing what things are (explicitly) versus what they stand for (implicitly).[8] This is the difference, cited above, between knowing that a line on a map is a boundary (in the depicted location and at the specified time) and what boundaries stand for in various political, community, or other contexts (i.e., what it means to establish and defend a political, neighborhood, or other dividing line). Connotation is a kind of indirect sign. Like indirect denotation, the sign has two (or more) levels of meaning. Unlike indirect denotation, however, connotation is not the explicit intention of the sign (Figure 7.12). With connotation, possibilities of indirect reference transform the map signs from containers that meaning is stored within to relations that meaning is brought to by the sociocultural context within which the map is created and used.

Guiraud (1975, p. 41) suggests that denotation and connotation represent what are actually two types of sign function: a *code* ("a system of explicit social conventions") and a *hermeneutics* ("a system of implicit, latent and purely contingent signs"). He views both codes and hermeneutics (or denotation and connotation) as conventional (or arbitrary), but sees hermeneutics as sign systems that are socially constructed in a "looser, more obscure and often unconscious way." This is an important distinction from Barthes's (1967) position that treats connotation as "insidious," "hidden," not "innocent." For Barthes, the assumption is that the

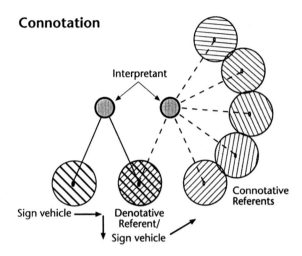

FIGURE 7.12. Connotation as a kind of indirect reference. Unlike indirect denotative reference, connotation is often not the explicit intent and, because the viewer is laregely responsible for "bringing meaning" to the sign relation, there are multiple possible connotative referents.

originator of the sign anticipates the connotations that will be associated with the sign (and that is precisely why one sign is selected over another). For Guiraud, the sign's connotations are truly "brought to the sign" by society, and therefore connotations can be "innocent" (i.e., unplanned). Signs will have multiple connotations and it is often difficult (and sometimes impossible) for the originator to anticipate which will be brought to the sign by particular interpreters in particular contexts.

The distinction between anticipated and unanticipated connotations is directly relevant to recent deconstructionist critiques of cartography. In addition to pointing out the vast connotative potential of both map signs and maps as signs, these critiques tend to condemn cartographers for trying to deceive the map-reading public into taking maps to be something they are not (i.e., reality rather than a representation of reality). Following Barthes, this is perhaps the only possible interpretation. Following Guiraud, "innocent" connotation is also possible.

Extrasignificant Codes

As a complement to their intrasignificant codes (discussed above), Wood and Fels (1986) propose five *extrasignificant* codes that delineate the possibilities of connotative meaning for map signs (Figure 7.13). These codes are:

A Lexical Approach to Map Representation 333

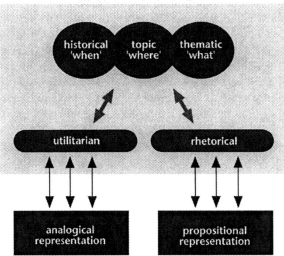

FIGURE 7.13. An interpretation of Wood and Fels's (1986) "extrasignificant" codes with their links outward to the society within which they are defined (in contrast to intersignificant codes that link inward to the individual).

Topic: Turns space established by the tectonic codes into place.
Historical: Turns abstract time into "eras" (which they refer to as "sequacious causal schemata" for organizing conceptions of time).
Thematic: Establishes the subject of discourse (e.g., "automobility," "nature subdues," etc.).
Rhetorical: Sets the tone—"The rhetorical code appropriates to its map the style most advantageous to the myth it intends to propagate" (p. 70).
Utilitarian: Affords the "real" uses of maps (e.g., state possession, monetary control, etc.).

In contrast to the intrasignificant codes that a map "exploits," the extrasignificant codes are those "by virtue of which the map is exploited." Wood and Fels (1986, p. 69) contend that these extrasignificant codes "operate at the level of myth," that they distort a map's meaning and "subvert it to their own use." They thus follow Barthes's lead in contending that no map connotation is innocent.

Wood and Fels (1986) first three extrasignificant codes again match our space, time, attribute distinction, as these fundamental components of mapping are extruded from the map into the social discourse of which the map is a part. Rhetorical and utilitarian codes seem to deal with the map as a whole more than with individual signs on the map. Their discussion, therefore, will be taken up in the section on maps as codes below.

Place, era, and subject of discourse are connoted through the cartographer's (or her client's) choices about what to denote. Issues of spatial and temporal scale, resolution, and partitioning will dictate the questions that can be posed and the issues that can be addressed via the map. As both Monmonier (1991a) and Wood (1992) so eloquently point out, the U.S. Geological Survey's choice to depict roads, railroads, glaciers, campsites, trails, fences, buildings, power lines, and so on, rather than military facilities, toxic waste dumps, impoverished neighborhoods, high crime areas, autos, trucks, dogs, cats, bicycles, trash, dead trees, litter, and so on, drastically limits the kinds of discourse that can take place through the map. Similarly, the choice to depict "enterprise zones" rather than "pedestrian crossings of the U.S.–Mexican border," or that to aggregate crimes against women by month rather than by temporal adjacency to televised football games, all act to conceal some issues while highlighting others.

One connotation always present with maps, and as a result never really innocent, is that the topic mapped is more important than what is not mapped. In fact, I suspect that this "connotation" is actually part of the general map schema for anyone who realizes that maps do not show everything. What is typically ignored, as Axelsen and Jones (1987) have noted, is that judging importance is not as intuitively obvious as it might often seem. Importance is determined by particular social structures and historical circumstances.

An element of map sign connotation associated with iconic codes, but one that Wood and Fels (1986) do not address explicitly, is that of color connotation. This is one feature of map signification for which cartographers have explicitly considered connotative as well as denotative meaning.[9] Several standard texts now contain sections in which typical connotations of color hues are listed. To Wood and Fels's denotative (i.e., intrasignificant code) iconic link between the hue green and lush, healthy vegetation exploited on many maps (whether or not the vegetation happens to be green, healthy, or lush), Dent (1993, p. 311) adds "immaturity, youth, spring, nature, envy, greed, jealousy, cheapness, ignorance, peace" as some likely connotations. To these Rader (1989) adds "life," "cool," "refreshing," and "proceed." Both Dent and Rader provide tables of connotative meanings associated with color terms.

Guiraud (1975) contends that over time connotations (his hermeneutic codes) can gradually acquire the status of a "technical code" or a denotative sign. The act of explicitly identifying connotations, as authors of cartographic texts have done with color connotations or as

Wood, Harley, and others have done in their critical analysis of maps, provides the mechanism by which this evolution takes place. As Barthes (1967) contends, connotations do not openly declare "their possession of a signification." As connotations are repeatedly and consciously exploited by cartographers they become explicit and conventionalized, and thus unambiguous denotations. Their place must then be taken by new forms of connotation.

Above, I argued that linguistic and presentational codes serve to link space, time, and attribute denotation with propositional and analog knowledge structures. Similarly, rhetorical and utilitarian codes can be viewed as linking space, time, and attribute connotation with broader social and cultural issues to which the map speaks. The utilitarian code, what the map's role is really intended to be, can often set the parameters for devising the rhetorical code. For Wood and Fels (1986, p. 70), the rhetorical code is about style: it "appropriates to its map the style most advantageous to the myth it intends to propagate." All maps use a particular style to connote a basic point of view or perspective on the world. As Wright (1942) noted in his classic paper, carefully executed maps with agreeable colors, neat lettering, and the like, will inspire confidence—and I would add, imply truth. The U.S. Geological Survey uses highly detailed, unadorned, visually unassuming maps to connote accuracy, impartiality, authority—thereby creating an impression that U.S. Geological Survey maps have no point of view. Similarly, maps by the Central Intelligence Agency have a clear, uncluttered style filled with light pastel colors. They connote, at the same time, the "executive summary" role they are designed for and an unassuming perspective that threatens to disappear from view. In contrast to these official government styles, the authors of the *State of the World Atlas* eschew the typical restraint (i.e., "good taste") of professional cartography. Their bold colors, dramatic symbology, and so on, connotes that they are different, socially conscious cartographers, that they have a point of view.

The intertwined utilitarian and rhetorical codes are evident in the following quote about an information designer: "He [sic] is, in fact somewhat like the poet or the ancient expert in rhetoric, whose oratory was both a tool of persuasion and a source of aesthetic enjoyment for the audience" (Krampen, 1965, p. 17). Although Krampen was referring to graphic designers working in advertising, he could just as easily have been referring to cartographers. The point is that there are similarities between map rhetoric and verbal rhetoric and that by considering some principles of verbal rhetoric we might learn something relevant to map connotation. Krampen (1965, p. 17), in fact, contends that "there is growing evidence that graphic persuasion and [verbal] rhetoric call on common underlying mediational processes." Krampen (1965) identifies five categories of rhetoric that might be considered mechanisms for prompting connotations:

> *Metonymy as a rhetorical figure (trick)*: Substitution of cause for effect, sign for thing signified, etc.
> *Synecdoche*: Use of part for the whole.
> *Hyperbole*: Exaggeration.
> *Zeugma*: Yoking together of customarily unrelated material.
> *Fusion or amalgamation*: For example, trademarks that combine effect forms with letter forms.

All are used to some extent on maps. While some are more obvious on advertising maps than on official government publications, it is the fact that they are there but not obvious that Harley, Wood, and others have recently condemned.

A Typology of Map Connotation

Just as there are several kinds of codes on maps for which connotative meaning occurs, there are several kinds of connotation that can occur with any map signs. To devise a typology of map connotation, we can start with Eco's (1976) typology presented in Chapter 5. As Nöth (1990) implies, however, Eco's typology actually mixes both indirect denotation and connotation. If we eliminate the denotative categories, the remaining categories suggest that connotation can be based upon metonymy (one thing standing for another to which it has a logical connection), ideology, emotion, metaphor, rhetoric–style, values, and intersemiotic translation.

These categories of connotation have an intriguing relation to Keates's (1984) discussion of map aesthetics. Keates lists five characteristics of art, or attributes that can distinguish objects (including maps) as artistic. These are:

1. *Imitative*: A re-creation of life, a faithful copy or likeness.
2. *Emotive*: Stimulates feelings or emotions.
3. *Expressionistic*: Makes a personal statement.
4. *Communicative*: Intellectually conceived idea communicated; produce an image of despair, hope, royalty, dominance, etc.
5. *Formalist*: Isolated from reality, art for art's sake (the opposite of imitative).

Emotional connotation and the emotive function of art are clearly linked, as are ideological connotation with communicative functions, and rhetoric–style (and/or values) with expressionistic functions. Since both metonymy and metaphor make use of representation or stand-for relationships, they are at least loosely associated with imitative functions of art. Formalist art is antirepresentational; therefore its goal is to not signify at all. Just as some have argued that nonrepresentational art is not possible (and therefore formalism is an unattainable goal), Keates points out that maps cannot be formalist. Wood and Fels (1986) contend that maps devoid of connotative meanings are equally impossible.

As Guiraud (1975) suggests, science tries to concentrate on referential functions (i.e., denotation) and to minimize interference from other functions (i.e., connotation). Art, presumably tries to do the opposite. It is not surprising, then, that Keates's discussion of the artistic properties of maps matches closely with concepts related to sign connotation. Maps, due to their melding of scientific and artistic approaches, always involve complex interaction between the denotative and the connotative meanings of signs they contain. When we consider the combined roles of maps in society, it is possible to isolate several connotative functions that can explicitly or implicitly be attributed to map signs. Drawing from the above discussion and ideas presented in Chapter 5, I offer the following typology of map sign connotation:

Connotations about the map

1. Connotation of veracity: Maps often suggest or imply accuracy—people believe maps because they think they are free from error. Precision of spatial location suggests accuracy, not only in location but in data representation, while precision in time (e.g., a specific date for a national census) suggests both temporal and attribute accuracy.
2. Connotation of integrity: Maps are thought to be unbiased. People believe maps, particularly those produced by the government or by scientists, because they think they are free from bias.

Connotations about the topic mapped

1. Valuative connotation: Maps proffer a value judgment, often through linguistic codes associated with tectonic, temporal, and/or iconic codes.
2. Incitive connotation: Maps are intended to arouse emotions, to convince and to persuade the viewer of the author's point of view, and at times to prompt particular actions.
3. Connotation of power: Maps often exert control over places or

people by establishing a claim to territory, to taxes, to minerals, etc., and/or to social–political control over those within the map's bounds.

Although this typology of map connotation is probably not exhaustive, it seems to cover a number of important categories of connotation that are at least partially independent from one another. Since connotations (or meanings of maps) are taken here to be brought to the map by the sociocultural context within which the map is made and used, it is unclear how much control the cartographer might have over connotation. Although Harley and others contend that cartographers have substantial control over map connotations, the lack of any attempt to assess the truth of this claim leaves it as nothing more than a claim.

The Map Itself as an Implicit Code

Connotation of Veracity: Truth and Reality

That maps are selective, that they represent a point of view, has been frequently noted by both cartographers and analysts of maps (e.g., Wright, 1942; Axelsen and Jones, 1987). In spite of this inherent selectivity and associated sociocultural bias, the tendency for maps to connote truth and reality is less often mentioned. Harley (1989) has been perhaps the most forceful proponent of the view that maps are typically accepted as truthful mirrors of nature (i.e., that maps connote veracity) and that the primary goal of map critique should be to break this assumed correspondence between reality and representation. He suggests that this connotation is explicitly fostered by cartographers' claims to scientific objectivity and by their tendency to condemn inaccuracy in maps by others (e.g., journalists, the Russians, etc.). The practice of consciously facilitating the map-as-truth connotation can be traced at least as far back as Renaissance cartographers who proffered a truth claim by emphasizing their support by (and access to information of) the state (Edney, 1993).

Making a similar point, Wood (1992) implies that for maps to work they must connote veracity. Specifically, he says that "our willingness to rely on the map is commensurate with our ability to suspend our disbelief in its veracity" (p. 20). He contends that cartography (and society in general) fosters this basic connotation by routinely pointing out cases in which maps do not match reality as if they were rare, extreme exceptions. The example given is the typical surprise (and outrage) voiced over finding a major blunder in inclusion or position of a specific map feature (e.g., a school building misplaced by about one half mile).

Among the best examples of the cartographic strategy of publicizing errors in order to foster a connotation of "good" maps as true was the "media watch" organized by cartographers in Great Britain (Balchin, 1988). This activity involved members of the British Cartographic Society in an active campaign to root out and admonish "poor" maps published by the media. It is Harley's contention that these "policing" efforts by cartographers have the effect (if not also the intention) of demonstrating to the public that cartographers take map quality seriously and that the public should expect high standards of veracity from professionally published maps.

Whether or not cartographers are consciously trying to dupe the public into interpreting "map" as a sign for "true," maps that connote truth (or even reality) are likely to work better than those that do not. A need to accept simplified models as if they were reality may be inherent in how the human mind makes use of mental models to grapple with a complex world. As noted in Chapter 4, Lakoff (1987) proposed ICMs as a basic mental organizing structure that we use routinely to comprehend experience "in an oversimplified manner." If we did not, at least for practical purposes, accept our simplified mental models as true, we might be incapable of functioning. At least in the context of maps as navigation aids, we must be willing to rely on the map as a source of information for specific decisions; therefore we must "suspend our disbelief in its veracity" during the time frame of our journey. Muehrcke (1974, p. 39), 20 years ago, suggested as much in his comments that "the human mind seems quite adept at letting the map serve as a surrogate for reality for convenience purposes, sometimes even to the extent that the symbol becomes, for most intents and purposes, the reality itself."

The idea that maps connote truth or reality, just by virtue of being a map, is by no means new to Harley (1989) or Wood (1992), or even Muehrcke (1974). This realization is quite clear in the following quote:

> The trim, precise, and clean-cut appearance that a well drawn map presents lends it an air of scientific authenticity that may or may not be deserved.... We tend to assume too readily that the depiction of the arrangement of things on the earth's surface on a map is equivalent to a photograph—which, of course, is by no means the case. (Wright, 1942, p. 8)

The difference in recent attention to connotations of truth in maps is an explicit realization that "good" maps as well as "bad" (or purposely misleading) maps connote a directness in representation which prompts users to overlook the fact that maps are representations. Wood and Harley both suggest that the connotation is often so strong that the map is taken to be the territory. Even this latter perspective, however, was noted at least as early as 1974 in Muehrcke and Muehrcke's analysis of maps in literature: "In direct contrast to map readers who do not allow themselves to see the reality that has been abstracted onto a map are those who actually substitute the map for reality" (p. 323).

A potentially interesting issue in relation to the rhetoric of map signs and maps as signs is the interaction between arbitrariness of the map as a whole and arbitrariness of signs within the map. If the map base is characteristic of the image end of the image–graphic continuum (see Figure 6.15), does that connote a highly realistic accurate representation of all information on the map? For example, if terrain shading or three-dimensional rendering is used to depict the landform structure, will a reader take that to mean that isolines representing surface ozone concentration are based on a highly comprehensive set of measurement sites for which accurate information is collected on a regular basis? Would they be less trusting of the map if the base on which isolines are drawn was a schematic one (e.g., does a schematic depiction of geography connote uncertainty or caution in relation to data represented on that base)? I offered this general suggestion in a discussion of uncertainty visualization, but it has yet to be tested (MacEachren, 1992b).

Connotation of Integrity: Map Ethics

Wood (1992) discusses what he calls the "myth" of a reference map's "dispassionate neutrality." He contends that the cartographic distinction between what are called "thematic maps" and what are called "reference maps" is an attempt to disguise the fact that all maps are about something. Labeling something as a "reference map," then, is interpreted to be a sign that has the likely connotation of "neutral" or "unbiased" because reference maps are presented as lacking a point of view.

Wood does not deny that cartographers do frequently make explicit statements about the map's imperfections and lack of ability to match reality precisely. However, he discounts cartographers' clear statements that maps have an undeserved aura of truth and exactness as something that they do not really mean or practice—and irrelevant because maps have a "ferocious power . . . to speak for themselves" (1992, p. 25). As an example, Wood cites the caution voiced by authors of *Goode's World Atlas*, that readers should be careful not to accept the maps too literally. He interprets this caution not as a literal denotation of "reader beware," but as a sign having the connotation of "integrity," "fair," "ethical." Thus, according to Wood, by admitting that their maps are not perfect, cartographers are not actually trying to neutralize the connotation that maps are true, but to prompt a complementary connotation that mapmakers are ethical, that they have integrity, that they are honest about their maps' weaknesses. Following this logic, we ultimately arrive at the (possibly absurd) conclusion that a blatantly biased propaganda map is more ethical than a conscientious attempt to map things fairly and to warn the reader about abstractness, because things cannot be mapped fairly and the propaganda mapper does not falsely imply that they can, by pointing to the potential lack of fairness.

The professed neutrality of maps is so ingrained in cartographic practice and the society of which it is a part that evidence for the connotation that maps are neutral can be found in ordinary literature (Muehrcke and Muehrcke, 1974). Based on evidence from Edney (1993), a case can be made that connotations of integrity and neutrality were firmly established for maps (particularly those by the government or produced with government sanction) by the 18th century. Edney contends that prior to this time European cartography was viewed as neither infallible nor neutral. It was with the development of systematic surveys and other government mapping efforts initiated in the late 17th century that maps began to present as single and apparently impersonal view of the world. It is this impersonalization of the map that allows it to connote neutrality, and neutrality has, in turn, come to connote integrity in all information media today. This use of an appearance of neutrality as a sign for integrity has become the hallmark of science and has even played a part in how integrity is judged in the news media.

A variety of explicit signs on maps are used to connote integrity. Pointing to a government authority is, perhaps, still the most common. Listing a government agency as a source, then, may denote where the data can be found if a reader wants to pursue them further, but at the same time, this citation connotes authority and a lack of bias in collecting the information. Following Wright's (1942) suggestion, reliability diagrams became a common feature of some maps, particularly those produced by the American Geographical Society from data that was known to vary in quality. A reliability diagram denotes those portions of the map about which the cartographer was less certain in her sources or interpretation from those sources (Figure 7.14). As Wood's (1992) analysis of a map supplement from the *Annals of the Association of American Geographers* demonstrates, however, reliability diagrams can connote integrity, even when what they denote is far from clear.

In scientific writing, citations to previously published work denote the sources of information and ideas that guided the author. Wood (1992) praises Conservation International for following this practice with their map of human disturbance of ecosystems. What Wood fails to note, however, is that (just as in scientific writing) copious citations con-

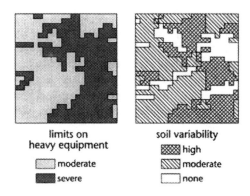

FIGURE 7.14. A variation on the traditional reliability diagram. On the left is a map indicating limitations (due to soils) on heavy forestry vehicle traffic. On the right is a form of reliability diagram that depicts the expected variability of soils within each cell, thus representing the certainty of the classes depicted on the "limits" map. *After Maclean et al. (1993, Figs. 2 and 3, p. 226). Adapted from* Photogrammetric Engineering and Remote Sensing *by permission of the American Society for Photogrammetry and Remote Sensing.*

note integrity, thoroughness, fairness, and the like, whether or not these are actually characteristics of the work. Just as a map author can use a reliability diagram for the express purpose of fostering a connotation of integrity, detailed citations can be used in the same way. Just listing them does not guarantee that they were selected or used in an unbiased way, that the sources were unbiased, or that they were used at all.

The dilemma we face is that any attempts to be ethical, to balance conflicting sources in a fair way, to make our point of view clear to the map reader, to work from the "best information available," can be interpreted as an attempt to disguise the flaws in our map, to obscure the inherent bias, and so on. Yet if we leave these items off the map, actual bias and concealment can prevail. The problem with trying to take into account connotations of map signs (and maps as signs) along with their denotations is that possible connotations are virtually unlimited and once understood they are always replaced by other, as yet unrecognized, connotations.

Valuative Connotations: Judgments

Harley suggests that there are basic "rules" of cartography that have nothing to do with the map as a functional representation or communication device (although cartographers might claim they do) but instead with "rules of social order" (Harley, 1989). We, for example, take for granted various hierarchies (e.g., the king is more important than the barons, who are in turn more important than artisans, who are in turn more important then peasants). Cartographic rules that match more prominent symbols to this implicit hierarchy are derived, then, from the social structure within which the map is produced, not from some fundamental principle of representation. Similarly, Harley suggests that cartographers follow a "rule of ethnocentricity" that dictates centrality and isolation of particular places on the map, which in turn sets them off as special, better, dominant, and the like. The single most important "rule" that leads to valuative connotation, of course, is the rule that "more important things" should be selected for inclusion on the map before "less important things." Presence on the map, therefore, connotes importance. A related "rule" is that important features should be at or near the map center. Centrality, then, also connotes importance.

Historically, medieval *mappaemundi* might be considered the most value-laden of all Western maps. On these maps the denotative function of locating places is subordinated to the map's connotations. Woodward (1985, p. 515) characterizes the main function of *mappaemundi* as "illustrated histories or moralized, didactic displays in a geographical setting." The primary connotation of these maps was the overriding importance of Christian ideals. While some have pointed to the centering of Jerusalem as a device for connoting the centrality of Christianity in the universe, Woodward provides evidence that this centering of Jerusalem was not consistently applied. Woodward cites evidence that, on some mid-15th-century *mappaemundi*, it was the center of humanity (the known world population) that was given prominence via a central location on the map. This change to a population-centered map, then, connotes that Christians (in Europe) rather than the historical origins of Christianity (in the Middle East) were the valued feature.

With contemporary maps, the U.S. Geological Survey's inclusion on topographic maps of roads, railroads, elevation contours, glaciers, campsites, township boundaries, fences, mines, power lines, and permanent versus intermittent steams, rather than toxic waste dumps, abandoned factories, high crime areas, drug-free zones, radon intensity isolines, water quality, shelters for the homeless, or school district boundaries does more than limit discourse. That the U.S. government considers the former items more worthy of inclusion on their maps than the latter connotes that infrastructure, mineral industries, natural landscape, political divisions, and so on, are more important than crime, natural or human induced environmental hazards, economic decline, or homelessness.

Valuative connotations of maps can be particularly disturbing in the context of underrepresented groups of society. As Muehrcke (1974) has noted, for example, most maps (such as U.S. Geological Survey topographic sheets) that depict "Indian reservations" provide no indication of pre-reservation extent of the particular tribes (Figure 7.15). These same maps typically include placenames selected by non-natives. Continued use on maps of names imposed on native society by white conquerors connotes that the taking of lands is still accepted as "right" (Harley, 1990). Similarly, inclusion of names such as "Niggerhead Creek" (in Alaska) and "Squaw-humper Creek" (in South Dakota) connote that the discriminatory racist and sexist attitudes prevalent when these place-names were coined remain acceptable to the U.S. government (Harley,

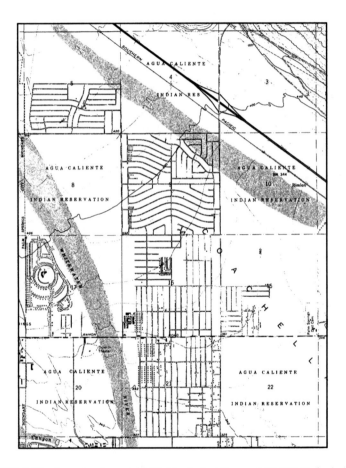

FIGURE 7.15. A section from the U.S. Geological Survey 1:24,000 Cathedral City, California, topographic sheet. The extreme dislocation of Native American groups is, however, strongly hinted at on this particular topographic sheet, with its clearly bureaucratic checkerboard reservation pattern. *Reproduced from Cathedral City, California, U.S. Geological Survey (1958, photorevised, 1972).*

1990). The U.S. Geological Survey has attempted to eliminate some of the most outrageous placenames, but many still exist.

It is, of course, not just what is selected for inclusion on a map nor how it is labeled that prompts valuative connotations. The way features are symbolized also influences these connotations. Symbol hierarchies have already been mentioned above (e.g., the king's castle gets a bigger symbol than the baron's). In addition, when U.S. Geological Survey topographic maps depict all places of worship identically, with a sign-vehicle having a cross on top, this connotes an official preference for Christian religions. When a cartographer chooses a "crossed tomahawk" or "person in a canoe" sign-vehicle to stand for "Indian reservation" on a map, the connotation of "savage" or "primitive" is probably implied (Muehrcke, 1974).

While Wood (1992), Harley (1990), and several other recent map critics imply that all valuative connotation is consciously implanted by mapmakers, Axelson and Jones (1987) allow for unconscious sign relations as well. They point out that cartographers are conditioned by their experience, cultural filters, and sources of knowledge and, as a result, may produce maps with connotations that can only be recognized by someone taking a view from outside the culture. Even if we accept Axelson and Jones's view, and that of Guiraud (1975), that connotations can be innocent, Barthes (1967) may still be right that these unconsciously (innocently) injected connotations are the most insidious and the most likely to perpetuate stereotypes, cultural bias, racism, and sexism.

Connotations of Power: Territorial Control

Muehrcke and Muehrcke (1974), in their analysis of maps in literature, identify what can be considered a connotation of power, or territorial control, associated with maps. They cite examples from fictional literature that changes *in the map* are often equated with changes *in the world*. As evidenced by Jincey in *Good Morning Miss Dove*, this confusion can extend to an expectation that if you can keep the map from changing you might keep the world from changing. As I write this passage, this connotation is being played out for high stakes on the international geopolitical scene. Member countries of the United Nations are taking sides in a dispute over the recognition of Macedonia as a sovereign country, with Greece leading the opposition under the assumption that if they can prevent a territory from being named Macedonia, the Macedonians do not exist.[10]

As the Macedonia example demonstrates, linguistic codes on maps can have a significant role in connotations of power or control over territory. The issues of placenames in general (discussed above), are equally important when connotations of power and authority are considered. The act of making an "official" decision on disputed placenames connotes that those with authority to make the naming decision condone a power relation in which those favoring the selected name exert control over others favoring the rejected alternative. The choice between Mt. McKinley and Mt. Denali is a case in point. In 1896 the peak known to the native population as Denali was renamed Mt. McKinley, and in 1917 a National Park was established surrounding the peak and also named McKinley. In 1975 the government of Alaska decided to change the name of Mt. McKinley back to Mt. Denali in response to pressure from Alaska's native population and their supporters. This name change recognized a shift in power relations that allocated slightly more power to Native Americans—power they had in effect won for themselves through a series of legal battles that awarded restitution for past territorial usurpation. In 1980 the U.S. government, responding to similar pressures, renamed the National Park Denali. The U.S. Board of Geographic Names has, however, never agreed to Denali as the name for the mountain itself and congressional legislation has been introduced to retain the name of McKinley "in perpetuity." This tension over naming (and the reticence of the federal government) can be taken as a sign of the unstable power relations between native and non-native interests in the state.

It is, of course, not only the linguistic codes on maps that connote control of territory or power over place. The act of delineating boundaries is replete with connotations of control, power, and obligation. That maps are often an accepted basis for litigation about ownership of territories—by individuals, by states, and by nations—makes the basic power connotation of maps obvious. As Wood (1992) notes, property boundaries can denote ownership while at the same time connoting numerous obligations that come with that ownership (e.g., paying taxes, mowing the lawn, paying for services such as garbage collection, not being a nuisance [as defined by the community nuisance ordinance], etc.). Exerting control over a piece of territory, then, often brings with it an acceptance of external control over what can be done with that territory. Hannah (1993) suggests that refusal to delineate property boundaries on the part of the Lakota Indians was a mechanism they used to retain control over their territory. Submission to Western practices of individual land ownership, rather than communal use with no "ownership" in the Western sense, brought with it an implicit acceptance of the white man's control of nature and of their lives.

As noted in Chapter 4, boundaries are associated with a fundamental kinesthetic image schemata. Bounded areas on maps imply (or connote) "containment" and control by an external force. While Western property owners may take the map delineation of boundaries of their property to connote possession and control of their property, the Lakota interpreted the same sign as a loss of control, a containment.

Harley (1988) provides the most comprehensive account of power relations in maps, primarily from a historical perspective. Much of his discussion involves direct use of maps to maintain power (e.g., for tax collection, military campaigns, etc.)—to denote features that were critical to the maintenance of power rather than to connote power or control. In relation to military map use, however, Harley points to map connotations of "socially empty space," connotations that make decisions of military commanders easier. This same point was made by Muehrcke and Muehrcke (1974) in their examination of maps in literature. They cite several literary examples of maps that allowed military decision makers to ignore the fact that "real people" would be affected by their decisions. The relatively bare maps, with terrain, water, structures, and so on—but no people—result in the connotation that it is just territory, not people's lives, being bombed and controlled.

As noted above, no map is unbiased. Just as omission from the map connotes a value judgment, omission is also a means of exerting power. Harley (1988), for example, points to surveyors, working in Ireland for English proprietors, who often omitted the cabins of the native Irish. With this exclusion, the resulting maps not only depicted a dominance of English proprietors, but also connoted a complete lack of opposition to their authority. As omission from the map signifies subjugation, overabundance on the map can signify domination. Monmonier (1991a) and others have discussed manipulation of symbol density on advertising and propaganda maps as a tool that implies control of territory (in a political or commercial sense). Harley (1988) notes a similar use of numerousness plus symbol size on 16th-century German maps to connote ubiquity of the church.

In addition to connotations of signs within the neatline of a map, maps connote through their marginalia. Most cartographers have treated marginalia (particularly that identified as "decorative") as irrelevant (at best) or something that obscures the real purpose of the map (at worst). Harley (1988), however, contends that it is often through the marginalia that maps acquire their political meaning and power. He notes the role of map decoration in connotations of land ownership, wealth, territorial control, and exploitation. Often, racial stereotypes are implicit in this decoration and act to connote the superiority of the mapmaker's culture over that of those being "civilized" by their conquerors. Similarly, the use of globes in the map margins carried a connotation of God's influence over the world. Harley suggests that this connotation was gradually transformed to the globe as an emblem of territorial power exerted by "divinely appointed" rulers. Carried to modern times, an argument can be made that globes, and world maps in various forms, are used as signs of global dominance. This is certainly the explanation for a preponderance of globes in company logos.

Recently, Crampton (1994) has extended this analysis of connotations of power to the power and control implicit in digital cartographic databases. He uses an analysis of digital cartographic databases used in geodemographic applications as a vehicle for discussing ethics in cartography. These databases are an example of Moellering's (1980) "virtual map—type 4" (nonvisible, nonpermanent maps). These maps contain spatially referenced data about individuals (including financial details) that companies use to target political information, advertisements for products, and so on. While most of us probably consider the junk mail deluge stimulated by access to these virtual maps an invasion of privacy, it is a minor annoyance compared to the powerful control that might be wielded by those with access to such maps.

Incitive Connotations: Persuasion to Action

Many maps are designed, either explicitly or implicitly, to persuade their viewers to accept a point of view, to support some decision, or to act on a decision.[11] Much attention has been given to maps that persuade in what is considered a deceitful manner (e.g., advertising maps, propaganda maps, etc.) (Tyner, 1974; Monmonier, 1991b). Less attention has been paid to subtler forms of persuasion inherent in perhaps the majority of maps produced. Maps try to persuade the reader that the map is accurate, that the value judgment made is appropriate, that the map is unbiased, or that the territorial claims are justified. Beyond the basic goal that the user accept a map author's perspective, incitive connotations stimulate action (or at least a desire that someone take action).

Among the most obvious attempts to persuade map readers to accept a point of view or to behave in a certain way are maps depicting boundaries. Representing a boundary as "disputed," for example, brings with it a connotation that someone should be doing something to settle the dispute (Figure 7.16). In addition, such a representation connotes an agreement on the part of those sharing the boundary that there is something to dispute—a connotation that both sides have a viable claim to territory.

Wood's (1992, p. 66) analysis of Van Sant's "global image map" provides a particularly good example of how incitive connotations can be identified: "The *suppression* of the clouds in the Van Sant [map] renders an earth so still it seems to be holding its breath, connotes an earth that is . . . *waiting for something*, an earth that is paper thin, that is delicate, frag-

FIGURE 7.16. A "disputed" boundary around which the United States was "persuaded to action."

ile." The connotation, then, is a call to action, a plea to save the earth before it is too late, before this beautiful "picture" can no longer be taken. Wood sees the Van Sant map as an emblem for environmental activism, a much more subtle emblem than the shattered crystal globe merged holographically with an intact globe on the cover of the December 1988 *National Geographic Magazine*. The latter directly *denotes*, rather than connotes, fragility of the earth's ecosystems.

As noted above, linguistic codes can be used to lay claim to territory. Changes in linguistic codes on maps connote a mandate by the government for local areas to change their attitudes and policies toward certain groups. Replacing Anglo-Saxon placenames on U.S. Geological Survey maps with Native American names or Russian names on U.N. maps with Lithuanian names empowers a subjugated and/or minority group to take action designed to increase control over their own lives. The decision by the National Geographic Society to depict Macedonia as an independent country connotes a view on their part that the issue is clear, with only one "correct" decision possible. The decision thus connotes that the United Nations should recognize Macedonia's right to exist.

Can Connotations Be Measured?

To consider whether or not a cartographer is able to dictate the connotative meanings engendered by a map, the issue of measuring connotative meaning needs to be addressed. There seem to be two approaches: public and private. By this I mean that connotative meanings can be discussed from the perspective of the society within which the meanings are formulated or from the perspective of individuals who actually derive (or do not derive) particular connotations from a sign. With the first approach, connotative meaning is deduced by reasoned textual analysis following the model of Harley's (1989) map deconstruction. The possible kinds of connotative meaning are identified (as above) and particular maps assessed according to the likelihood that such connotations are prompted. Much of this chapter emphasized this approach. The last several years has seen a dramatic upsurge within the social sciences in general and geography in particular of the application of principles of critical theory to analyzing the multiple meanings of texts of various kinds. For cartography, Harley has taken the lead in adapting the principles of this form of analysis to maps and he has been followed by an increasing number of scholars delving into the cartographic void he opened up.

An alternative to addressing connotation at the public–social level is to attempt to measure connotations developed by individuals and arrive at an overall assessment of the connotative meanings of particular maps through an inductive process whereby many individuals' connotations are amalgamated to a common model. Here, I offer one suggestion of how this individual–inductive approach to identifying connotative meaning in maps might be addressed.

Since connotation is a secondary, indirect, implicit meaning, it seems reasonable to treat connotation as something that can occur in degrees. There can be strong and weak connotations, and particular signs can have several connotations that vary in strength. The degree to which a connotation holds for a particular sign, then, is analogous to the degree of iconicity of a denotation. This variable degree of applicability suggests that connotations have much in common with attitudes; both are dependent upon first knowing what the object or concept is, then deciding the strength of relation between this concept and another. It may, therefore, be appropriate to apply attitude-measuring principles to the assessment of connotative meanings. Perhaps the most appropriate such measuring device is the *semantic differential*.

Semantic differential is a technique in which subjects are asked to judge the relative applicability of bipolar word pairs to a thing or concept, usually on a 7-point scale. An intermediate rating implies that the words (or the referents they signify) are not associated with the thing or concept being judged. The more strongly one end of the scale is selected over the other, the more clearly that word (or referent) is linked with the object.

In an attempt to assess overall reader attitudes toward specific maps, Gilmartin (1978) used a semantic differential analysis composed of word pairs such as good–bad, legible–illegible, hard–soft, ordinary–innovative, and so on. She found a high degree of consistency among her subjects on particular word pairs as well as for logical groups of word pairs that were interpreted as measuring different attitude categories. Gilmartin identified three categories: evaluative, activity, and potency measures. These categories were predicted prior to analysis, but the specific word pairs assigned to them was determined through a factor analysis of actual subject scores. This factor analysis resulted in several word pairs being associated with a different attitude category than expected, and with four word pairs judged to be independent measures.

A similar application of semantic differential might be used to assess map meaning at the secondary level of connotation. Usefulness of the method would depend (as it does when assessing attitudes) on being able to select appropriate word pairs. Possibilities might be word pairs such as powerful–weak, controlled–free, accurate–inaccurate, careful–careless, true–false, convincing–unconvincing, socially conscious–socially negligent, ethical–unethical. To apply semantic differential, however, we also need to have some general agreement on the kinds of connotations that might be typical for maps (and of the proper opposites of these connotations). The typology of map connotations presented above offers a starting point. One advantage of the semantic differential method is that it provides a mechanism to assess the typology of map connotation suggested above. Factor analysis of semantic differential ratings could be used to determine the number and characteristics of connotative meaning categories occurring in any given map or set of maps.

SYNOPSIS AND DIRECTIONS

This chapter has explored the "terrain" of map sign meaning. A dichotomy of denotative (or explicit) and connotative (or implicit) meaning was presented along with the contention that this dichotomy is not as discrete as often implied. Through repeated use, connotation can evolve toward denotation. Looked at historically, as Harley has done, it becomes difficult to determine the degree of complicity of past cartographers in what now appear as perfectly obvious examples of intentional connotations. These connotations may seem more intentional now than they were at the time a map was produced because the intervening years have transformed an unconscious, perhaps accidental, connotation to the status of explicit denotation.

What I have attempted here is to demonstrate that a "traditional" cartographic approach to map meaning and a "critical social theory" approach to map meaning are not incompatible. In isolation, either is prone to bias for or against maps and to disregard for important aspects of the power of maps. While fusing work on explicit denotation with that on implicit connotation of map signs is perhaps like bringing the United

States and the former Soviet Union to the arms negotiation bargaining table, failure to do so (in either case) prohibits comprehensive "global" understanding.

Parts I and II of this book propose a fundamental dichotomy in approaches to the study of how maps work. In Part I, I considered issues in the private realm of map percipients with emphasis on perceptual and cognitive processing of sensory (largely visual) information. In Part II, I have balanced this approach with a two-pronged analysis of map semeiotics on functional and lexical grounds, an analysis of the map's public realm. My overall argument is that we cannot hope to fully comprehend the complexities of how maps work by limiting ourselves to either perspective in isolation. I conclude the book, in Part III, with one example of how the integrated perspective advocated here might be applied to a rapidly emerging area of cartographic concern: how maps work as visualization tools.

NOTES

1. The distinction was inspired by Wood and Fels's (1986) identification of two aspects of sign function: intrasignificant and extrasignificant codes.

2. A similar distinction can be found in Nyerges's (1991a, 1991b) application of a semiotic approach to understanding geographic information abstractions that underlie the digital data structures of GIS and in Peuquet's (1994) efforts, also in the context of GIS and their associated data structures, to provide a unified representational framework that combines "location-based," "time-based," and "object-based" categories of representation.

3. Technically, it might be argued that all aspects of world space are denoted but that the denotation may not be easy to interpret visually. For example, all map projections depict the distance between places, but most depict those distances at a continuously changing scale. Thus, while it might be easy to calculated distances between places using the map-projection formulas, it is difficult to judge them visually.

4. Dynamic maps simulate continuous time by making the interval between discrete frames displayed smaller than is perceptible.

5. While this may seem trivially obvious, consider the case of virtual type III maps (digital databases) for which the codes (coordinate pairs) defining rivers might be grouped in a separate data file and be sorted according to the longitude of their initial coordinate pair (for ease of plotting). The names linked to the rivers through pointers, on the other hand, might be sorted (in another file) using alphabetical order (for convenience of indexing the map).

6. See the "Directness of Reference" section below for more details on this point.

7. Familant and Detweiler (1993) follow Peirce in identifying discrete symbolic and iconic categories of reference rather than a continuum.

8. Barthes's (1967, 1977) work has popularized a distinction originally due to Hjelmslev.

9. Following from Wood and Fels's (1986) arguments that cartographers use omission of various feature categories to connote that these features are less important than those included, we must question why they chose to leave discussion of color connotations out of their analysis. Is it because with color connotations we find an example of cartographers explicitly talking about connotations of their sign-vehicle choices? To acknowledge that cartographers are (at least *sometimes* with *some* issues) concerned with the connotations of their design choices would weaken Wood and Fels's argument that cartographers consistently try to obscure the true meanings of maps by inordinate claims to objectivity. Wood and Fels, therefore, did what most cartographers (and writers) have always done: they ignored evidence that did not happen to support their point of view.

10. The fact that I did not put "Macedonia" in quotes above, of course, connotes that it is a country rather than just a proposed label for a not-yet-recognized country. Putting it in quotes would connote the opposite—it is impossible to avoid an appearance of taking sides.

11. In the case of geodemographics, virtual maps are used to identify the people who might be persuaded. See Axelsen and Jones (1987).

PART THREE

How Maps Are Used: Applications in Geographic Visualization

The goal of Part III is to explore how social, cartographic, and visual–cognitive spatial representations mesh in the context of geographic visualization (GVIS). There has been a dramatic resurgence in the role of visual analysis within the natural and earth sciences, particularly in the early stages of research. Computer systems are being used to depict spatial information in a rapidly increasing variety of cartographic and graphic forms. We know little about how maps work in these contexts or about the "appropriate" tools to provide. It is clear, however, that cartography's "optimal map" goal of the past 40 years becomes less relevant in an environment where neither the person making the map nor the person using it knows what she wants to *see* until she *sees* it. Developments in "scientific" visualization offer a particularly exciting opportunity to apply our evolving concept of "how maps work" to a rapidly expanding area of graphic and cartographic depiction.

"Visualization" is a term with many meanings. In the most general sense of "to make visible," it can be treated as a superordinate category to which cartography belongs. The term "scientific visualization" has taken on the narrower meaning of advanced computer technology to facilitate "making visible" scientific data and concepts. Research on scientific visualization has emphasized creating technology that allows scientists to turn nonvisible (digital) data, collected via measuring devices or generated by computer models, into visual representations. In contrast to visualization as creation of concrete displays (whether via computer or otherwise), the term "visualization" can also refer to making visible in the sense of mental images. Great insight by scientists like Einstein or Kekule is often linked to a purported ability to generate elaborate mental images through which connections and relationships critical to solving a problem are "seen."

One justification for the development of scientific *visualization tools* (computer technology for production of concrete visual representations) is that visualization tools facilitate exploration of both data and the general problem context. As Beyls (1991, p. 313) has noted, "Exploration gives birth to side effects. Very often, these side-effects are more interesting than the initial ideas that triggered them." Most scientists are not as good as Einstein at mental visualization, and therefore their ability to mentally visualize data and relationships will benefit from concrete prompts provided via visual display tools. In particular, data-intensive problems driven by a rapid increase in remote sensing capabilities (or problems approached through computational modeling) cannot be addressed by just thinking—concrete visual displays must replace or supplement mental visualization when information volume and problem complexity makes mental visualization alone impractical. As Friedhoff and Benzon (1989) note, some of the most significant inventions of the 20th century are instruments designed to extend vision. Like the Hubble telescope or an electron microscope, visualization tools make concrete and manipulable what was previously only imaginable. Visualization tools thus extend our means for Gestalt perception and for using embodied image schemata. As a consequence, they allow basic-level categorization processes to be applied to abstract concepts that have taken on some of the characteristics of the human-scale environment.

As alluded to above, visualization has many possible meanings including concrete physical representations, mental images, and the processes involved in generating either. For some writers, it is not visualization unless a computer is involved, for others any drawing tool can produce a visualization or a display that prompts visualization. Some cartographers have approached cartographic–geographic visualization as primarily a kind of map use (DiBiase, 1990; MacEachren et al., 1992). For others, it is a map display technology that can be treated as a user interface to spatial or space–time data (Makkonen and Sainio, 1991; Taylor, 1991; Dorling, 1992b; Monmonier, 1992). Cartographic visualization has also been described as a mental process facilitated by maps (MacEachren and Ganter, 1990). Still other researchers have emphasized visualization as a process (or environment) for manipulating information (Monmonier, 1989a; MacDougall, 1992; DiBiase et al., 1994b). As is obvious from the citations above, the critical aspects of visualization as a concept are not even fixed for individual researchers.

The multiple possible definitions of, or approaches to, "visualization" as a category are similar to those cited in Chapter 4 for "map" as a category. How people define a concept determines the way they confront potential instances of that concept. As a subcategory of a discipline, the interpretation of "visualization" by practitioners within the discipline will determine what research is treated as within the category, what critical questions are identified as central to visualization research in cartography, and what research methods are considered acceptable for answering those questions.

"GVIS," like "map," is probably best viewed as a fuzzy (probably radial) category. Any attempt to place rigid bounds on the category will place artificial barriers between camps within the discipline. We will be better served by considering various maps, map uses, and map production processes to be good or poor "exemplars" of GVIS. If we treat GVIS as a fuzzy category, we can identify characteristics of the prototype. Being able to characterize the prototype will help in determining which cartographic representation issues covered in Parts I and II are most critical to GVIS applications. Like all concepts, GVIS can only be usefully defined in relation to other concepts with which we are already familiar. Due to past dominance of the communication paradigm in cartography, the most useful approach to characterizing GVIS may be in terms of how it contrasts with (or subsumes) cartographic communication.

In another context I offered a three-dimensional model of human–map interaction space that defines what might be considered prototypes for visualization and communication (MacEachren, 1994c). The dimensions of the interaction space are defined by three continua: from map use that is private (tailored to an individual) to public (designed for a wide audience); map use that is directed toward revealing unknowns (exploration) versus presenting knowns (presentation); and map use that has high interaction versus low interaction (Figure PIII.1). There are no clear boundaries in this human–map interaction space. All visualization with maps involves some communication and all communication with maps involves some visualization. The distinction made is in emphasis. Geographic visualization is exemplified by map use in the private, exploratory, and high interaction corner. Cartographic communication is exemplified by the opposite corner.

Cartographers delineate categories and map syntactics for organizing and structuring the concrete visual representations we are charged to create. When we do this in either a visualization or a communication context, we impose structure on the world. The extent to which a map works

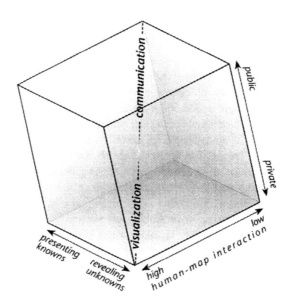

FIGURE PIII.1. Cartography depicted as a cubic map-use space in which visualization and communication occupy opposite poles. Reproduced from MacEachren (1994c, Fig. 1.3, p. 6); see this source for a more detailed discussion. Reprinted by permission of Elsevier Science Publishing Company.

(if at all) will depend upon how the structure we impose matches with the cognitive structures brought to bear by the map reader and upon how the imposed structures relate to social structures within which the visualization tools are used.

If cartographers are to extend their attention from communication to visualization, with the concomitant emphasis on the research process and use of interactive exploratory visualization tools by scientists, it is essential that our view of representation be broadened. At the level of the individual, processes inherent in perceptual "representations" of visual scenes are critical to the perceptual organization of maps. The cartographic "representation" choices concerning fundamental visual variables and their dynamic visual and audio counterparts (e.g., location in space, color hue, size, duration, pitch) have as yet unknown consequences for the groupings and patterns seen by a researcher exploring data for the first time. To play a part in developing visualization tools that prompt insight we must also begin to understand the conceptual "representations" and knowledge schemata that researchers/analysts bring to the visualization environment and that constrain and direct what they can see. At present, we have only limited understanding of how to develop mapping structures that match the perceptual organization tendencies and schemata humans have been demonstrated to use. We have not even begun to explore the potential of using multiple cartographic representations of data sets to prompt changes in schemata that allow (or force) researchers to see their data from different perspectives, nor have we investigated the implications of tools that expedite interactive restructuring of cartographic information. In addition to using our multirepresentation approach to design better GVIS tools, attention should be directed to explicating the kind of structure we impose and helping map readers see the implications of that structure and how it might differ from the everyday structures (i.e., schemata) they routinely use to deal with the world around them (e.g., when we re-express time and space to place emphasis on particular features).

There are many specific aspects of designing and using GVIS environments that have received little, if any, attention in the cartographic literature. Among these are questions of how visualization tool users identify spatial–geographic features and anomalies in data displays and the appropriate representation tools to foster this identification, how humans compare spatial features and the role of interactive display environments in depicting multiple variables and their interrelationships, and how dynamic interaction with a system can be used to best advantage in helping an analyst understand dynamic processes. Answers to all of these questions depend upon how the various levels of map representation (from those of vision to those of society) interact.

The three questions posed correspond to increasing levels of problem complexity that visualization tools might be useful in addressing. *Feature identification* (either recognition of known features or discovery of previously unknown features) deals with the processes by which human information-processing and visualization tools might interact to make spatial entities or patterns emergent features of displays. When multiple variables are considered, feature identification remains an important stage in visual analysis. The additional variables, however, mean that an analyst must be able to "see" both individual features and the relationships between them (where the "relationships" might exhibit further features or patterns). The task becomes one of *feature comparison*. A concern for process adds not just another variable (in the form of time), but creates a qualitatively different kind of information-processing task. Not only are spatial features important, but temporal features and spatiotemporal features as well. Using map representations to understand a spatial process probably involves both feature ID and feature comparison, but in a spatiotemporal (rather than just spatial) domain. This visualization task will be termed *space–time feature analysis* (STFA).

Part III will apply the multifaceted approach to cartographic representation detailed in Parts I and II to the unique issues posed by GVIS. The emphasis will be on individuals using highly interactive specialized virtual maps to search for patterns and relationships that are not known a priori. While there are certainly other issues relevant to GVIS, the three visualization task levels suggested here are typical of those that cartographers must attend to as our role shifts from that of producing individual concrete representations to producing systems that provide custom maps in response to user queries. The first, feature ID, provides a basis for all GVIS, and will be considered in Chapter 8. The remaining two task levels both involve using GVIS to find relations among multiple pieces of information and will be considered together in Chapter 9. This pair of chapters emphasizing how to make GVIS environments that work will be followed (in the final chapter) by a brief look at how to judge whether GVIS does work, along with some broader implications, for both science and society, of what "work" might mean.

CHAPTER EIGHT

GVIS

FACILITATING VISUAL THINKING

As presented in the introduction to Part III, GVIS is typified by the search for unknowns, real-time interaction with spatial information, and individual control over the map display process. GVIS has the potential to help us cope with the flood of information that technology increasingly provides and to stimulate insightful discovery—but only if those creating the GVIS environments understand how visual–cognitive representations interface with the sign systems of cartographic representation. At this point, there are few certain answers about this interface in a dynamic interactive environment, or therefore about how to design GVIS environments. The most ambitious goal for this (and the next) chapter, then, must be to offer some suggestions about where to start and to provide a structure for framing the necessary research questions. This structure will be based upon aspects of vision, categories, schemata, and logical denotation considered in Parts I and II.

This chapter will emphasize the interaction of visual–cognitive representations with representations at the functional level used by cartographers for explicit signification. The core GVIS task of feature identification is our specific focus here. Feature identification will be defined to include both *recognition* of anticipated features and *noticing* of unanticipated features. Visualization should facilitate both, but it is the potential for noticing the unanticipated that is the most compelling reason to develop visualization tools. While feature identification is treated as the lowest level visualization task, as such it is also a fundamental step in all visualization. In Chapter 2, the pattern ID model of cartographic visualization that John Ganter and I developed was introduced (MacEachren and Ganter, 1990). Here, I will elaborate on this model, then extend it as the framework for integrating ideas on vision, schemata, and functional representation presented in Parts I and II.

The original feature ID model was labeled a "pattern-matching" model. The term "pattern" was used in the sense of schema or mental construct. As part of the model extension undertaken here, the term "pattern" is being replaced by "feature." The reason for this terminology change is that pattern (in the context of geography and maps) has a connotation of global scale (i.e., a spatial or geographical arrangement for a mapped region as a whole). It is related to entire schemata that might be applied to a display interpretation/analysis task. Feature, in contrast, has a somewhat broader connotation that can include geographic entities at both local and global scale. Global-scale "features" match with a complete schema (and include spatial patterns) while local-scale features match with slots within a schema. By discussing features rather than patterns, the visualization model clearly includes both recognition of global phenomena such as a drumlin field, a central place hierarchy, and the like, but also local-scale geographic entities such as a gentrified neighborhood or isolated disease infested trees in a forest.

A MODEL OF FEATURE MATCHING

The feature ID model of individual interaction with visualization tools and David DiBiase's (1990) research-sequence model of scientific visualization as a private-to-public process were developed concurrently. An active exchange among the three of us led to a productive blending of ideas. In particular, John Ganter and I built on the research sequence idea, putting particular focus on private exploratory–confirmatory analysis that involves a high level of interaction with a map or other information display. As detailed in the introduction, an emphasis on private exploration for unknown patterns facilitated by high levels of interaction eventually resulted in what I have termed the (CARTOGRAPHY)³ approach to delineating visualization and communication as complementary aspects of map use (MacEachren, 1994).

A primary source of inspiration for the original feature ID–pattern-matching model was Howard Margolis's (1987) *Patterns, Thinking, and Cognition: A Theory of Judgment*. In this monograph, he posits that scientists (and humans in general) make decisions by matching present situations against a collection of patterns (or schemata) representing past experience and "knowledge."[1] Margolis gave only cursory attention to visual displays as the situations to be assessed, but his visual metaphor of pattern matching stimulated us to adapt his perspective to visual (particularly cartographic) analysis. A fundamental contention of Margolis's argument is that there is an evolutionary advantage to reacting quickly, in a manner that is *usually* right, rather than slowly in a manner that is *always* right. Gregory (1990) more recently made the same point specifically in relation to vision:

> If perception were *driven* by knowledge it would be almost impossible to perceive anything highly unusual. But unusual objects and events do occur; and may threaten danger, or be potential opportunities requiring action. Separateness of perception allows sensed novelty to overcome the inertia of wisdom. (p. 329)

Margolis went on to suggest that, due to survival-based tendencies for making quick decisions about whether or not a pattern is real or about what the pattern represents, we need powerful higher level processes to evaluate the initial reactions. These top-down processes can then prompt further interrogation of the situation. Margolis's view, adapted to visual display, seems to be a direct complement to Pinker's (1990) theory of graph comprehension (discussed in Part I). Pinker argues that precognitive processes can have ambiguity at various stages of information processing and that the context provided by the schema allows the viewer to infer which of several alternative interpretations is most likely. This schema-based approach provides a formalization of the iterative cycle of "seeing that" and "reasoning why" first proposed by Margolis and subsequently incorporated in the original feature ID model.

The feature ID model, as depicted in Figure 2.16, added the distinction between *recognizing* and *noticing* to Margolis's conception of human pattern-matching behavior (MacEachren and Ganter, 1990). Rather than using the term "notice" to refer to seeing something expected due to familiarity or practice (as Pinker, 1990, does), we followed Perkins (1981) in equating noticing with a process of sudden realization of something *unanticipated*. In relation to schemata, as described below, noticing will often require application of several alternative schemata, thus providing alternative perspectives on the problem.

The "model" as originally proposed provides only a rough sketch of how human–map interaction might proceed during a visualization task. The perspective on human information-processing, categories and schemata presented in Part I provides the conceptual underpinning for elaboration of this model (Figure 8.1). One step in this elaboration is to define the links between terms. In addition to substituting "feature" for "pattern," "sensory input" can be interpreted as Pinker's "visual array" and "seeing that" can be matched with his "visual description." The orig-

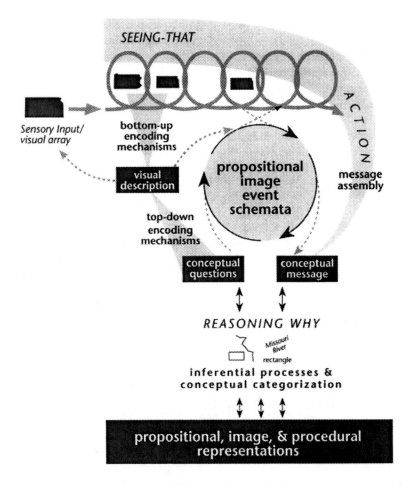

FIGURE 8.1. The feature ID model for map-based visualization, an extension of the original pattern-matching model to incorporate schema and category theories.

inal model emphasized a module in which stored patterns were matched to what was seen. This stage is interpreted here as the instantiation of a schema (or filling the slots of a schema). The "stored patterns" of the original model, then, can be viewed as specialized kinds of schemata (or components of schemata) adapted to spatial–geographic feature matching. The "reasoning why" stage of the original model is linked here to Pinker's "inferential processes."

A distinction between spatial and geographic feature matching is useful here. Spatial feature matching is treated as the recognition or noticing of various spatial (or geometric) arrangements of map sign-vehicles independently of what those sign-vehicles signify (e.g., noticing that a pattern has a predominantly linear trend). Geographic feature matching, on the other hand, implies application of geographic knowledge (e.g., many small shallow lakes scattered across an area produce a pattern recognizable as glaciated terrain). Spatial and geographic feature matching, then, represent two levels of cognitive processing and two levels of schemata (a general all-purpose schema to identify spatial features and highly specialized, domain-specific schemata for geographic features).

Features (whether spatial or geographic) can involve the "state" of things or "processes" by which things occur. Margolis referred to patterns associated with "context" versus those associated with "action." In relation to the approach to schemata outlined in Chapter 4, state features (or Margolis's context patterns) are clearly associated with image schemata, process features (his action patterns) are associated with event schemata, and propositional schemata are likely to provide elaboration for either and the links needed to form schemata complexes.

Human mental categorization occupies a critical role in the amplified feature ID model. Sign-vehicles (through which the map represents) are linked not to individual referents, but to conceptual categories (e.g., forest). The features "seen," due to the relative arrangements of sign-vehicles, are also categories rather than individuals. For GVIS, what is critical is that analysts "see" a geologic lineament, an urban income sector, karst topography, a hole in the ozone, a concentric ring structure for cities, or whatever, not that they discriminate among the unique "spatial signatures" of every individual topographic feature, climate anomaly, or urban area they have ever encountered. For each identifiable component of the visual display, schema instantiation involves categorization. Features of the entity identified are matched with the possibilities allowed by the particular schema applied. The instantiation process results in some or all of the entities specified in the visual description being judged as members of specific feature categories that can occupy "slots" in the schema or that represent sets of slots and their relationships.

The graphic model proposed (for practical reasons) shows the simplest case of schema instantiation, in which the feature identified is judged to signify an individual discrete geographic entity, a state of the United States (specifically the state of Kansas). For most visualization applications, it is categorization of groups of sign-vehicles (rather than individual ones), along with the matching of these visual categories to conceptual categories, that is the critical step in data exploration (Figure 8.2). Chase and Simon (1973), for example, provide evidence that chess

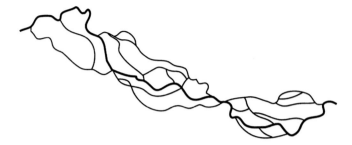

FIGURE 8.2. Given an appropriate geographic schema related to geomorphology, the hydrologic "pattern" depicted here cannot be mistaken. In the absence of the appropriate geographic/cartographic schema, however, the pattern may still be identified, but as a drawing of a rope rather than as a braided stream.

grandmasters can recall about 50,000 game patterns to which they are able to compare an observed game situation. Larkin et al. (1980, p. 1342) describe use of similar patterns by physicists in problem solving: "Large numbers of patterns serve as an index to guide the expert in a fraction of a second to relevant parts of the knowledge store. This knowledge includes sets of rich schemata that can guide a problem solver's interpretation and solution and add crucial pieces of information."

Once initial "seeing" has arrived at a "hypothesis" about apparent map features (a stage roughly equivalent to Pinker's "message assembly"), the "reasoning why" process takes over (a stage analogous to Pinker's "application of inferential processes"). This stage compares conceptual representations of what is seen on the map (as structured by the initial schema and stored in short-term or working memory) to representations in long-term memory in the form of analog, propositional, or procedural representations. This matching of what is initially "seen" with stored knowledge brings top-down mechanisms into play. These top-down mechanisms can have several effects. Separately or in combination they may alter the schema being applied, prompt new sensory input through shifts in attention or fixation, and/or stimulate the visualization tool user to alter the parameters of the display to suppress or enhance particular features. The overall process is considered to be iterative with the possibility for several potential schemata being tried until one provides an acceptable number of logical matches with parts of the visual description.

The idea of an iterative process of comparing observations with knowledge (through the mechanism of schemata) finds support in a variety of literature about how scientists work. Judson (1987), for example, cites a conversation he had with Joshua Lederberg, a Nobel Prize-winning geneticist. Lederberg is quoted as commenting:

> You go back and forth from observation to theory. . . . You don't know what to look for without a theory; and you can't check the theory without looking at the fact; and the fact is only meaningful in the light of some theoretical construction. . . . I believe that movement back and forth occurs thousands, even millions of times in the course of a single investigation. (p. 229)

A similar perspective is offered by Perkins (1988) in relation to Beethoven's "working out" of brilliant ideas through "numerous cycles of generation and selection" of ideas. This generation–selection cycling is analogous to the seeing that–reasoning why cycle proposed here for visualization.

The following two sections consider the role of the "seeing" and "reasoning why" modules of the feature-matching process independently. While the linear nature of language forces such a sequential presentation, it is important to stress that the visualization process proposed is a complex interconnected one in which there is a continual give-and-take between vision and visual cognition through the intermediary of knowledge schema.

LINKING PERCEPTUAL ORGANIZATION AND MAP SYNTACTICS

This section addresses several aspects of perceptual organization, attention, and categorization as they relate to functional representation in a GVIS context. A few recent attempts by visualization researchers to combine what we know about seeing perceptual units of spatial displays with the capabilities of interactive computer technology to facilitate feature noticing are highlighted. These are considered in relation to issues of map syntactics detailed in Chapter 7.

There are probably few research situations in which data are examined independently of a problem context (where not even a general schema is suggested a priori). Situations in which there is an identified threat to humans or to the environment are, however, often encountered well before spatial (or temporal) features have been identified or viable hypotheses formulated concerning the cause of the situation (e.g., the AIDS epidemic). It is at the stage of initial problem exploration, before conceptions of the problem context become so strong that what we see is discounted in favor of what we think we know, that the issues of perceptual organization discussed in Chapter 3 are most critical to GVIS.[2]

At the early stages of visual thinking, particularly when large volumes of data must be sifted, relatively low-level processes of vision may have a role in data exploration that rivals that of prior knowledge applied through context-specific schemata. Bottom-up perceptual processes may even dominate in some cases, particularly before an explanation of a pattern, or hypothesis about its cause, has been clearly worked out. At this stage, general schemata (e.g., such as that which might distinguish between random and regular spatial distributions, among regular geometric features, between a map and a graph, etc.) are most likely to come into play.

When visualization tools act as a catalyst to early visual thinking about a relatively unexplored problem, neither the semantics nor the pragmatics of map signs is a dominant factor. On the other hand, syntactics (or how the sign-vehicles, through variation in the visual variables used to construct them, relate logically to one another) are of critical importance. What we "see" independently of specialized knowledge schemata will be what the syntactics allows us to see—what Bertin (1967/1983, p. 151) has called the "image," or the "form perceptible in the minimum instant of vision."

Indispensable Variables

In Chapter 3, the concept of "indispensable" variables was introduced. According to Kubovy (1981), position in space and time have a dominant role in perceptual organization. In relation to visual information displays, the contention for space as an indispensable variable seems indisputable. For centuries, humans in all cultures have "mapped" the nonvisible to space in order to facilitate understanding. GVIS usually follows the logical route of using display location to depict geographic location. In addition, GVIS environments typically can extend representation to 3-D, thus providing an added dimension (although generally a simulated one) to this indispensable graphic variable. Time as a tool of information display has a much shorter history. The evolutionary advantage of noticing things that move or change, however, leaves little room for doubt that change in a map or other graphic display will provide a powerful cue to early visual processes about the structure inherent in the scene. As Movshon (1990, p. 122) notes, "Practically everything of any interest in the visual world moves." The importance of time for feature identification is dramatically illustrated in the context of camouflage

(Friedhoff and Benzon, 1989). Objects that are virtually invisible on individual static scenes will pop out of an animated time series display.

2-D Space

In GVIS as in other applications of mapping, of course, space is typically used to depict space. As a result, it is not available as a variable open to use and manipulation as part of an interactive visualization tool kit. Linking space to space is, however, not obligatory. Dorling (1992c) has argued quite forcefully that when dealing with information about people, we should treat all people equitably by mapping people, not geographic space, to map space. Such a mapping results in a population cartogram, which Dorling asserts is the ideal map base for representing any social, economic, or political data collected by contiguous enumeration unit. His argument can be linked to the concept of indispensable variables of perceptual organization by translating it to say that we should pay more attention to the power of space in what the user of a GVIS environment sees. Space as one of the indispensable variables should represent the map theme. When the theme is people, space should represent people.

Dorling's argument for cartograms seems particularly valid in a GVIS context in which the features being looked for are spatial–geographic patterns covering extensive map area. In this case, the standard mapping method (choropleth maps) puts visual emphasis on large political units (usually the places where people are not). Any spatial pattern noticed is likely to be one associated with the large units, and thus involving small numbers of people. Patterns associated with small units (e.g., districts within and around cities) will be hidden from view. Following Dorling's arguments, for cartograms as a base for visualizing social phenomena, would involve using the indispensable visual variable of space (combined with the visual variable of size) to depict the key data variable of population, while at the same time retaining a partial mapping of geographic space to map space. His cartograms retain approximate spatial location of enumeration units, and are therefore nearly a topological mapping of space to space.

While most cartograms retain at least partial world-space to map-space mapping, some geographic representation problems are more productively addressed by using display space to directly signify an attribute. A common alternative space used in geographic problem solving is that derived through multidimensional scaling (MDS). Gould and his colleagues (1991), for example, displayed geographic places (cities in Ohio) in AIDS space to emphasize the fact that interaction among locations in terms of AIDS transmission is as dependent upon aspatial links among cities as on their relative geographic location. While AIDS has a definite pattern in geographic space, that pattern can be misleading concerning the interaction among places.

Simulated 3-D

Efforts in GVIS, along with those in scientific visualization in general, have included considerable attention to three-dimensional representation. An assumption behind much of this enthusiasm for 3-D is that expanding the dimensions of an indispensable variable will make representing higher dimensional data more successful. For GVIS, an added impetus for 3-D representation is that the geographic environment, at least when considered at certain scales, is three-dimensional. Considerable effort, then, within GVIS has been given to techniques for tricking vision into seeing three dimensions on a two-dimensional display screen. Much of this work has been with 2½-D perspective displays in which the surface of a three-dimensional solid is represented but locations within the solid are not accessible (e.g., reminiscent of the fishnet plots developed in the late 1960s). An alternative to such perspective views is to use the dynamic nature of most visualization environments to produce depth from motion. Depth cues (as discussed in Chapter 3) are derived from the same graphic or dynamic variables that are used to represent data being mapped. A critical issue to be considered here is the implication for pattern (feature) identification, of interaction between depth cues and data representation.

Both 2½-D and true 3-D displays have two problems that may interfere with feature analysis. These are hidden areas and scale changes across the map. Although scale change can be eliminated by using isometric rather than perspective projection, isometric projection contains none of the depth cues cited in Chapter 3 except frame of reference.[3] The angles formed by the edges of the map with the boundary of the display device cue us to interpret a 2½-D map as an oblique view. Although intellectually, we can interpret an isometric projection map as three-dimensional, it looks "wrong" because the shapes, orientations, and positions of things are not as they would be if we observed a real three-dimensional figure. We are not used to seeing oblique views of objects that do not also have linear perspective and size disparity from front to back.

Although there is at least some evidence that oblique-view maps (whether using isometric or perspective projection) are effective in suggesting a three-dimensional scene (with perspective projection more effective), for many map applications the hidden features that result are unacceptable. One solution to the problem in a GVIS environment is that the oblique view can be dynamic rather than static. With dynamic display, not only is the problem of hidden features addressed, but depth cues provided by motion are added.

Motion provides an important depth cue termed "motion parallax." Motion parallax will result when an observer's position (or hypothetical position) is smoothly changed relative to a true three-dimensional scene. A key to simulation of three dimensions through motion parallax is that apparent shape of objects change as the apparent viewpoint changes.

Animated flybys (e.g., *L.A. the Movie*) typically make use of motion parallax. All dynamic computer displays involve change in the display, not in the observer's position. In flybys, therefore, motion parallax is achieved by generating the impression that the observer is moving. This impression results because of changes to where things are in the scene and to their projected geometry. For each scene, displayed in perspective, a slightly different viewpoint is selected and the position/shape/orientation of all objects in the scene is calculated as they would be observed from that vantage point. The individual perspective views alone provide strong depth cues and appear three-dimensional (particularly with the realistically rendered images such as those of *L.A. The Movie*). When hundreds (or thousands) of these scenes are linked in temporally displayed sequence, the result is particularly dramatic. The sensation is of observing a truly three-dimensional world from a moving aircraft.[4] I am aware of no research concerning the ability of viewers to recognize features (or associate Z-height with other features) on a map movie (or on an actively controlled flyby). Both passive and actively controlled flybys, however, have been advocated as a tool for presenting topics such as the results of forest management practices (Bishop, 1992).

The idea of actively controlled terrain flybys (with GVIS applications in mind) was first implemented by Moellering (1980a) and (due to both hardware development and innovative software techniques) is quickly becoming possible in real-time on desktop workstations (Kraak, 1993). As yet, little or no empirical analysis has been done concerning how viewers cope with dynamic change in point of view or whether the ability to interactively manipulate the viewpoint on the map representation enhances the impression of depth. As Goldberg et al. (1992) point out, however, whatever advantages might ensue from dynamic control of viewpoint, there is a potential for generating confusion about orientation. Kraak (1993) has addressed the orientation issue to some extent by incorporating a dynamic directional indicator along with his dynamic perspective views, but has not yet done empirical testing to see if this method is effective. Evidence from a study with helicopter pilots (using a dynamic, but plan view map) indicates that, at least for the task of target location, a continuously reorienting "track-up" map impedes performance (Harwood, 1989). Further empirical work is needed to see whether a simulated 3-D view makes the continual reorientation of track-up displays easier to cope with.

Kaiser and Proffitt (1992) describe motion parallax as composed of two components, an observer-relative transformation (overall movement of the scene relative to the observer) and an object-relative transformation (movement of scene components relative to each other). The latter they label the *stereokinetic effect* (SKE). A good example of the SKE is provided by considering the impression we would have of the dots in Figure 8.3. The figure at the left appears to be a flat display of randomly located dots. The linear-perspective cues in the middle give a sensation of depth and we can begin to imagine the dots contained within a cylinder (even though there is not size disparity among the dots). If we add a perspective impression of an internal cylinder (on the right), with some effort we can imagine dots scattered within two concentric cylinders. If we animate this last scene so that interior dots rotate clockwise while the remaining dots are stationary or rotate in the opposite direction, the impression of depth is dramatically enhanced.

For contourlike displays of a conic shape, Kaiser and Proffitt demonstrate that the object-centered transformations are the key to depth-from-motion, with motion parallax and SKE producing virtually identical sensations of depth. They describe an application of their SKE technique to contour lines on dynamic air navigation maps. On the SKE display maps, contours shift in the X and Y directions by an amount proportional to their altitude. Contours translate relative to one another, but do not change shape. The method results in an appearance of contours being layered in depth. Based on informal evidence that the depth percept created by the SKE transformation persists for some time, the system presents short, cardioidal depth-inducing translations followed by static intervals. The result is claimed to be stable, natural depth perception that does not interfere with eye movement translation across the surface of the display.

When we simulate the third dimension in a physically two-dimensional map, whether through perspective, motion, or alternative depth

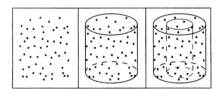

FIGURE 8.3. Random dot distribution (left), with linear perspective cues indicating an enclosing cylinder (middle), and with further perspective cues suggesting an interior cylinder (right).

cues (e.g., shading, color, etc.), the impact of these cues on viewer ability to interpret the resulting map must be considered. Wanger et al. (1992) looked at this issue from the perspective of simple perceptual judgments that might be made for objects in perspective views generated with sophisticated computer graphics techniques. Specifically, they considered six depth cues (or variations on cues) as these cues affected accuracy with which position, orientation, and size could be judged. The depth cues considered were motion of an object in the scene (controllable by the user), linear perspective, shadow, elevation in the reference frame, texture gradient of the background, and texture gradient on the objects being judged. The depth cues as they appeared in the scaling task are depicted in Figure 8.4. Although the experiment described included only 12 subjects, there are a few fairly dramatic results that should be investigated in more detail (Figure 8.5). Both shadow and perspective resulted in substantial increases in positional accuracy. Shadow also resulted in a substantial increase in size estimation accuracy and a minor improvement in orientation accuracy. Perspective, however, had negative impacts on both orientation and size accuracy (with the effect on orientation being greater than the positive influence of perspective on position judgments). Motion had a generally positive effect, while elevation in relation to the reference frame decreased performance on size tasks. Somewhat surprisingly, texture had virtually no significant influence on task performance.

In related research, Kraak (1988) combined a semiotic approach with a concern for visual processing of three-dimensional maps. Specifically, he investigated the question of "whether the perceptual qualities of the two-dimensional graphic variables are still valid in three-dimensional maps, since there is a strong link between them and the psychological or pictorial depth cues" (Kraak, 1988, p. 61). He hypothesized possible rela-

FIGURE 8.4. A sample display from the Wanger et al (1992) experiment in which all depth cues tested are included. This figure was used in their positioning tasks. Subjects interactively adjusted the position of one ball relative to the other two (whose positions remained fixed). The task was to position the moveable ball such that it appeared to lie on the midpoint of an imaginary line adjoining the two fixed balls. *Reproduced from Wanger et al. (1992, Fig. 13, p. 49); original in color. Reproduced by permission of the author and IEEE Computer Graphics and Applications. Copyright 1992 by IEEE.*

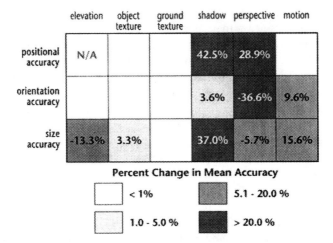

FIGURE 8.5. A summary of the interaction of depth cues and task performance found by Wanger et al. (1992). *Reproduced from Wanger et al. (1992, Fig. 14, p. 49); original in color. Reprinted by permission of the author and* IEEE Computer Graphics and Applications. *Copyright 1992 by IEEE.*

tions between graphic variables (as originally defined by Bertin) and his own taxonomy of depth cues (Figure 8.6). Kraak conducted a series of experiments to determine the relative advantages of two-dimensional maps in which graphic variables represent Z-values versus three-dimensional simulations in which Z-values are represented by apparent height of oblique views (both linear perspective and stereo pairs were tested). These 2-D and 3-D options were labeled GEO2+C versus GEO3 (where "C" stands for a graphic component or variable). In addition, when two-dimensional graphic variables were used to depict an additional attribute on an oblique view seen with or without stereo (and identified as GEO3+C), the potential interference of the different depth cues in interpreting these graphic variables was evaluated. These issues were addressed for four categories of map that can be depicted with perspective (or oblique view) symbolization: an urban building plan, a DTM, a prism map of contiguous enumeration units, and a three-dimensional point symbol map (using spheres or cubes).

When GEO2+C was compared with GEO3 Kraak found no difference. This led him to advocate three-dimensional simulations over standard two-dimensional maps (for reasons that relate to the indispensable nature of space for perception). Most of Kraak's test cases dealt with GEO3+C maps. Results were summarized graphically (Figure 8.7). Two general findings are worth special note. First, there was not a single case

GVIS: Facilitating Visual Thinking 375

GRAPHIC VARIABLES to display spatial distributions	DEPTH CUES to display the third dimension
size	retinal image size
value	shading
texture	texture
color	color
orientation	line perspective
shape	perspective
-----	area perspective
-----	detail perspective
visual hierarchy	overlapping/ interposition

FIGURE 8.6. Kraak's proposed links between graphic variables and depth cues. (To compare this table with discussion of depth cues in Chapter 3, we can match his texture with texture gradient, color with chromostereopsis, line perspective with linear perspective, perspective with oblique projection, area perspective with areal perspective, detail perspective with detail, and obstruction/overlap with interposition.) As Kraak notes, visual hierarchy is not a graphic variable, but an aspect of map design. *Derived from Kraak (1988, Fig. 5.6, p. 60).*

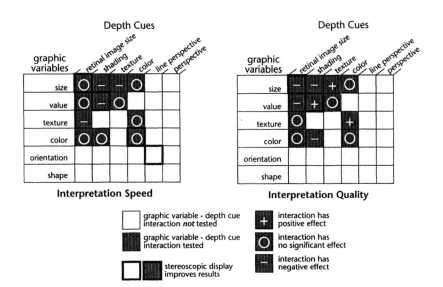

FIGURE 8.7. The relation between graphical variables and depth cues tested by Kraak. He labeled these matrices "understanding" and "quality" to refer to results from reaction time and accuracy of response measures, respectively. Here, the more explicit label of "interpretation speed" is substituted for "understanding." *Derived from Kraak (1988, Fig. 9.1, p. 103).*

in which nonstereo depth cues speeded interpretation of graphic variables (but four cases in which interpretation was interfered with). In relation to accuracy of interpretation, there was a nearly even trade-off between enhancement and interference. Overall, the most likely negative impact of representing geographic information in simulated three-dimensional form, with a fourth (data) dimension added via graphic variable variation, is for judgment of feature size. Finally, in at least two cases, a negative impact of three-dimensional representation seems to be counteracted if the map incorporates retinal image disparity and is viewed with a stereoscope. As is clear from the two matrices, Kraak's experiments leave several unanswered questions. Most significantly, the effect of oblique views was not tested at all, since all maps in this portion of the study were oblique views.

Studies by Kraak (1988) and by Wanger et al. (1992) leave open many issues concerning how features will be seen on three-dimensional maps. Both studies, however, point to potential problems encountered when we try to extend the information content of visualization displays through simulation of a third spatial dimension. Although viable arguments can be made for making displays that fit what evolution has adapted vision for (see Bishop, 1992), one of the main strengths of maps for centuries has been their ability to abstract the world to make it more manageable. It is likely that by adding to the visual processing load by including cues to a third dimension or simulating motion, we will decrease the ability of vision to process pattern that relies on variables other than space or time.

Time

Maps (at least visible ones) always use space to signify something. In contrast, although time has frequently been an attribute to signify on maps, it has only recently become a potential sign vehicle. Due to the indispensable nature of time as a variable, however, it may prove to be the most powerful tool for feature identification in GVIS. Except for circumstances in which the problem context makes it acceptable to signify geographic space as something other than space, time is the only indispensable variable at our disposal. Like mapping space to space, the use of time to depict temporal data is an obvious application of this powerful variable. Like mapping population to space, however, time can also be used to prompt the noticing of atemporal features.

In a dynamic cartographic environment, time adds several new variables to our static set of visual variables (see Chapter 7 for a detailed discussion of these variables). Perhaps the simplest use of time as a feature identification cue involves "blinking," the repeated change of a sign-vehicle or set of sign-vehicles between two states (e.g., on–off, light–dark, etc.). As noted in Chapter 7, blinking is an application of the dynamic variable duration to a repeated cycle. Monmonier (1992) demonstrated this technique as a way to draw attention to individual extreme values of a data set. Time can be similarly used to visually group any entities that have been assigned to a class or category. Once categorized, the entities (whether symbolized identically or not) will appear as a group if some common attribute of their symbolization is changed in synchronization. This technique employs the Gestalt principle of common fate (entities that move or change together will be seen as grouped together). Monmonier suggested using this kind of blinking to highlight the spatial pattern (or lack of it) exhibited by successive classes of a choropleth map. DiBiase et al. (1992) used the same idea to highlight the global pattern of earthquake occurrence.

The success of blinking as a visualization tool will depend upon understanding how vision reacts to temporal change. Cartographically, we know virtually nothing about the role of cycle duration, frequency, or any other dynamic variables in relation to figure–ground (or to other aspects of perceptual organization/ feature identification). Some clues, however, are provided by noncartographic research into the influence of temporal frequency on figure–ground segregation. In early vision, temporal frequency is considered to be as potent a determinant of figure–ground organization as spatial frequency (Klymenko and Weisstein, 1989; Klymenko et al., 1989). In one experiment Klymenko and Weisstein (1989) had subjects view a bipartite circle in which sine wave grating patterns having identical spatial frequency but different temporal frequency were juxtaposed (Figure 8.8). As in their similar experiments with varied spatial frequency patterns, subjects were required to fixate on the center of the pattern and continually indicate (by moving a joystick in the appropriate direction) which side appeared as figure. In all cases, the area with higher temporal frequency was seen as ground. The static pattern was figure to all temporal frequencies tested. In a similar study using dot patterns that

FIGURE 8.8. A sample test stimulus from Klymenko and Weisstein. The actual test pattern was dynamic. After Klymenko and Weisstein (1989, Fig. 1, p. 629). Adapted by permission of the authors and Pion Limited, London.

flickered at specific frequencies Wong and Weisstein (1984) had similar results. Specifically, they found that frequencies of 6 to 8 Hertz were optimal for organizing the pattern of flickering dots in depth behind an area of stationary dots. In situations where both spatial and temporal frequency are combined, their effects appear additive. High temporal frequency combined with low spatial frequency produced the strongest ground in one experiment (Klymenko and Weisstein, 1989), and when these factors are opposed (e.g., low spatial frequency with low temporal frequency) they cancel each other out resulting in ambiguity (Klymenko et al., 1989).

Based on the perceptual research cited, effectiveness of map sign-vehicle blinking will depend on duration of the blinking cycle and frequency of change within each cycle. Blinking at high temporal frequencies may even produce deemphasis rather than draw attention because patterns with high-frequency flicker tend to result in the flickering area being seen as further in depth—thus as ground rather than figure (Wong and Weisstein, 1984; Klymenko and Weisstein, 1989)! At the other extreme, the relatively low frequency (temporally coarse) blinking used by Monmonier and Gluck (1994) was disliked by viewers of their graphic narratives. The problem seems to have been that blinking at the frequency used was harsh and disturbing to look at. Clearly, to use blinking effectively in a GVIS environment, we need to determine the relevant thresholds for enhancing and suppressing features.[5]

Only one indispensable variable, time, is used in blinking. As time passes, a nonindispensable visual variable applied to a stationary feature is changed (e.g., the color value of a point symbol switches from red to yellow). An alternative that makes use of both indispensable variables, and that derives more directly from the Gestalt grouping principle of common fate, is to change location rather than a visual attribute of the features of interest. For lack of a better alternative, I will label this technique "quivering." Quivering as envisioned here involves simultaneously shifting a group of features back and forth. The goal is to allow vision to group the shifting features while not producing an irritating display. The key, as with blinking, is to determine an appropriate distance and cycle duration for movement to and from the starting location.

To consider how quivering might be applied, imagine an astronomical image/map containing thousands of stars. You see what look like regular geometric features (e.g., lines and circles) formed by sets of stars. These features may be just an accident of projected geometry due to our observation point (Darius, 1990). If lines or circles of stars exist in the same plane, however, the alignment would be considered a phenomenon in need of explanation. Assuming that distance to earth of a sufficient number of stars is known, we can visually examine the pattern of stars in stereo to see whether the lines and circles still appear.[6] Alternatively, we could retain the monocular plan view of the original image, but quiver the position of sets of stars within narrow distance bands.[7] True star strings or rings will remain together and be highlighted by grouping through common fate. A user controllable visualization tool might allow the analyst to step sequentially through overlapping depth bands progressing out from earth and to halt the progress in order to repeat any section that caught her attention.

If visualization tools are dynamic, not only visual change and movement are possible, but aural change as well. Sound has been investigated as a data "auralization" tool in a number of contexts. Speeth (1961) developed a data analysis tool to facilitate identification of earthquakes on the basis of seismic recordings. The key problem he was trying to solve was how to distinguish between earthquakes and bomb blasts, which were visually very similar on seismic recordings. Seismometers, of course, record sound, but at frequencies outside of human hearing. Speeth's approach was to play the sound recordings of seismometers at speeds high enough to be audible (a compression of 24 hours to about 5 minutes). Based on the audio representations, subjects in an experiment correctly classified seismic events as bomb or earthquake 90% of the time.

Human abilities to locate objects by sound is much less than that to locate objects visually or to distinguish among sounds. We probably should not expect sound to be particularly effective for visualizing spatial patterns. It will be best for localizing events in time (as with the seismic data discussed above) or for classifying individual entities that can be selected interactively. Krygier (1994) produced a cartographic example of the latter. His implementation allows an analyst to use a mouse to point to counties on a map of Pennsylvania. Each time the analyst clicks on a county, a pitch is sounded at high, medium, or low register to represent high, medium, and low distance for the mean journey to work of county residents. By itself, this map sonification would not be very useful, except possibly if modified to allow visually impaired analysts to retrieve information that a sighted person might obtain by observing differences in color value or symbol size. The technique, best described as a *sonic probe*, is intended as a way to increase the number of variables that an analyst can consider simultaneously (see Chapter 9 for more details). Although the process of querying points for information may seem far less efficient than visual (potentially parallel) processing of information, it is a technique that can allow us to quickly narrow in on important local features, much as knocking on a wall is an efficient way to quickly locate studs (Buxton, 1990).

Scale and Resolution

Features are scale- and resolution-dependent, not only in space, but in relation to attributes and time as well. Anything we depict on a map results from a decision about the portion of obtainable information to be represented, how much area (e.g., a state), what range within the data (e.g., only people over 65 years old), and over what time span (e.g., 1980–1990). Similarly, we must specify how much detail to portray in space (e.g., divisions at the county level), attributes (e.g., high, medium, and low percentage), and time (e.g., yearly).

Space

The size of "features" on a display will interact with the geographic scale of features in the world. If the map scale for analysis is too small (the display covers too much of the world), small geographic features will be invisible. If the scale is too large, features bigger than an individual display screen are likely to be missed. To anticipate whether particular kinds of features will be noticed by a visualization tool user we must consider the likely size of feature and the proportion of the visual scene occupied by features when they appear on the display.

Based on the empirical evidence discussed in Chapter 3, one hypothesis that might be proffered at this point is that global–local precedence will be a more significant issue for exploratory visualization with detailed maps than for any other map-use context. If, for example, the apparent size threshold for global precedence identified by Antes and Mann (1984) and by Kimchi (1988) holds on maps, we might anticipate that display screen size will have an impact on results of exploratory visualization. This could happen due to the apparent influence of initial global patterns on subsequent processing of local detail and the fact that different patterns will be global ones with different display sizes. We might take advantage of this seemingly unfortunate circumstance by purposely altering scale of a data display as a technique for prompting analysts to "see" different global patterns, and therefore make different interpretations of what appears local at that scale.

It is, of course, not just the size of the display area that determines whether or not an analyst is able to identify a feature. Geographers have long known that geographic features are likely to be resolution dependent, with different features identifiable at different resolutions of analysis (MacEachren et al., 1992). The idea of multiresolution analysis is, therefore, not new to geographic research—it forms the basis of central place hierarchies, theories of crop yields (Haining, 1978), and applications of fractal analysis in geomorphology. It is also a factor prompting the development of spatial analysis techniques such as quadrat analysis and variable lag spatial autocorrelation. What is new is the possible identification of underlying psychological factors that may interact with the natural scale-dependent nature of real-world patterns.

A geographic feature that exhibits high frequency spatial variation can be unrecognizable on a low resolution display, but in contrast a feature with low frequency change will be enhanced. Visualization tools that change the scale of geography depicted (by zooming in and out), the resolution of the display, and the spatial frequency of change across the display (by manipulating display contrast or by spatial filtering) are therefore likely to facilitate feature identification.[8] The search for global patterns, therefore, can be facilitated by altering either of two new graphic variables proposed in Chapter 6, resolution or crispness of the scene. If the data underlying a map or image is organized in a grid structure, either procedure is quite easy to implement. Fine-resolution grid data can be aggregated to larger and larger grids (producing a resolution change) or a spatial filter can be applied in which each grid cell value is replaced by a function of itself and surrounding values. This kind of spatial filtering was presented in Chapter 2 as a fundamental stage in visual processing. It is not surprising, therefore, that a method that human vision has evolved to enhance the identification of features in the visual scene also works as a user-controllable visualization tool.[9]

In the context of meteorological/atmospheric simulations Berton (1990) suggested a strategy for implementing multiresolution analysis of model-derived information. The method involves dynamic specification of regions of interest that are visually and computationally zoomed in on. Once the bounds of a region of interest are delineated, zooming can be linked back to the model (or a separate regional scale model) that calculates details only for the specified region, leaving adjacent areas at their original level of detail. A perceptual–cognitive problem yet to be addressed with such a system is how the human visual information-processing system will cope with a scene that varies in detail. Will the coarse-resolution background initially appear to be background, or will the larger visual units trick vision into assuming that they are closer (and therefore that they are figure rather than background)? Will knowledge that the coarse textures are coarse due to data manipulation rather than to the presence of large homogeneous features be able to overcome the perceptual evidence to the contrary?

To consider how resolution and crispness might interact, a good place to start is with Margolis's (1987) analysis of how human vision copes with blurring of an image. He uses a raster image of Abraham Lincoln as an example. Here, a different image at the same resolution is substituted (Figure 8.9; see Figure 2.2 for a discussion of the Lincoln image in relation to Marr's theory of vision). Human vision is particularly good at edge detection, so the initial impression generated of the image is probably of an irregular arrangement of squares. If the squares were much smaller, even at the same crispness, vision would be much less likely to isolate them as independent features. A reduced version of the figure therefore enhances the global features at the expense of the local (Figure 8.10). As Margolis points out, however, a similar effect is achieved, perhaps more dramatically, by "defocusing" the image (i.e., changing its crispness). Simply squint your eyes (or if you are far-sighted, remove your glasses). The image is immediately identifiable. By depriving vision of detail information, you seem to have actually enhanced the image.[10]

Reducing crispness of the image acts to break up the appearance of distinct edges forming blocks or squares. Margolis (1987, p. 38) contends that the dramatic change in interpretation of the scene results because "your brain has engaged in a process that amounts to seeking out some pattern in its repertoire that seems to fit—that satisfies ... the set of

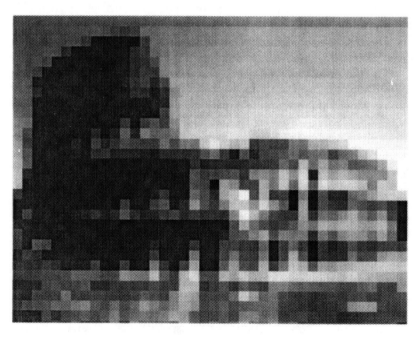

FIGURE 8.9. A low-resolution image that may be difficult to recognize, unless looked at from a distance or with defocused eyes.

FIGURE 8.10. A reduction of the previous figure to 25% of the original size makes the structure depicted identifiable.

cues." In other words, vision selects the best-fit schema at its disposal in order to cope with the loss of stimulus information. That vision can sometimes come up with a more accurate interpretation from partial than from complete information is strong evidence for the role of schemata in visual information processing. "In Darwinian terms, it makes sense that the brain has a bias favoring seeing something rather than nothing, so that it tends to jump to a pattern that makes sense of a situation" (Margolis, 1987, pp. 38–39). Margolis goes on to argue that this process of scene interpretation is a bottom-up process; even when you now know what the coarse grid represents, it is difficult to see it as a building (or as a face, in the case of the Lincoln figure) without defocusing or a change in scale (without changing crispness or resolution).

The power of vision to impose global patterns when local detail is suppressed suggests that a visualization tool that moves a scene in and out of focus could be a quite powerful prompt to feature noticing. Gershon (1992) developed an animation technique that does just that. He based his approach, in part, on evidence that vision will ignore small details of objects in successive scenes if a global object is perceived, and that vision will "conserve" objects through a series of changes (an idea similar to the Gestalt principle of objective set) (Ramachandran and Anstis, 1986). Gershon applied the technique to a set of gridded global sea surface temperature (SST) data (specifically a monthly blended average SST as a function of geographic position and time, and derived from a combination of satellite radiometer and in situ measurements). Maps used a red to blue color scale for high to low temperature. A series of six increasingly blurred images were created by applying a local moving average filter. The images were then displayed in an animation loop in which the appearance was of successive blurring and sharpening. Gershon contends that the animation enhanced the visibility of structures embedded in the data and also that it drew viewer attention to the structures. He also compared blurring done with image filtering (what I have called a change in crispness—see Chapter 6), with "blurring" by subsampling the grid (what I have called resolution change). Filtering was considered the superior technique, although no formal test of either method was conducted.

Attributes

Just as manipulation of scale and resolution can be used to uncover scale-dependent features and isolate global from local patterns, manipulation of range and classification of data will isolate general to specific features. Selecting subsets of data from what is available has been labeled "focusing" in the EDA literature (Buja et al., 1991). Rather than relying entirely on vision to group features, we can make use of interactive tools designed to isolate subsets of data, thus enabling vision to concentrate on seeing and interpreting the spatial pattern exhibited by those subsets.

Focusing is, in essence, a special case of interactive data classification in which single classes are highlighted. Several implementations of interactive classification for mapping have been presented (Yamahira et al., 1985; Plumb, 1988; Ferreia and Wiggins, 1990). Interactive classification is a concept that was (is) a bit hard to get used to for those of us who pursued our graduate education at a time (and in a place) when (and where) "accurate" grouping procedures were considered to be the ultimate goal of any quantitative classification. At the time, George Jenks's (1977) program for "optimal" data classification was seen as the end to research on data classification for choropleth mapping. A few years ago, I even wrote a rather critical evaluation of the first of these interactive classification systems I came across (that by Yamahira et al., 1985) (MacEachren, 1989). There were two issues I did not consider in writing that critique. First, while a case can be made for fixed "statistically optimal" class choices if a single map is to be printed (i.e., when only one view *can* be seen), single perspectives are at odds with the whole concept of exploratory data analysis. They impose a single statistical criteria on what we might see. Just as importantly, statistically optimal classification takes into account only how similar geographic places are to one another in terms of data values, not in terms of location. If mapping the data is a good idea in the first place, techniques that get at the spatial groupings should be considered just as viable as those that get at attribute groupings (see MacDougall, 1992, for a recent implementation of spatial clustering in a GVIS context).

Of the interactive data-classification techniques presented in the literature, all are based on the idea that classification (by grouping data into a small number of categories) can highlight spatial patterns or the lack of them (by directing attention to clusters of similar data values), but can also hide some of the similarities that might span class break points. By providing analysts with the ability to change the number of classes data are parsed into, and to shift class boundaries around, the chance to uncover patterns should be increased (Figure 8.11). At this point, no empirical assessment has been made of the impact of focusing through interac-

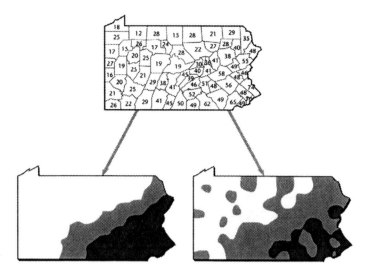

FIGURE 8.11. The bottom pair of maps depict the same data (shown at the top). The only difference is in how data breaks are chosen. If an analyst is given an interactive data classification tool, then these extremes could both be identified, thus providing a measure of the confidence about apparent pattern. The issue of visual confidence limits will be taken up in more detail in Chapter 10. *Reproduced from MacEachren and Ganter (1990, Fig. 6, p. 77). Reprinted from* Cartographica *by permission of University of Toronto Press, Inc. Copyright 1990 by University of Toronto Press, Inc.*

tive data classification on the propensity of analysts to find patterns (or on the verity of patterns uncovered).

Time

Like space, time is continuous but is treated as discrete in order to measure or represent it. The search for geographic features depends not just on what is measured where and at what resolution, but on when measurement is done at what frequency and what temporal resolution. Space–time patterns will be evident only when an appropriate temporal interval is selected with appropriate bounds. If the pattern is cyclic, the frequency with which we sample is critical as well.

Monmonier's concept of *geographic brushing* (discussed in more detail in Chapter 9) includes the idea of a temporal as well as a geographic brush (Figure 8.12). His example showed a time line with a temporal slide bar that could be dragged with a mouse to focus on a particular year (Monmonier, 1989a). It is a simple extension of the idea to envision a

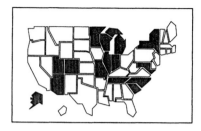

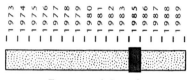

Temporal Brush

FIGURE 8.12. Monmonier's concept of a "temporal brush." Analysts can change the year for which data are displayed by moving the slider with a mouse. *Reproduced from Monmonier (1990, Fig. 23, p. 42); only a portion of the original figure is shown here. Reprinted from* Cartographica *by permission of University of Toronto Press, Inc. Copyright 1990 by University of Toronto Press, Inc.*

slide bar in which both a beginning and an end time (the bounds of a temporal interval) could be set dynamically. Depending on the topic represented, data might be totalled or averaged over the interval, or represented as a repeating animation loop.[11]

Questions of temporal scale and resolution are primarily relevant to space–time feature analysis. Further temporal issues will, therefore, be taken up in the next chapter.

Spatial Feature Enhancement through Graphic Variable Manipulation

Even though the above approaches to GVIS emphasize indispensable variables and/or manipulation of detail, they also depend upon use of the static graphic variables. If there is no variation of size, color value, shape, and so on, there is no sensation. Just as for communication-oriented map use, therefore, attention should be given to the role of the remaining graphic variables in the context of GVIS. This section focuses on ways in which these graphic variables influence apparent spatial pattern.

For monochromatic arrangements of sign-vehicles, Caivano (1990) (as described in Chapter 6) defined the term "pattern" as the result of combining three graphic variables, what he termed directionality (orientation), size, and density (texture). In the sense of "features" as defined above, however, manipulation of any graphic variable for a set of spatially distributed point or line sign-vehicles will alter global and local features that are seen. We will, therefore, consider pattern enhancement through variation in color attributes (value, hue, or saturation) and/or size, shape, orientation, texture, and arrangement (crispness and resolution were covered above and transparency is only relevant to situations in which multiple variables occupying the same geographic coordinates are to be viewed in conjunction.). Spatial pattern on maps and other information displays, then, is made up of repeated positional, linear, or areal sign-vehicles. These sign-vehicles typically hold constant several graphic variables while varying one, two, or three.

Using Monochrome Variables

As discussed in Chapter 3, choice of graphic variables and variable combinations will determine what (if anything) a viewer will "see" at first glance in a map display. Bertin's (1967/1983) concepts of selectivity and associativity (with appropriate modification to match knowledge obtained from empirical research on selective and divided attention) are quite relevant to the design of interactive tools for facilitating spatial pattern recognition (and noticing).

Perhaps the most elaborate attempt to make use of what is known about links between fundamental graphic variables and perceptual organization in the context of exploratory visualization is the EXVIS (Exploratory Visualization System) project underway at the University of Massachusetts at Lowell (Grinstein et al., 1992). In all EXVIS displays, data is depicted by an array of positional sign-vehicles termed EXVIS "icons." Grinstein et al., (1992) state their goal as

> to create displays in which the regions of contrasting texture "pop out" at the analyst. A considerable body of evidence and perceptual theory has accrued to guide the design of icons that induce the visual system to segment the displays into regions of different texture automatically. (p. 338)

An underlying principle of many EXVIS displays is that vision is highly selective for orientation. In the earliest version of their system, EXVIS icons consisted of line segments whose orientation, length, and thickness could be varied. In the example shown, the characteristics of the icons allow an analyst to identify key locations in a plasma flow. Most applications of the EXVIS approach have been to multivariate analysis. In an effort to depict more variables, the "icon" was initially extended from one to five "limbs" (Figure 8.13). More recent implementations combine the five-limbed stick figure, sound, and color to represent more than 20 variables at once. Further details will be considered in Chapter 9.

At Pennsylvania State University, my students and I have been

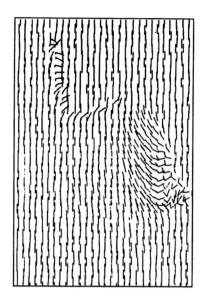

FIGURE 8.13. An EXVIS display in which stick-figure icons are used to represent five (of 22) parameters in the solution of a partial differential equation that describes a reaction inside an inductively coupled plasma chemical reactor. *Reproduced from Grinstein et al. (1992, Fig. 3, p. 640). Reprinted by permission of John Wiley & Sons, Inc., from* International Journal of Intelligent Systems. *Copyright 1992 by John Wiley & Sons, Inc.*

working on a visualization environment designed to facilitate analysis of uncertainty in data values estimated for a surface at various points in time (MacEachren et al., 1993). Again, much of the system is designed to deal with multivariate analysis (and will be considered in Chapter 9). One tool developed, however, is as relevant to univariate as to multivariate analysis. This tool uses sign-vehicle size to depict a variable at numerous points across a surface. Initially, the analyst sees narrow bars scaled by height. These "bars," however, are really triangles in which the base is very narrow. While triangle height represents the data values, the analyst can interactively control triangle width. As the analyst increases triangle width, two things occur simultaneously. First, the phenomenon as a whole becomes more visually dominant in relation to the background (because triangle area, a redundant variable, becomes a visually dominant factor in the scene). Second, a redundant (highly selective) cue to data values is created, slope (or relative orientation) of the triangle sides. It is anticipated that analysts will easily develop a schema that matches short/flat slope to low values and tall/steep slope to high ones.

The use of size, orientation, and dynamic manipulation to facilitate feature noticing builds upon principles of perceptual organization related to selective attention (and Bertin's [1967/1983] concept of selectivity of certain graphic variables). Another goal in some visualization tool development is to prevent the emergence of patterns that are artifacts of the data-processing method (and are thus false patterns).[12] Because many simulation models (particularly for atmospheric phenomena) generate data values in a regular grid, patterns produced when the output data are displayed often have an artificial appearance. Similarly, multiple layers in a three-dimensional representation tend to result in moire patterns (a clear artifact of the grid nature of the models). Papathomas and Julesz (1988) encountered this problem when using the "point-cloud" technique to display dynamic changes in cloud water content. Their solution (although they did not use the term) was a manipulation of the graphic variable arrangement. Papathomas and Julesz (1988) manipulated arrangement of sign-vehicles (round point symbols) by slightly randomizing location of dots in the point clouds. This change from regular to irregular arrangement enhanced the realism of the display and eliminated spurious features of the moire patterns produced when layers of the three-dimensional cloud of points aligned.

Using Color Variables

In virtually all GVIS environments, color is an option. Brown and Hershberger (1992), for example, point to some simple manipulations in which color can enhance feature identification dramatically:

> Color also displays global patterns very effectively. For example, perceiving global patterns when a monochromatic group of small triangles changes to a monochromatic mixture of circles and squares is much harder than seeing the patterns when a group of black dots changes to a mixture of red and blue dots. (p. 60)

To use color effectively in GVIS, we need to make use of what we have learned about color vision combined with semiotic approaches for matching color with categories of referent. The color assignment problem is quite different for visualization uses than for communication uses of maps. Dynamic interactive displays do not require a single "optimal" color mapping to be selected. Imposing such a limit could impede feature ID. For GVIS, we must provide users with tools that allow dynamic selection from a set of logical choices, each of which might enhance a particular aspect of the data.

A good example of the kind of color assignment tool that can be incorporated in a GVIS environment is Brewer's (1993) hypermedia tutorial on color selection (see Chapter 7). Users of this tool are able to select one of several map topics (e.g., land use, soil permeability, etc.), then interactively change among alternative color schemes. Although the hypermedia tutorial was created to illustrate Brewer's approach to color (originally conceived for presentation maps), it also illustrates the analytical power provided by a tool that ensures logical matches between data variation and color variation, while at the same time providing flexibility in highlighting different aspects of the same data set.

Among the most important color issues in GVIS design is how to make use of color for the representation of data that are ordinal or numerical. As discussed in Chapter 4, hue (though inherently ordered in terms of wavelength) does not appear ordered to most viewers. Vision easily identifies order in color-value ranges and in part-spectral ranges (particularly those starting or ending at yellow (the highest value hue). Many GVIS applications are complicated by a need to treat a numerical range as bidirectional or diverging (e.g., mapping temperatures above and below freezing), thus requiring both qualitative and quantitative distinctions on the same map. Color connotation often comes into play here, for example, with assignment of reds to high temperatures and blues to low. I was told by a researcher trained in oceanography, for example, that a blue to red scheme is so standard in oceanographic mapping that researchers assume red means warm and blue means cold, without the need for a legend (William Peterson, personal communication, March 1993). This blue–red scheme is so commonly recognized for temperature maps, it is often applied even when there is no specific intention to suggest that data are bidirectional. As indicated in Chapter 7, the barrage of full-spectral temperature maps on television weather reports and in major newspapers (with red and blue extremes) may be producing a similar association between temperature and full-spectral hue ranges for a broader public. An exception to this trend was uncovered when discussing color meaning in an interdisciplinary seminar on visualization. I was surprised to learn that for at least one group, astronomers, the meaning of red and blue was exactly the opposite of that used on many temperature maps (James Leous, personal communication, March 1993). This reversal is easy to comprehend if one realizes that the hotter stars emit radiation in the blue part of the visible spectrum.

Full-spectral schemes are common in scientific visualization. They are, however, problematic for reasons beyond the lack of an apparent order (or the need to apply top-down processes to interpret that order).[13] Adjacent colors of the spectrum have vastly different levels of discriminability. Edge detection within the blue–green range, for example, is poor (Mollon, 1990). The result for visualization, at least in early exploratory stages where preattentive processes are critical, will be that differences among adjacent locations will be exaggerated in some places while differences in other places are hidden. What is emphasized will be an artifact of the particular match of data values with hues, not of any characteristic of the data. Beyond this problem, red and green (colors applied to opposite extremes in this scheme) can be indistinguishable for people with the color deficiency.

For communication with single published maps, the discussion of color would probably stop here. For visualization, however, we need to provide analysts with the flexibility to adapt displays to unique exploration tasks. This can at times call for conscious violations of standard cartographic syntactics. In the context of fluid dynamics, Berton (1990) compared advantages of the red–blue color set, a value range of a single hue, and what is termed a banded palette. The banded color palette (designed to create maximum contrast between adjacent hues), is cited as particularly good at highlighting "artifacts in the interior of the dataset." The banded palette provides no cues for early vision to sort out order in the data. As one among several interactively selectable palettes, however, it can complement a simple color-value scheme by highlighting the shape of features at various places in the data range.

In addition to establishing dynamic color syntactics for GVIS, the nature of the environment (under user rather than cartographer control) suggests a need for tools to help users see the structure imposed on their data by particular color choices. An approach similar to that developed by Brewer (1989) in the context of manual color selection on printed maps is needed. The Tektronix *Color-picker*™ is a step in that direction, but, like Brewer's system for printed maps, it requires a color to be selected individually for each display category. To be effective for exploratory GVIS, a color selection tool must allow selection of an entire scheme, and it should provide feedback to the analyst about how colors in the scheme are related.

One attempt to link theories of color vision, strategies for structuring color for information display, interactive color selection tools, and feedback to users on the color structure selected is a tool called Calico developed by Rheingans and Tebbs (1990). Their system deals with selecting sequences of colors from a *color space* defined as a three-dimensional solid. Any color space model (e.g., RGB, HLS, CIE LUV, etc.) can be used (HLS was selected for their initial map example). The basis of color selection in Calico is definition of a one-dimensional curve, a *color path*, through the color space. Rheingans's (1992) own review of how Calico might be used for color selection on a map of median family income in the United States provides the best (nonvisual, noncolor) illustration of how their system works:

> As the path curves through the color space it completely describes the sequence of colors used in a mapping from a set of values of a single scalar variable to a set of colors ($f(s)$,s = distance along the path). For

example, if median family income for U.S. counties is mapped to a combination of hue and lightness using a rainbow scale in the HLS model, the color path runs through the hues in an ascending spiral from black to white.... Counties with a low median family income would be displayed in dark reds, those with an average median income in medium greens and blues, and those with a very high median income in pale purples. (p. 253)

Although the particular color choices described do not match with general cartographic guidelines or Brewer's (1994) more precise color syntactics, the example does produce a logical order due to changing value. More importantly, the color selection described is one of a vast range that the analyst can choose among. Calico, or systems like it, should not be evaluated in terms of a single color assignment, but in terms of the tools it provides for manipulating the color assignment and giving feedback on that manipulation.

Users of Calico see the color space they are working with on the display (represented by colored sample patches) along with their map. When they fit a color path to the color space, they see the path's direction and shape through the color space. As a color path is selected, its corresponding impact on the map is seen immediately. The color paths are implemented as spline curves through a set of control points. Parameters of the spline equation can be tied to slider bars. One example given is for saturation attached to a slider bar. As the analyst moves the slider to full saturation, colors selected (and used on the map) are fully saturated colors of varied value spiraling around the outside shell of the color space. As the slider is moved toward zero saturation, the spiral gradually moves in toward the central axis of the color space, ultimately straightening to a line specifying a value range of gray tones. An additional feature of the Calico interface is that it allows users to manipulate individual control points defining the spline curves as well as the curves as a whole. This kind of manipulation is likely to be particularly helpful in enhancing ranges within a data set where the analyst needs a bit more contrast in order to explore a pattern.

THE ROLE OF CATEGORIES AND SCHEMATA

Whatever the tools used to enhance features, visual evidence will be interpreted differently if approached with different schemata. For GVIS, new schemata are likely to be needed, both for system users and for their designers. From a designer perspective, Friedhoff and Benzon (1989, p. 84) draw an analogy between what they call visual schemata and computer algorithms for three-dimensional representation: "As with schemata, a successful algorithm becomes part of an ever expanding vocabulary among computer graphicists and profoundly influences both the conception and appearance of their images." This view seems to match well with Margolis's (1987) arguments concerning the role of pattern in science as a whole. Schemata, then, are like computer algorithms in that they provide a structure that controls possibilities and creates commonality among unique situations.

When designing maps or graphs to communicate, we should strive toward depictions that allow existing schemata to be used in the most automatic way (Pinker, 1990). For exploratory visualization, in contrast, Ganter and I argue that the goal should be to "permit, indeed perhaps demand" that users are given multiple perspectives on information (MacEachren and Ganter, 1990). Based on the overall thrust of this book, I contend that visualization tools can only prompt dramatic insight when they succeed in breaking through existing schemata so that new perspectives are brought to bear on a problem. Design of effective GVIS environments will depend upon understanding the schemata that a viewer or analyst brings to the display-reading process. This is the case whether the goal is to match with a schema to facilitate recognition of expected patterns or to change the schema so that alternative patterns can be noticed.

A key issue to consider in the context of GVIS is that there will be significant differences in schemata available to domain specialists (experts) versus novices. That application of appropriate schemata requires learning and practice is emphasized by Giere (1988). To make the point, Giere presents one example of Wason's *selection problems* in which subjects are provided with a set of partial information and they must select from a set of choices those pieces of further information that are essential to solve a particular problem (Figure 8.14). Giere (1988, p. 177) contends that failures on selection tasks such as these are not due to deficiencies in

a

b

c

d

FIGURE 8.14. Giere's depiction of one of Wason's selection tasks. Subjects see four cards with half of each hidden (the lined areas). They must determine the card(s) which it is absolutely necessary to see in order to determine the truth of the statement: If there is a circle on the left, there is a circle on the right. Typically, few subjects answer correctly that cards *a* and *d* must be uncovered. *After Giere (1988, Fig. 6.8, p. 177). Copyright 1988 by the University of Chicago Press. Adapted by permission.*

reasoning ability, but are instead "indicators of ignorance of the most appropriate models [schemata] for the situation." This view is nearly identical to that proffered by Margolis (1987) in discussing similar Wason selection problems. It is empirically supported by Cheng and Holyoak (1985) who demonstrate that it is possible to dramatically improve performance on selection tasks by prompting subjects to adopt particular general schemata (that they term *pragmatic reasoning schemata*).

Some problems, of course, require learned domain-specific schemata. Edwards (1991) provides an example and presents an argument about how domain-specific schemata are activated.[14] In a limited empirical study, Edwards applied an image segmentation algorithm to a remotely sensed scene, then asked a pair of individuals to describe two small sections of the result (Figure 8.15). One of the individuals "knew" the image well, the other did not. Edwards found that the interpreter who knew the image applied a more analytical approach emphasizing the "agricultural landscape," while the other interpreter's description seemed to derive from associative cues and emphasized analysis of the scene as "a drawing" rather than a landscape. Edwards contends that the ability to take the landscape-analysis approach depends upon being able to access labels for the features from memory. Experienced map readers are said to have "a whole dictionary of names for grouped entities which he or she carries around her head, allowing her to read and understand maps more quickly and effectively" (Edwards, 1991, p. 298). Knowledgeable map reading is thought, by Edwards, to differ from interpretation of abstract drawings primarily as a result of the "vocabulary" available for labeling complex entities. For the experienced map reader, an extensive vocabulary exists that defines and labels complex entities. For abstract drawings, interpreters "appear to chunk visual information into entities which are named through association with a set of prototype objects of known shape characteristics" (Edwards, 1991, p. 300).

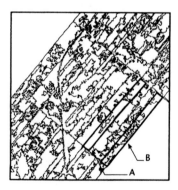

FIGURE 8.15. The automated scene segmentation result from Edwards's experiment (derived from a synthetic aperture radar image acquired over an agricultural site near Drummondville, Québec). The test regions are highlighted and labeled. *After Edwards (1991, Fig. 3, p. 302). Adapted by permission of Kluwer Academic Publishers.*

As discussed in Chapter 6, Chang et al. (1985) proposed that expert and novice topographic map readers differ in the number of learned patterns (or categories of landforms) they have at their disposal. It would follow from Edward's arguments that learning of these landform patterns will not happen in the absence of labels to attach to each distinct pattern. In addition, Chang et al. link pattern expertise to more general pattern-recognition aptitude. This finding suggests that some analysts will benefit more from GVIS tools than others.

Chang et al. (1985) related their concepts of knowledge schemata used in topographic map interpretation to empirical results of a map-reading task. Eye-movement analysis was used to evaluate differences in how expert and novice map readers approached the task of studying a map to determine patterns of absolute and relative height. As expected, the experts performed better on questions requiring them to locate high and low points on each of 10 test maps as well as tasks requiring them to determine which of two map locations was higher. For maps with a definite landform pattern, experienced readers had more fixations, concentrated on the high and low points of the map, and had fixations that were shorter in duration. These findings were interpreted to mean that experts had a better schema which allowed them to quickly convert the 2-D map to a 3-D mental representation. They contend that experts had shorter fixations and were better at height judgments because they could process larger "chunks" of information (i.e., patterns of contours) in the available time, and therefore could more easily integrate information into their schema. For the experienced reader, visual search was apparently guided by "familiar patterns of contour lines" acquired through experience. An interesting finding was that expertise seemed to get in the way for interpretation of maps that did not have a clearly identifiable landform pattern, that is, landscapes for which the experts had no stored pattern match. As Chang et al. point out, this result matches with that of Chase and Simon (1973) for chess experts, who could recall coherent (i.e., typical) chess-piece arrangements far better than novices but were equal to novices in memory for random arrangements.

Where Chang et al. (1985) concentrate on expert–novice differences in schemata that facilitate spatial feature recognition on topographic maps, Crampton (1992a) considered expert–novice differences in "mental models" that guide the *process* of wayfinding based on information presented on orienteering maps (a specialized topographic map). Mental models, as defined by Crampton, are a combination of mental representations and procedures for problem solving based on those representations. They can, therefore, be considered to be a schemata complex dominated by subordinate event schemata. Crampton's empirical evidence deals with map-based navigation through the environment, not with GVIS. It seems particularly relevant here, however, because understanding what constitutes a successful event schema for wayfinding with maps in the environment is likely to provide some clues about how to best structure the process of navigation through geographic data spaces accessed via GVIS.[15] The similarity between environmental and database navigation is also likely to increase as visualization tools follow a trend toward Cyberspace (artificial computer information environments within which an analyst can travel, manipulate information entities, etc.).

Using protocol analysis, Crampton (1992a) compared seven expert orienteers with seven novices, all of whom were provided with an orienteering map and presented with the task of explaining how they would go about negotiating a route between a specified origin and a destination. Experts (compared with novices) were found to have a refined ability to "abstract terrain from the map" and to talk about a wide range of morphological features. The latter finding seems to lend further support to Edwards's contention that the ability to label features and spatial patterns is a critical component of being an expert. Labeling, of course, requires categorization and the ability to assign what is encountered to a category.

Expert wayfinders were also found to differ from novices in sharing two main strategies related to the process of spatial problem solving. They were adept at "attacking," a method in which an easily found location near the goal destination is identified and used as a subgoal which, once achieved, leaves the task of locating the final goal easy. None of the novices employed the strategy of attacking. In addition, experts used an active self-analysis/error-prevention strategy that Crampton refers to as "meta-wayfinding." Meta-wayfinding is analogous to "reasoning why" in the feature identification model of GVIS described above. Crampton's expert wayfinders adopted an iterative strategy of quickly achieving approximate answers followed by self-evaluation, with progressive repetition to focus in on a solution. Experts were able to quickly distinguish between promising routes and unsuitable routes and could "switch on" precise terrain inference at critical stages. This process suggests a hierarchical schemata that initially facilitates global pattern seeking but provides links to lower level subschemata that deal with making detailed decisions.

Based on his analysis, Crampton (1992a, p. 63) concludes that "expert wayfinding is 'enabling' and seeks good routes, performs plenty of self-monitoring and error-prevention, and depends on finding a decent attack point with which to approach the destination in some detail." Successful data exploration is likely to follow a similar model. GVIS environments that facilitate this kind of iterative narrowing in on a problem (perhaps through scale and resolution controls on space, attributes, and time) should be investigated, as should the role of "attacking" as a strategy in non-wayfinding problems. As should be clear from Crampton's work with expert wayfinders, however, success of a GVIS environment will be as dependent upon the schemata brought to analysis by the user as upon tools that we might incorporate in the system.

In an empirical study that comes much closer to dealing with the role of expert versus novice schemata in GVIS, McGuinness et al. (1993) compared strategies of information used by experienced and inexperienced GIS users. In one of two scenarios, 18 subjects were presented with a database of nine variables relevant to machine peat cutting in the Sperrin Mountains in County Tyronne, Northern Ireland (the subjects all resided in Northern Ireland). Variables were available as layers in ARC-PLOT and included incidence of cutting, type of peat, contours, roads, rivers, rainline, rain stations, areas of natural beauty/scientific interest, and density of wading birds. Subjects were asked to use the database to consider how environmental and social variables were related to the incidence of machine cutting, giving particular attention to potential conflicts between economic and environmental issues. Subjects participated for two to three hours overall, with the peat scenario (described here) first. They were free to plot and replot any individual or combinations of the nine variables during the session.

Experts made use of most of the available information, at least at a cursory level, with all experts looking at either eight or all nine data layers. Experts, however, considered fewer layers at the same time (an average of only 2.2) than did novices (with a mean of 2.6 and one novice who overlaid all nine variables at the same time). While experts looked at fewer variables simultaneously, they were much more likely to go back and re-examine combinations (an average of 30 vs. 9 re-examinations for experts vs. novices, respectively). These findings seem to be in keeping with the idea that a successful procedural schema for spatial problem solving involves an interactive process of setting subgoals and narrowing in on a solution through a process of seeing potential solutions and then applying a self-assessment procedure (engaging in reasoning why) to determine whether they are viable. From the task performance data and the summaries subjects were asked to write, there is clear evidence that experts obtained more information and were able to provide more adequate assessments of the situation. Experts made more frequent reference to spatial relationships, and protocol analysis suggests that they approached the task in a more systematic and structured way. "Experts were more likely to make systematic searches through the databases and to explore them in a planned and careful way . . . ; they consistently tested hypotheses, checked their interpretations . . . and reached comprehensive conclusions" (McGuinness et al., 1993, p. 484). Overall, it seems clear that the experts imposed domain-specific schemata that allowed them to categorize what they saw and that provided a structure for organizing the process of data exploration. Such schemata were apparently unavailable to the novices who tended to engage in more local and superficial hypothesis testing.

The few studies of expert–novice differences in map interpretation suggest significant differences in conceptual categories that experts bring to a problem context and schemata they can apply to organize these categories and/or the process of exploring data. If a potential GVIS user can be no more precise concerning landform types than to distinguish among mountain, valley, and plane, they are unlikely to recognize "glacial till" unless the system provides them with visual prototypes to match against observations. Similarly, unless an analyst can rely on an appropriate event schema for logically structuring analysis, the GVIS session may become a purposeless random walk through the data that results in feature identification only through chance. We have an opportunity to facilitate visualization tasks (particularly for nonexperts) by making systematic analysis the default option.

CONCLUSIONS

This chapter has presented a selection of GVIS issues for which our multiple-representation approach to maps can be evaluated. An elaborated cognitive model of GVIS as a process of human–map interaction is presented. In this context, attention was directed to the ways in which principles of perceptual organization, attention, and categorization relate to functional representation concepts of map syntactics. The ways in which schemata and mental categories can control what we "see" through GVIS tools were then explored. Together the cognitive and semiotic structure provided in Parts I and II seems to offer a framework for extending cartography from emphasis on static displays to dynamic information-processing tools. We will continue this foray into interactive analytical aspects of GVIS in the next chapter, where our emphasis moves to looking for relationships across attributes in space, time, or space–time.

NOTES

1. Although Margolis chose the term "pattern" over schema, it is clear that his concept of pattern was similar to that of schema, as developed in Part I of this book. Margolis, in fact, lists patterns, frames, schemata, and attunements as equivalent terms.

2. Letting knowledge overcome what we see is, of course, not always bad. Situations in which visual evidence should be carefully considered in relation to other sources of knowledge, theoretical constructs, and so on, will be taken up in more detail in the final chapter.

3. Within cartography there has been considerable confusion of terms related to different kinds of projection of three-dimensions onto two. Lobeck (1958) uses the terms "isometric diagram" and "perspective diagram" (distinguishing between depictions in which scale is or is not commensurable). Jenks and Brown (1966) use "parallel perspective" and "angular perspective." These terms are particularly confusing because what they call "parallel perspective" does not involve any perspective depth cues, and in art, the term "parallel perspective" is used to mean one-point perspective (Lobeck's isometric diagram). I have opted for "isometric" and "perspective projection" here because these terms both imply a process of projecting three dimensions to two, while making clear that the isometric version does not involve perspective as it has been defined in Chapter 3.

4. This sensation is enhanced when the scene is projected to a large curved screen of the sort in the Smithsonian Air and Space Museum, where I had an opportunity to view the Venus flyby produced by JPL.

5. Based on rather disappointing experience with blinking, Mark Monmonier (personal communication, November 1993) suggested that there are better alternatives. Specifically, "Any temporary, prominent visual contrast might well be sufficient, especially if under the user's control."

6. Stereoscopic analysis, of course, requires specialized hardware. Some techniques, such as using a binocular stereoscope, do not work well for all people.

7. As noted above, temporal frequency at which static patterns are blinked can result in the changing pattern being seen in depth behind a stable pattern. It seems unlikely, however, that quivering (because it includes not just change, but movement) will result in anything other than the moving elements being seen as figure.

8. Spatial and space–time data structures that support an effectively seamless transition among geographic scales are a major research effort within GIS. Although resolution change can probably be implemented with current technology, significant scale change will require further developments in both data structures and cartographic generalization. It is unlikely that the near future will provide storage capacity to retain multiple representations of the world at all potentially required scales. As a result, attempts are being made to formalize the steps of cartographic generalization to the point that large-scale databases can be successfully adapted to use at a range of scales.

9. One area of geographic research in which both resolution and crispness have been given considerable attention is image analysis (where the latter is referred to in terms of the process by which crispness is adjusted—spatial filtering).

10. As I learned when trying to illustrate this argument in a graduate seminar, simply seeing the image through the back of the page (when I was passing out copies) was enough to make it immediately identifiable.

11. Although he did not implement the idea as an interactively control-

lable feature, Monmonier's (1992) prototype graphic scripts include sequences in which the temporal interval used for data aggregation is systematically increased.

12. See Chapter 10 for detailed consideration of this issue.

13. See Hoffman et al. (1993) for a discussion of the problems of full-spectral colors in the context of meteorological displays.

14. Although Edwards does not use the term "schema," he discusses hierarchical structures in which chunks and their relation are organized into a framework that guides the interpretation of visual scenes.

15. Crampton (1992a) suggests as much in a comment about potential application of his findings for teaching expert database navigation.

CHAPTER NINE

GVIS

RELATIONSHIPS IN SPACE AND TIME

In the previous chapter, a model for GVIS-based feature identification was presented along with a variety of issues related to feature-enhancement tools. For the most part, discussion was limited to recognition/noticing of individual features. Most applications of GVIS involve spatial comparisons across multiple attributes (feature comparison) or multiple times (space–time feature analysis). A goal for GVIS in this context is to enhance the likelihood that an analyst will see not only features, but relationships among features. From the perspective of tool development what is needed are tools that take advantage of human vision, our propensity to categorize, and standard schemata for judging similarity and difference along with a semiotically robust sign system that facilitates the perceptual/cognitive processes involved. In this chapter, the GVIS approach outlined above is extended to deal with these more complex tasks.

FEATURE COMPARISON: LOOKING FOR RELATIONSHIPS IN MULTIDIMENSIONAL DATA

Most of the feature-identification techniques discussed in Chapter 8 have been extended to feature comparison, and several (e.g., the EXVIS iconographic display system) were developed specifically for feature comparison with multidimensional data. The basic GVIS principles (that have been proposed and/or tested thus far) were outlined above in the context of schema theory and a model of how an expert will approach a visualization session. The goal of this section will be to highlight a few of the attempts to build tools that facilitate feature comparison. Efforts that draw upon concepts of perceptual organization and attention, as well as those that are grounded in semiotic principles, are considered.

Discussion is divided on the basis of which representation device, tool, or variable is emphasized in the method (with space, orientation, color, time, focusing, and sound considered separately). It should be made clear from the outset, however, that the division is for organizational convenience only and that most efforts have incorporated two or more of the methods discussed here.

Space

Both Bertin (1977/1981) and Tufte (1983) advocate use of "small multiples" over composite maps as a way to represent multiple variables (see Figure 2.11 for an example). There is little doubt that selective attention will be superior for individual variables represented spatially on separate maps. What is as yet untested is whether variable conjunctions can be derived by mentally merging patterns seen on two or more maps.

Small multiples represent an extension of simple side-by-side comparison of map pairs as a method to investigate relationships. Cartographic research dealing with map comparison supports some scientists' resistance of small multiples (see below). Taken as a whole, map comparison research suggests that vision is not particularly well suited to judging spatial correspondence between two variables represented on side-by-side maps (Olson, 1972; Muehrcke, 1973; Lloyd and Steinke, 1976; Steinke and Lloyd, 1983b). Most of this research has focused on comparison of maps using darkness and/or hue of area fills as the symbolization technique. Bertin (1977/1981), in contrast, argues for reliance on space alone as the sign-vehicle (with location of symbols in space or the amount of space they cover allowed to vary, but not both at once). He contends that perception of pattern is immediate for identical sign-vehicles arranged in space (such as those in Figure 2.11). Although he does not cite any psychological research, this contention is consistent with psychological research on selective and divided attention (see Chapter 3). Varying sign-vehicle size at fixed locations, however, is comparable to varying color value at fixed locations as on a choropleth map, in that both involve a variable conjunction (location with size or value). While the locations of all sign-vehicles of the same size or color value can be visually grouped (and these groups compared from map to map), vision will resist visually grouping all color values or all sizes in a particular region. Bertin, in fact, points out that neither size nor value are associative variables—that we cannot attend to all sizes (or values) at once. Comparing regions from map to map (or entire maps) will, therefore, not be an immediate visual process (Figure 9.1).

Although interpretation may not be immediate, we might expect viewers to be more successful in judging map similarity using size or color value variation at fixed locations than they will be at comparing isarithmic maps (in which size, shape, and orientation of features will vary). Also, a map with five size or color value categories contains information that would require five small multiples if we restrict the representation to a single repeated sign-vehicle. It seems, therefore, that the ability to identify variable relationships should be compared for small multiples using only location of a single sign-vehicle, small multiples in which one additional graphic variable is allowed to vary, and composite maps in which dynamically adjustable graphic variable conjunctions are used in an effort to visually sort out the individual variables and their combinations (e.g., the EXVIS icons in which up to 20 variables can be combined).

Bertin (1977/1981) derived a method of matrix manipulation designed to help find pattern in any set of variables that could be cross-tabulated. The method was a precomputer interactive visualization technique in which rows in the matrix were printed on cards that could be reordered by hand. Bertin demonstrated how the technique could be extended to geographic visualization (although he did not use this term) by iteratively changing the row–column location of small multiples arranged in matrix form. The goal of this rearrangement was to group maps according to dominant spatial pattern (Figure 9.2).[1] Clearly, given the flexibility offered by computer processing, Bertin's matrix-analysis procedures provide a potentially powerful strategy for making use of small multiples in a GVIS environment. It seems particularly well suited to analysis of data sets with large numbers of variables (probably a minimum of 16 for a 4 × 4 matrix).

Despite the claims of Bertin and Tufte for their utility, scientists in Pennsylvania State University's Earth System Science Center (ESSC) rejected small multiples when some colleagues, students, and I began to explore problems of multivariate analysis of global model-derived data. They demanded composite maps in their GVIS tools. They seemed to feel that pairwise comparison is too difficult and imprecise (and the number of variables to be considered initially did not approach the threshold of 16 at which the advantages of small multiples begin to be obvious). What the ESSC scientists seemed to want was a technique by which two or three variables could be superimposed while retaining the ability to examine each variable individually.

Our research group devised a trivariate display system that uses spa-

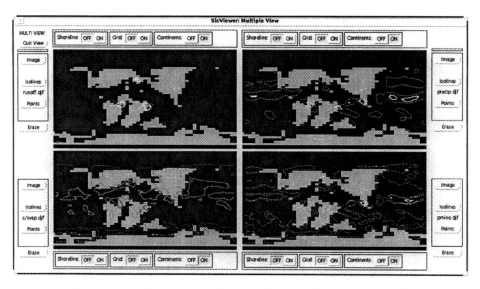

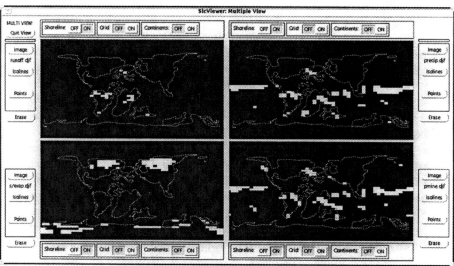

FIGURE 9.1. Sample screens from a GVIS toolkit (named SLICViewer) designed to facilitate analysis of data generated by a global climate model. The top panel depicts a screen from the SLICViewer interface. Four variables from a global climatological model (runoff, precipitation, evapotranspiration, and precipitation minus evapotranspiration for mid-Cretaceous summer) are represented with isolines. It takes a good deal of effort to sort out relationships among these relatively simple four-category maps (the original is, of course, larger and in color, thus making the task somewhat less difficult than it might seem here). The bottom panel depicts one method to interactively focus on particular relationships. If we reduce the visual processing task to that of using space alone, by representing a single category (runoff is shown here) with small multiples (with cells above a specified value highlighted), pattern comparison becomes a virtually automatic process.

tial dimensionality as a way to separate the different variables (DiBiase et al., 1994b). Data to be represented was model-derived for grid cells in a global matrix. This regular arrangement of data (and the techniques available in IMSL/IDL, the visualization development environment at our disposal) led to image (shaded grid cell) maps as one logical data representation. Rather than trying to come up with a trivariate color scheme that would signify individual distributions and their relationships, we decided on a system in which one variable was matched with color value on image maps (an area symbol), a second was matched with isolines (a line symbol), and a third was matched to graduated point symbols (a point symbol) (Figure 9.3). The image maps were produced with gray tones so that the point and line symbolization could be superimposed in other hues. Isolines made use of the concept of weighted isolines in which isoline width is scaled to data values, thus eliminating the need for labels and producing a perceptually salient cue to location of high and low values (DiBiase et al., 1994a). The data matched to dots were treated as diverging (above and below average) with the positive–negative regions coded as red and blue, respectively.

We, as yet, have done no empirical evaluation of spatial dimensionality as a trivariate mapping method. Initial reaction from the scientists the system was designed for, however, has been positive. The discipline specialists working with us on system design (two graduate students) seem to rely most heavily on isolines, probably the least "stable" symbolization in the system.[2] This tendency implies a dominant schemata complex, among paleoclimate researchers, adapted to isolines. That such a dominance exists is supported by an examination of graphics in the *Journal of Climate*. DiBiase et al. (1994a) examined all graphics in a single volume of that journal and found that more than half, 1,435, used isolines. The tendency to rely on isolines also suggests the lack of appropriate domain-specific schemata for dealing with the alternative area and point symbolization. Effective use of symbol dimensionality as a trivariate mapping tool, therefore, will probably require a learning phase in which the needed schemata can be developed. Use of similar displays in related applications may, however, signal a willingness to learn new symbols in an effort to observe increasingly complex relations (Treinish, 1993).

Plans for the future include empirical comparison of spatial dimensionality, tridimensional color coding of grid cell maps, and EXVIS-like linear point symbols that rely on orientation change (see the section below). In addition, it would be useful to evaluate possible extension of the spatial-dimensionality approach to four variables by projecting the current trivariate representations onto a perspective (or stereo) view, the height of which represents a fourth variable. In all such empirical evaluation, care must be taken not to rely exclusively on positive or negative re-

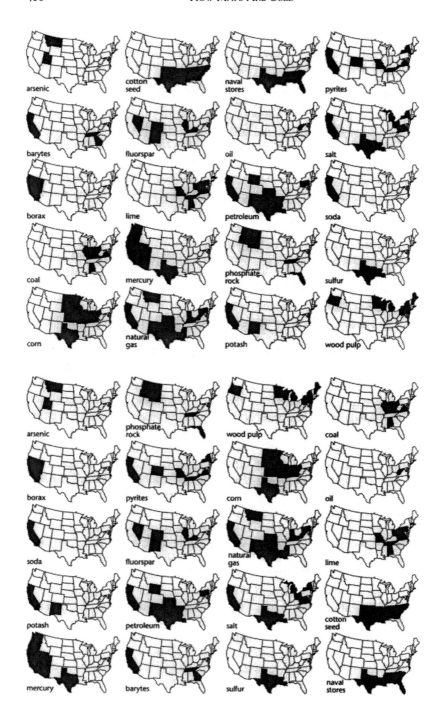

sults by subjects who have little experience with a particular symbolization type. Some experience is needed with any symbolization (even those matched well to general visual-processing capacities) in order to develop an interpretation schema that can link the visual input to domain-specific knowledge.

Monmonier (1991b) presents a somewhat more abstract use of space as a direct tool in seeing geographic relationships that he terms the *geographic biplot*. A standard biplot is a two- or three-dimensional information graphic used to depict the structure of a multidimensional data set. It is typically based on a principal-components analysis, with the two components that explain the most variation defining the plot's axes. On a biplot, locations near one another are similar and those apart are different in terms of the components on which the biplot is constructed. On a geographic biplot, the eigenvector coefficients for the variables used in the principal-components analysis can be treated as coordinates for plotting location of these variables in the principal-component space. In the example Monmonier provides, six socioeconomic variables (metropolitan population, black population, persons younger than 18 years, persons 65 years or older, population change for 1985–1988, and population density) are reduced to two components. Component scores for geographic regions on the two main components can be used as coordinates for plotting regions in this same space. The result is that display space is used to express the relationship between geographic location and socioeconomic attributes. For data in which more geographic locations are represented, labels would make the biplot confusing and it would be difficult to mentally link biplot points to geographic positions. Monmonier suggests replacing the labels with a rectangular or circular "brush" that can highlight locations on the biplot and their corresponding locations on an adjacent map (Figure 9.4). In addition, he illustrates the potential use of class-sequenced linking in which states with data values in each category are highlighted on both the map and the biplot.

As Monmonier points out, proximity on the biplot can reveal relationships such as whether or not groups on the biplot are spatially groupd or dispersed. Separation on the plots makes differences obvious. A biplot,

---

FIGURE 9.2. The unstructured small multiples from Bertin's analysis of raw materials used by the U.S. chemical industry is shown in the set of maps at the top. The result following matrix manipulation is shown in the bottom set of maps. This arrangement organizes raw material maps geographically with north-dominant distributions in the top row grading to southerly distributions in the bottom row. Similarly, distributions favoring the west are located in left-hand columns with those favoring the east at the right. *After Bertin (1977/1981, Figs. 3 and 4, pp. 158–159). Adapted from* Graphics and Graphic Information Processing *by permission of Walter de Gruyter & Co.*

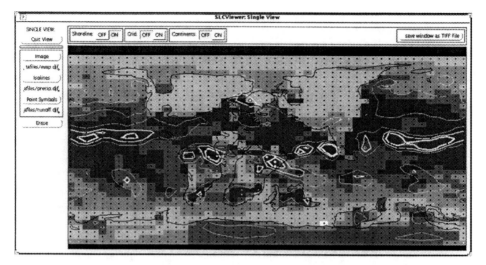

FIGURE 9.3. Use of spatial dimensionality to simultaneously represent three variables from Figure 9.1. Evapotranspiration data are represented with area symbols (as a gray scale image), precipitation is represented with weighted isolines (in white), and runoff is represented with point symbols scaled by magnitude (again, white above the mean and black below).

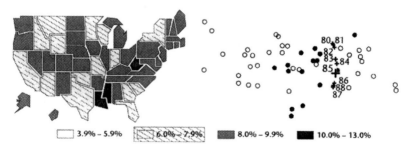

FIGURE 9.4. A geographic biplot for yearly unemployment and mean annual rates by state (1980–1987). Representations on the map, scatterplot, and legend are dynamically linked. In the example, a legend category has been selected, resulting in the corresponding states and scatterplot points being highlighted. *After Monmonier (1991b, Fig. 9, p. 46). Adapted by permission of the* Journal of the Pennsylvania Academy of Science.

because it is based on the indispensable variable of space, should result in similar entities grouping immediately in vision. A simple spatial schema is all that is required to interpret the *near = similar* logic of the plots. Conceptually, however, the biplot is quite abstract. A fairly complex domain-specific schema is likely to be required in order for the analyst to determine what "similar" means.

Orientation

As discussed in Chapters 2 and 3, orientation is a powerful variable in contexts where multiple data points are arrayed in space and the goal is to identify spatial patterns. GVIS tools making use of orientation have been suggested by Carr et al. (1992) and Dorling (1992b). In both cases orientation of lines connected to data locations was linked to a key variable. Carr et al. used symbol conjunctions in which pairs of lines representing variable pairs were joined at the data location (see figure 3.39). This symbolization should result in both individual attributes and their relationships being easily extracted from the display (see Chapter 3 for more details).

The most sophisticated application of orientation to multivariate analysis is the EXVIS visualization environment described in Chapter 8 (Grinstein et al., 1992). The system makes use of the powerful human capacity to notice orientation (direction) differences. A critical advantage of the EXVIS system over other geographic applications of orientation as a variable for multivariate spatial analysis is the interactivity built into the system. The designers contend that this interactivity has been a key to successful exploratory data visualization. The system is now implemented on a massively parallel processor, a Thinking Machines Corporation CM-2 Connection Machine to make real-time interaction with these complex displays possible.

In EXVIS, each data element is assigned to a position in display space, with one variable mapped to X and another to Y location (space can, of course, be mapped to these spatial dimensions in the case of geographic data). At each location, up to 15 additional attributes can be represented by one of several five-limbed stick figures (Figure 9.5). If color hue is used, additional variables are at least representationally possible (if not perceptually and cognitively functional). Patterns produced (even with only 5–7 variables rather than 17) are necessarily complex. Stick-figure parameters that highlight relationships in one application are not always the best for other applications. The analyst needs freedom to dynamically alter parameters until an apparent pattern emerges (if one does), then to examine characteristics of this pattern in more detail to see

FIGURE 9.5. The 12 standard EXVIS stick figures. These figures vary in base limb location (shown with heavy lines) and the way limbs are joined (with two or three joins possible at each connection point). The slope, length, and thickness of each limb can be matched to a variable, which together with X and Y position yields 17 variables potentially represented. Only one of the 12 stick figure forms is used on any one display. *After Grinstein et al. (1992, Fig. 2, p. 639). Adapted by permission of John Wiley & Sons, Inc., from* International Journal of Intelligent Systems. *Copyright 1992 by John Wiley & Sons, Inc.*

if they are real and what might be responsible for them. Not only does the EXVIS system allow the analyst to alter the mapping of variables to EXVIS icon attributes (e.g., different variables can be matched to X and Y and/or to the icon limbs), but several global controls are available (Smith et al., 1991). These include:

1. *Orientation of the base limb*: This parameter can be related to evidence that horizontal and vertical orientation tend to dominate human perception.[3] By rotating different sets of limbs (representing different variables) to the horizontal and vertical positions, subtle changes in emphasis can be achieved that sometimes result in a previously hidden relationship "popping out" of the display.

2. *Bounding box size*: Bounding box size controls the apparent density of icons on a display. In a GVIS application, this size could correspond to the spatial resolution at which data are examined. In EXVIS, this box size is linked to size of icons via specification of the range in limb length allowed. If the resulting icons are small relative to the box size, individual details of each icon will be discriminable, but overall patterns of relationships will be less noticeable (i.e., local detail will dominate global pattern). On the other hand, if icons fill the box, an overall surface texture will be generated, thus emphasizing global over local relationships.

3. *Limb length*: Although the system allows data to be mapped to limb length, the scaling factor that determines maximum length (or length of all limbs if this variable is not used to represent data) is under user control. As indicated in paragraph 2, the interaction of global limb length and bounding box size determines the relative emphasis on global versus local patterns. Manipulation of maximum length of individual limbs would allow specific variables to be emphasized or de-emphasized (but it is not clear whether EXVIS incorporates this particular global control feature).

4. *Jitter*: As found by Papathomas and Julesz (1988), regular spacing

of regular sign-vehicles can result in features (and relationships) being identified that are a function of the regularity of display method rather than of the data characteristics being represented. In an effort to check for potentially false patterns, users are given a global control over "jitter" (the amount of randomness imposed in the mapping from data to X–Y location).

5. *Limb angle*: By default, EXVIS maps the range of any variable to a 180° span of icon limb orientation with an alternation between clockwise and counterclockwise directions for subsequent variables. The maximum limit of 180° (rather than 360°) is imposed because human vision cannot distinguish between orientation of pattern elements that differ by 180° from each other (see Figure 3.34). The user can select consistent direction for angular change from limb to limb and limit the maximum change to less than 180°.

Smith et al. (1991) provide some compelling examples of the power of the EXVIS orientation-based approach to representation of relationships in multivariate information. They generated test patterns having predetermined spatial relationships that are not apparent with typical monochrome representation of the data. In one example, they demonstrate that two high-resolution gridded displays (similar in appearance to many remotely sensed images or GIS-produced maps) can have perfect positive correlation in one region and negative correlation in another with no perceptible difference in the regions (if the displays are compared side by side). When the variables are mapped to a single EXVIS display, however, the region boundary is immediately obvious. For two variables, this effect could also be achieved with color hue (if colors were carefully selected using one of Brewer's [1994] diverging and bivariate schemes). For three variables, the ability of color to highlight regions of positive and negative association becomes questionable. Smith et al. provide a second example in which three distributions are generated for which the top half of the display has paired correlations of approximately +0.5 and the bottom half has correlations of –0.5. Standard gray tone representations of the three variables does not even hint at these relationships (Figure 9.6). The default EXVIS display does no better (Figure 9.7). Global change of limb angle parameters, however, causes the previously hidden relationship to emerge (Figure 9.8).

Color

Assigning different colors to data variables in a multidimensional display is an easy thing to do. As a result, most multivariate efforts in GVIS (and scientific visualization as a whole) have attempted to make use of color.

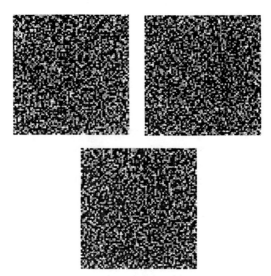

FIGURE 9.6. The three sample distributions generated to produce the +0.5 and −0.5 paired correlations in top and bottom halves, respectively. *Reproduced from Smith et al. (1991, Fig. 7, p. 200). Reprinted by permission of the Society of Photo-Optical Instrumentation Engineers (SPIE) and the author.*

FIGURE 9.7. The default iconographic composite of the three images from the previous figure. The difference between top and bottom halves of the figure is no more evident than in the gray tone displays of Figure 9.6. *Reproduced from Smith et al. (1991, Fig. 9, p. 202). Reprinted by permission of the Society of Photo-Optical Instrumentation Engineers (SPIE) and the author.*

FIGURE 9.8. In this display, angular change from limb to limb is all in the same direction with the maximum change set at 45°. The distribution between top and bottom halves is now visible. *Reproduced from Smith et al. (1991, Fig. 9, p. 202). Reprinted by permission of the Society of Photo-Optical Instrumentation Engineers (SPIE) and the author.*

That observers can sort out at least three hues in a complex display of positional sign-vehicles was demonstrated cartographically by Rogers and Groop (1981) (in a noninteractive context). They had subjects perform a region delineation task. Their subjects did as well (actually slightly better) at extracting univariate regions from a three-color dot map as at extracting the same regions on three univariate maps. When asked to identify regions of commonality for two or three variables, subjects were again as successful with the tricolor map as with side-by-side black-and-white maps.

Designers of the EXVIS system have attempted to extend their original model by adding both color and sound as additional attributes assignable to the stick-figure icons. The Rogers and Groop results suggest that up to three additional variables might be practical on an EXVIS display while preserving an ability to "see" individual variables as well as their combinations. On both the Rogers and Groop dot maps and the EXVIS iconographic displays, relationships among variables become visible due to spatial proximity of sign-vehicles with contrasting hues. Proximity in space is the primary visual cue, but the task is a conjunction task because the viewer must try to attend to proximity and hue contrast at the same time. Two issues that were not addressed by Rogers and Groop are the impact of color hue choice on this kind of regionalization task and whether regions on the color maps were emergent for their subjects or simply possible to delineate through careful point-by-point analysis.

Bertin (1977/1981) offers an opinion relevant to both questions. He argues that human vision can achieve nearly instantaneous synthesis of a trichromatic map in which three variables, represented by regularly arranged dots of varied size (Figure 9.9), are printed on top of each other using cyan, magenta, and yellow ink. He contends, however, that the map will not work given any change from the three process printing colors. Brewer (1994), in the context of her overall conceptual system of color matching, agrees with Bertin on hue choices for trichromatic printed maps. Other starting colors would result in less than a full spectrum of possibilities and locations at which all variables correspond would be a hue rather than black. Bertin does not consider the implications for his trichromatic approach if colors are displayed rather than printed. In a computer display environment, for example, Bertin's approach should probably be modified to use the additive primaries of red, green, and blue. A problem that arises in this context, however, is that blue and green are not very discriminable.

Unlike dot maps or an iconographic display (also constructed from discrete positional sign-vehicles), many distribution maps represent contiguous areas (e.g., choropleth maps of socioeconomic data or image maps derived from remote sensing or GIS). Although Bertin's repeated graduated circles could be used as fills, most analysts seem to prefer the less coarse appearance of solid color fill. In this case, representation via color is in the form of combinations of color attributes defining the fill of contiguous areas. Proximity, then, is no longer the primary grouping variable. The ability to visually sort out spatial correspondence from lack of corre-

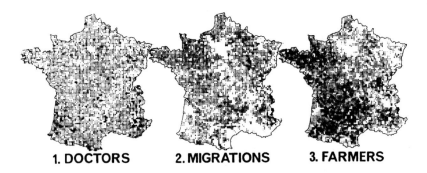

FIGURE 9.9. The three variables of Bertin's trivariate analysis. *Reproduced from Bertin (1977/1981, Figs. 1–3, p. 162). Reprinted from* Graphics and Graphic Information Processing *by permission of Walter de Gruyter & Co.*

spondence will depend (more so than with positional symbols) upon the logic of color matches and the extent to which those color matches allow vision to select out various combinations of spatial and attribute relationships in the data. With a poor color scheme, variables might be correlated but exhibit no identifiable spatial patterns, as in the EXVIS examples described above.

As noted above, Brewer's (1994) approach is the most comprehensive system of color logic for maps used in geographical analysis. Included in her system are color choice guidelines for multivariate maps (primarily emphasizing bivariate maps, but also addressing trivariate maps). A key issue that she addresses is specification of the relationships in the data that particular color matches can bring out on the map. She distinguishes among the following bivariate situations and provides guidelines on how color hue, value, and saturation should be manipulated in each case (see Figure 6.46):

1. *Qualitative/binary:* Both variables are nominal, with one being limited to a binary choice (because no color scheme can be successful for a bivariate map of two nominal variables if both variables have more than two classes).
2. *Qualitative/sequential:* One variable is nominal and the other represents unidirectional ordered or numerical data.
3. *Sequential/sequential:* The typical bivariate example, with two unidirectional ordered variables.
4. *Sequential/diverging:* Both variables are ordered, with one variable unidirectional and the other bidirectional (e.g., temperature above and below zero).
5. *Diverging/diverging:* Both variables are ordered and both are bidirectional.
6. *Binary/diverging:* One variable has a two-class distinction and the other is ordered and bidirectional.

Brewer's color assignment system can be applied to the problem of locating regions of high correspondence between two (or even three) variables. One of my students, Bill McLay (1987), investigated use of maps in this context. Specifically, he considered how color bivariate maps can be used to locate soil suitable for use in lining sanitary landfills. The key map-use task was to locate soil that was relatively impermeable, but did not have high clay content (because clay soils are too difficult to work with). McLay demonstrated that color bivariate maps (depicting soil permeability and clay content) were considerably more efficient tools than standard soils maps in which the sign-vehicles consist of one to three character abbreviations of soils types (several of which might fit the low permeability–low clay criteria, depending upon the thresholds used for each). The superiority of a color bivariate map for this task is not surprising, because standard soil maps provide no visual cues at all. The result does, however, demonstrate that a logical combination of two sequential color ranges can be produced that works for applications in which one corner of the space defined by the legend must be emphasized.

Although a logical syntactics for color matching should enhance the possibility of GVIS to yield insight, part of the power of computer-assisted visualization is due to dynamic interaction. Fixed color choices are seldom the best approach. The "optimal" multivariate analysis tools will probably be those that define general syntactics governing selection of potential color scheme types but leave specific choices within a category for the analyst to experiment with (e.g., a general guideline that quantitative sequences are best represented by color sequences using value change leaves a plethora of potential color ranges to apply). These might be accessed by simply selecting different color tables, as in Brewer's (1993) hypermedia tutorial. In addition, different color combinations might allow an analyst to weight one variable over the other or to treat them equally.

On the issue of manipulable color assignment for multivariate maps, Rheingans (1992) has extended Calico, the interactive color selection system described in Chapter 7 (Rheingans and Tebbs, 1990), to bivariate color assignment. The concept of a color path is replaced by that of a *color sheet*, a smooth continuous, but not necessarily planar, surface through the color space. At any point, the color sheet indicates the color to be assigned to a particular combination of two variables. Rheingans demonstrates the system with a bivariate choropleth map of the United States by county in which median income is mapped against mean education level. Although her example map matches education with hues and income with values, the system could just as easily be used to select one of Brewer's more logical color combinations. It would also provide a powerful tool for comparing Brewer's suggestions.

Time

If the spatial distribution of two (or more) variables are to be compared, comparison can be side by side (as with small multiples) or can involve overlay or merger of the variables into one illustration (as with the EXVIS iconographic display or the Pennsylvania State University SLICViewer overlay technique using spatial dimensionality). Time (in the form of temporally varying symbolization) can facilitate visual comparison in either case. For a pair of adjacent maps (e.g., choropleth maps of income and education), the technique of sequencing (Slocum et al.,

1990) might be applied to the maps simultaneously to direct attention to commonality, or lack of it, in successive categories from high to low or the reverse. A related strategy, suggested by Monmonier (1992), involves dynamic scatterplot brushing linked to a pair of maps representing the two variables. Alternatively, blinking rather than sequencing might be used. At this point there is no conclusive evidence concerning the ability of analysts to attend to pairs of blinking or changing scenes. Much of the research on attention, however, suggests that this will be a difficult task regardless of dynamic symbolization choice.

A promising alternative to the extreme abstraction required for bivariate maps is application of an "alternating syntagm" (see Chapter 5). In film, an alternating syntagm is used to deal with events taking place at the same time but at different locations. Presenting the events as alternating scenes helps the viewer to link them together. On a pair of maps in which time is fixed, similar alternation could be used to depict the same place but different attributes. Place (rather than time) is held constant, and attributes (rather than places) alternate. This use of alternation to represent correlation was suggested by Monmonier (1992). The technique was implemented as part of a graphic script developed to provide a narrative on the relationships between percent of women in the work force and percent of local politicians who are women. The two variables are depicted as four-class quartile maps using "signature" hues of red and blue. These are alternately displayed in register (13 cycles in 12 seconds). What vision notices the most is change in color value. States that are ranked similarly on the two variables appear to be stable but those ranked differently create a "visual dissonance in the form of a jerky, throbbing pseudo-motion" (Monmonier, 1992, p. 252). My students and I have adapted this alternating syntagm idea to comparison of data–metadata maps related to pollution in the Chesapeake Bay (MacEachren et al., 1993). The procedure adopted (more concisely called "flickering") involves alternating between a dissolved inorganic nitrogen surface (derived by interpolation) and an uncertainty estimate concerning that interpolation. The technique allows an analyst to look for relationships between derived values and the confidence that should be put in them.

Gershon (1992) proposed a similar use of an alternating syntagm for exploring geographic correlation. To enhance the feature-identification process, Gershon suggests creating an illusion of motion to supplement the likelihood that preattentive processes will detect correlation. In his method, the motion illusion would be created by shifting one image relative to another. If this concept was combined with the SKE technique for depth from motion discussed above (Kaiser and Proffitt, 1992), the anomalous regions would float to the top with correlated regions forming a background in depth. The process would be to shift those geographic units that differed the most by the greatest distance. Based on the success of dot map and maplike representations in various perceptual tasks (Julesz, 1975; Rogers and Groop, 1981; Papathomas and Julesz, 1988), and the arguments concerning space as an indispensable variable, I suggest the hypothesis that flickering (i.e., an alternating syntagm technique) will be particularly effective if dot rather than choropleth (or other area fill symbolization) is used.

Focusing

Blinking, sequencing, and flickering are feature-enhancement processes that require dynamic display but not real-time interaction (although they are probably most effective in an interactive setting). An alternative exploration strategy that becomes available with a dynamic system is to provide focusing tools designed to facilitate pattern comparison through interactive classification. The best known implementation of the concept of focusing (discussed above) is termed "brushing" and was initially designed as a technique for comparing multiple potentially interrelated variables on a set of linked scatterplots (Becker and Cleveland, 1987). The technique allows an analyst to select a subset of points from a scatterplot of two variables with the data entities represented by those points highlighted on all other scatterplots. Monmonier (1989a) extended the brushing concept to geographic information (and labeled the technique *geographic brushing*).[4] With his implementation of brushing, when the analyst selects points on a scatterplot, not only are points on other scatterplots highlighted, but their map locations are as well (Figure 9.10). Alternatively, the analyst might select a particular region (either with a tool to draw a boundary around the desired region or a tool that allows geographic units to be individually selected) and the corresponding scatterplot points will be highlighted.

If variables to be compared are merged into one map, we eliminate the potential problems of attending to spatially separated features. As discussed above, one method of variable comparison is to produce color bivariate maps. Monmonier (1992) proposed a dynamical bivariate map as one of several GVIS tools for investigating correlation. His implementation consists of a four-class map (two classes of each variable) called a "cross-map" (Figure 9.11). As part of the narrative script, the class break point (the center of the cross) was moved around the scatterplot space, alternately focusing on values that were high on both scales, high on one and low on the other, and so on. The cross-map is particularly amenable to interactive manipulation because a single point on its legend controls the bounds of all four classes. In contrast to this advantage, however, it

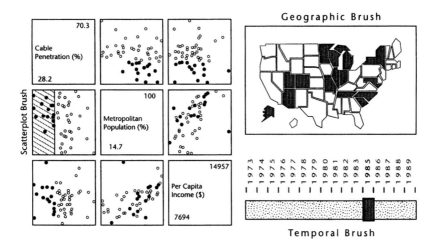

FIGURE 9.10. An illustration of geographic brushing which includes a scatterplot brush to select attributes, a geographic brush to select places, and a temporal brush to set the temporal bounds of analysis. *After Monmonier (1990, Fig. 23, p. 42). Adapted from Cartographica by permission of University of Toronto Press, Inc. Copyright 1990 by University of Toronto Press, Inc.*

requires extreme data abstraction (with only two categories for each variable).

Sound

As the number of variables to examine increases, human vision begins to have increasing difficulty sorting out relationships. Not surprisingly, a number of researchers have investigated sound as a way to add to the variables that can be compared simultaneously. Sound, like time itself, is suited to representing temporal information. Changes in sound over time, like changes in visual variables, will be noticed, and temporal patterns will be apparent.

An intriguing concept concerning recognition of temporal patterns is proffered by Neisser (1987). Based on a variety of evidence, he suggests a neurophysiological basis for pattern identification that depends not on which particular cells are firing, but on which pattern of firing is prevalent. If pattern identification worked this way, "a temporal rhythm expressed in one perceptual system might be recognized elsewhere without any new learning or recoding, and an information structure appearing in any of several modalities could produce the same sense of where you are" (Neisser, 1987, p. 300). Such an independence of modalities would allow

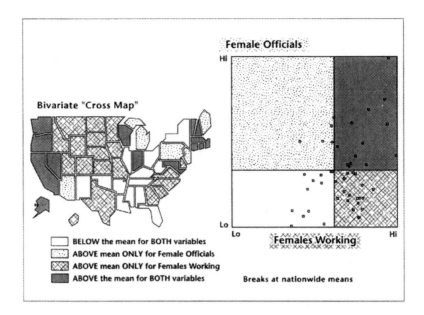

FIGURE 9.11. A monochrome representation of a snapshot from an interactive session using a bivariate cross-map to explore correlation. *After Monmonier (1992, Fig. 4, p. 272); original in color. Adapted by permission of the American Congress on Surveying and Mapping.*

both vision and hearing to be used for temporal pattern representation, with patterns learned in one modality equally identifiable in the other. In addition, similarity of patterns could be noticed even when one is presented visually and the other aurally. In fact, temporal pattern similarity might be more easy to recognize in a dual-modality representation than a single-modality representation (where both patterns would in essence be competing for the same brain cells).

Sound has also been investigated for applications with nontemporal multivariate data. An example is provided by Rabenhorst et al. (1990). They devised a system to facilitate semiconductor design. The system displays the semiconductor as a solid and allows two of three variables (electrostatic potential, electron concentration, and hole concentration) to be examined simultaneously (one visually and one aurally). The analyst is able to select a focal point of the representation. Three cross-sections based on that focal point are displayed in which value of a hue (blue, red, and green) is varied to depict one of the variables (e.g., electron concentration). At the same time, the three-dimensional vector gradient of another variable (e.g., electrostatic potential) is sonified. The analyst can then dynamically change the focal point, resulting in appropriate change in sounds. Sounds are three-note chords in which a combination of "detuning," stereo balance, and voice overtones are used to separately represent the three-dimensional vector. The analyst (after learning the system) can use this "sound probe" to quickly zero in on local maxima and minima of one variable while visually concentrating on another variable.

As mentioned above, Krygier (1994) implemented a geographic sound probe to demonstrate the potential of simple sound-generation techniques to embed additional variables in a multivariate map. His example begins with a choropleth map of population percent not in the labor force as a base overlaid by graduated circles representing median income. A third variable, a drive-to-work index, is added by matching three levels of the index to pitches at three different octaves. Finally, percent of poor in each county is matched to another sonic variable, pitch within each octave. When a user clicks on a county, she hears two pitches in sequence, first a pitch at one of three octaves, then another pitch within the octave. A signal of high initial octave followed by low pitch within the octave, for example, indicates a county having long drives to work and low poverty. Having experimented with the tool myself, I agree with Krygier that it is possible to extract "quad-dimensional" information from the display. It is currently unknown, however, whether this sonic probe can provide an understanding of variable relationships not available from a purely visual representation (in which users can examine anything from single variables to all variables superimposed on one map).

Sound has also been incorporated in the EXVIS system described above (Grinstein et al., 1992). In this case, the implementation might be thought of as a sonic brush or stylus. In addition to being able to probe a point for which a sound is returned, the user can move the mouse cursor across the screen to generate a "sound texture" composed of a series of individual sounds. The authors describe the resulting display as a "two-dimensional musical score" that users can "play" using any exploration strategy they desire (e.g., a slow spiral out from a point, profiles along or across visual features, etc.). Their system uses stereo sound generation (as well as a simulation of depth away from the users), thus making possible a spatial "sound texture" (for at least limited regions of a display/map). It is unclear from their description, however, whether the dynamic stylus can be varied in size (to allow the sounds attached to groups of icons to be heard at the same time). That the sonic representation is interpretable is demonstrated empirically through an experiment in which system users were presented with a pattern categorization task. They compared task performance for categorization using only visual representations with categorization using visual representations plus added sonic cues. The experiment used simplistic iconographic displays of single-limb stick figures. Users were given two prototypes representing the two categories into which they were asked to group a series of examples. Subjects classified the examples using only visual information, then using visual plus sonic information. Classification performance was reported to be significantly better for visual plus sonic representation than for visual alone.

SPACE–TIME PROCESSES

A third level of both cognitive and computational complexity for GVIS is that associated with explicit attention to dynamic space–time multidimensional processes (STFA). A large proportion of scientific-visualization tool developments have been directed toward use of dynamic computer environments to facilitate processes of change through time. It is in this realm that the power of computers is expected to offer the most benefit in relation to traditional static graphics used for exploratory analysis. While dynamic tools have demonstrated their usefulness in providing visual cues to feature identification and comparison, their application to representation of temporal processes seems as natural as the application of static maps to spatial structure.

If a problem has both spatial and temporal components, visualization tools that do not take both into account have the potential to do more harm than good. Although this contention has not been formally tested, some support is offered from a study of mental visualization. Antonietti (1991) presented subjects with a problem called the high-tide problem. The problem is stated as:

> A rope ladder was hanging from a boat so that the ladder had six rungs above the sea. The distance between any two rungs was 30 cm. At high tide, the sea rose 70 cm. How many rungs were above the sea at high tide? (p. 213)

Performance of three groups of 35 subjects each was compared: a verbal group that was given the problem only in verbal form, a picture group that received both the verbal problem statement and a drawing of a boat with ladder down the side to the water line (Figure 9.12), and a mental visualization group that received the verbal statement and instructions to mentally visualize the situation. Less than one-third of the subjects correctly realized that the boat would move with the tide. There was some difference across groups, with the picture group having only 5 correct responses, the verbal group having 7, and the mental visualization group 10. This study provides support for Antonietti's hypothesis that static pictures provide a "frozen" representation that inhibits the correct problem solution, as well as weak support for the hypothesis that mental visualiza-

FIGURE 9.12. A typical visual representation to accompany the "boat problem."

tion offers an opportunity for a dynamic representation that facilitates the correct approach to the problem. What seems most clear, however, is that none of the three problem presentations is particularly effective in prompting the appropriate schema. A dynamic representation, in contrast, could hardly fail to result in a dramatic improvement.

Categorizing Space–Time Phenomena

In general, map syntactics structure the manner in which categories of things in the world can be matched to categories of sign-vehicle variation. In previous chapters I have delineated the aspects of time that might be coded in or derived from a space–time database (Chapter 7), the dynamic cartographic variables that become available when computer tools allow display changes in real time (Chapter 6), and some principles of filmic syntax that might be used to structure how space–time phenomena are represented (Chapter 5). Attention to time in these chapters was directed to how maps or map sets might be imbued with temporal (and spatial–temporal) meaning. What remains before considering applications of GVIS to STFA is consideration of what temporal aspects of the world are to be represented. This issue is primarily one of how temporal concepts are conceptually categorized. Whether images in a display of small multiples, dynamically selectable individual maps, or frames of an animation, representations of space–time are inherently discrete. To represent temporal processes, we must first decide what the basic units of analysis are to be. The following tabulation is one parsing of this complex issue.

There can be at least two primary structures for conceptually organizing time:

- *Time series*: Time treated as a linear sequence.
- *Time cycle*: Time treated as a repeating cycle linked to earth's rotation or revolution.

with at least four temporal entities (see Chapter 6 for more detail on these):

- *Scene*: A time period with no change.
- *Episode*: An identifiable stage in a sequence (a coherent set of scenes).
- *Time phase*: An identifiable stage in a cycle.
- *Event*: An identifiable break point or transition between discrete scenes, episodes, or phases.

Our conceptualization of an entity as a scene, episode, phase, or event will depend on the way we bound time and the detail at which time is considered:

- *Time frame*: The overall time span being considered (e.g., hour, week, season).
- *Time resolution*: The temporal precision used to record, retrieve, or display information.

Together, time frame and resolution set the bounds and precision for time series and cycles (e.g., a cycle spanning a year, divided by week, and running for a century).

As with spatial representation, there are scaling issues when we try to conceive of an entity in time or to represent our conceptualization of time:

- *Conceptual time scale*: The ratio between nonexperiential and experiential time frames, such as that between a geologic epoch and a day or a week (a comparison we often use in order to comprehend a time frame that is so much smaller or greater than experienced time that it is otherwise beyond comprehension).
- *Representational time scale*: The ratio between chronological time and display time.

Mapping Temporal Entities to Display Variables

When we move from three-dimensional space to a two-dimensional map, we lose the ability for one-to-one mapping between world space and map space. Abstractions such as contour lines have therefore been developed to deal with this loss. If we consider time to be a fourth dimension, static representation of space–time processes must face the more difficult problem of collapsing four dimensions to two (analogous to collapsing three spatial dimensions to one and representing a city with a dot on a map). By taking advantage of technology that allows us to use binocular parallax (and some type of viewing aid such as red–green or polarized glasses), we can achieve at least a close equivalent to three-dimensional representation. This 3-D display has the advantage of allowing us to enhance contrast by manipulating the spatial Z-value (just as we might manipulate X and Y in constructing a cartogram). At the same time, a 3-D display has the disadvantage of producing a scene we can no longer see at once. With time, on a dynamic map, the situation is similar. We can achieve a representation of time that is matched to our experience of it and can enhance temporal features by manipulating display time (usually by compressing it or leaving out uneventful pieces). While gaining a more intuitive (and manipulable) sign-vehicle for time, however, we lose the ability to see everything at once and have to rely on spatiotemporal memory to link what we are seeing now with what took place before.

In relation to molecular chemistry, Dennett (1990, p. 300) proposes the concept of a space–time "microscope" that can "enlarge and slow down such happenings by a factor of a million and then go on to clothe the invisible features with visual metaphors!" For GVIS, we want essentially the same thing, but usually acting in reverse both spatially and temporally. World time can be mapped one to one with map time, but like the one-to-one map about which Lewis Carroll's farmers complained (because it blocked out the light), a temporally one-to-one dynamic map offers mostly disadvantages (although it can allow us to experience a time that is distant from us just as a one-to-one map could allow us to walk through a distant place).[5] Maps in general are useful, in part because they transform entities of interest to human scale. Temporally, some geographic space–time processes (e.g., an earthquake tremor) are fast enough that we need to slow them down to understand them (as when we "map" a molecule, cell, or computer chip, for which an increase in scale makes visible a pattern that would otherwise remain hidden). Most temporal geographic phenomena, like spatial ones, have a time span too large to be grasped at once, so therefore we need to compress time as well as space.

In the case of both microscopes and maps, the goal is to bring phenomena to human scale (both spatially and temporally) so that we can examine them using a sensory system particularly adapted to this scale. To be able to see (and hear, and manipulate) "objects" as if they were a flower, or rock, or bird in our hand, allows us to use our most highly refined mental-processing abilities, to which more than half of the brain is devoted. As noted, for most GVIS problems, we need a space–time tool that compresses both space and time. This process alone, in contrast to Dennett's space–time microscope, however, increases the detail per unit space and unit time rather than decreasing it. For most space–time problems, we have a temporal reduction in scale comparable in severity to the spatial scale change. GVIS, then, requires not just scaling and symbolization but extreme abstraction to meet the resolution limits of human sensory systems.

It is not scale change alone that makes space–time GVIS tools powerful; it is the ability to manipulate both spatial and temporal scales in the search for pattern (at various scales) that provides the key. As Dorling emphasizes in his comments on, "stretching space and splicing time," non-Euclidean transformation and an ability to pick, choose, and reorganize are critical to the power of GVIS. When temporal episodes are spliced (in a dynamic representation), how we identify events in the continuity of time can be as critical as the resolution and scale at which world time is mapped to display time. Gombrich (1990) provides a number of compelling examples in the context of pictorial instructions, particularly those that must work for people of many languages (e.g., illustration of the process of donning a life jacket). He suggests that there are prototypic processing "chunks," as identified by human factors engineers, that the illustrator must isolate for the diagrams to be successful. In a visualization context, we can give an analyst dynamic tools that help to isolate these chunks. Space–time manipulation tools can include ones that allow the analyst to consider spatial features and temporal features independently and in combination.

An additional aspect of temporal resolution to deal with is the difference between treating time as a series and time as a cycle (see Chapter 7). Temporal aggregation can involve grouping adjacent points in time (in a manner similar to that used for spatial aggregation). Alternatively, however, it can be applied across cycles, grouping all times at the same position in the cycle over some time span. Moellering (1976), for example, presented traffic accidents for "composite time." He aggregated across three years of data to produce a composite week by quarter-hour interval. This temporal abstraction de-emphasized individual daily events in relation to temporally global patterns. Dorling and Openshaw (1992) report an analysis of similar data, telephone-line faults. They applied space–time filtering (aggregating adjacent spaces and times) and found no space–

time pattern. It would be interesting to see whether treating their data as a cycle rather than as a series and compositing a day or week from the data would uncover the same lack of pattern.

Exploring Space–Time Processes: Kinds of Interaction

In relation to simulation modeling, Marshall et al. (1990) proposed a categorization of visualization applications that include process *tracking*, *postprocessing*, and process *steering*. Although their categorization was proposed in the context of numerical models, it seems applicable to how we might interact with space–time information of any kind. Considering differences in how their taxonomy applies to measurement- versus model-derived information can help to highlight important differences in visualization based on phenomenon- and concept-representations (i.e., the key differences between using visualization tools to explore process by measuring real-world variables over time and examining their relationships vs. exploration of process through manipulation of model parameters).

Tracking in the context of P-reps allows an analyst to track real-time processes (e.g., a hurricane or other meteorological event for which the representation is driven by a real process occurring in the present). The analyst working with C-reps is offered the added ability to "run" a modeled process multiple times with changes to various model parameters, thus using tracking to identify features that change as a result of parameter changes. Postprocessing is similar for the analyst working with measured or simulated data. Once the information has been turned into data, the application of GVIS tools is the same. The only real difference is in the potential for the analyst to generate new data if an interesting phenomenon is uncovered during initial analysis. While possible in both cases, from a practical point of view it is almost always easier to run a model again than to collect more sample data.

Process steering uses visualization tools to provide feedback concerning the impact of various parameter changes, with the feedback used to make decisions about adjustment of additional parameters. Its application has been thought of almost exclusively in the context of simulation models where the "parameters" are values used as inputs to model equations. The concept of process steering with GVIS, however, might be productively extended to include attempts to control real-world processes (e.g., manipulating resources used in fighting a forest fire, dynamically controlling traffic lights to react to and manipulate traffic flow, etc.). Perhaps the most important distinction between real-world and simulation process steering is related to the implications of steering decisions made (a topic to be taken up in the final chapter).

Process Tracking

Although process tracking, by definition, prohibits changes in the process, it can still involve considerable interactive exploration on an analyst's part. It is possible to put any parameters used to generate the display under user control so that the user is free to chose the features to be emphasized while the process plays out. Major changes during the process (e.g., sudden directional change for a hurricane being tracked or a switch in direction of temperature trends in a global climate model) might be linked to particular visual and/or sonic cues designed to attract the user's attention to critical features. In an actual or model hurricane depiction, color palettes might be manipulated to emphasize different wind speed zones at different times, or the analyst could have the ability to move among atmospheric levels to explore the three-dimensional structure of the storm as it progresses.

One (aspatial) application area in which techniques for process tracking via visualization have been investigated in some detail is computer algorithm development. When the logic and efficiency of an algorithm is to be assessed, a visual depiction of the algorithm as it progresses over time has proved quite effective (Singh and Chignell, 1992). One particularly effectual technique for algorithm process tracking is color highlighting as proposed by Brown and Hershberger (1992). This technique is analogous to blinking (described above) for drawing attention to aspects of an otherwise static scene. One of the drawbacks of blinking, encountered by Monmonier and Gluck (1994), is that viewers find it irritating. The color-highlighting method developed by Brown and Hershberger uses a clever trick with color tables to give the effect of "painting with transparent color." This method of highlighting produces a much less harsh impression (than that achieved by blinking) because it, in essence, changes the graphic variable of transparency rather than hue plus value (as most implementations of blinking do).[6] Although developed for process tracking, the Brown and Hershberger color-highlighting technique also seems suitable as an alternative to blinking for highlighting important aspects of static scenes.

In the context of GVIS, the effectiveness of process tracking will be particularly dependent upon both spatial and temporal scale/resolution decisions. Unlike postprocessing, real-world process tracking allows just one pass. If recording is not begun at the right time or measurement resolution is too coarse, a feature can be missed entirely. Even when the process being tracked is simulated, and can be run repeated times, an inappropriate time frame or resolution can cause critical features to remain hidden regardless of how other parameters might be manipulated for each model run.

Postprocessing

Postprocessing provides the opportunity for more flexible interaction with information than does either process tracking or process steering. All of the feature identification and comparison principles and methods discussed above can be applied. Perhaps more importantly, time can be used to represent not only chronological time but other ordered attributes of the data. The way time is parsed can also be manipulated (if the original data are comprehensive enough), thus allowing the analyst dynamic control over how events are defined or how cycles are specified.

The advantage of postprocessing (over tracking) for understanding a space–time process (through application of some of the feature identification and comparison methods discussed above) is illustrated by Dorling and Openshaw (1992). They explored 20 years of childhood leukemia data for Britain using animated maps to look for space–time trends. Because the entire data set consisted of approximately 680 cases over the 20 year period, analysis required "the careful exaggeration of any space–time trends in the points displayed" (p. 646). Their analysis demonstrates that in a search for global patterns, filtering of fine-resolution data is often a better strategy than aggregation of that data to coarser resolution. Aggregation can eliminate features that occur at more detailed time or space scales. Filtering can act to exaggerate global features rather than hide them.

Dorling and Openshaw were able to use GVIS (although noninteractively due to the size of the data set in relation to the power of the computer used) to experiment with a variety of smoothing functions by manipulating the size and shape of a weighted moving average applied to the space–time lattice. After each manipulation, they "played" the map series as an animation. A spherical 25 kilometer × 12 month weighted average was found to be most informative. The high resolution of the original data, apparent motion produced by the rapid (25 frames per second) sequencing of static scenes, together with global pattern enhancement through blurring, resulted in identification of a space–time feature (what might be called a "hot spot") in the leukemia data. This hot spot, in the Middlesborough area, had been repeatedly missed in previous analysis due to coarse temporal resolution and the visually noisy data set. A second unexpected feature "popped out" of the animation, a feature Dorling and Openshaw describe as a "peculiar oscillation" between cases in Newcastle and Manchester. A slightly out-of-synch rise and fall of cases in Newcastle then Manchester cycled four times over the 20-year time span examined. They suggest that this oscillation is similar to space–time oscillations observed for many viral and infectious diseases that are sometimes linked to rarer diseases.

In Chapter 8, efforts by Gould et al. (1991) to model and visualize geographic factors of another disease, the AIDS epidemic, were discussed. In the same article, they provide anecdotal evidence that map movies showing AIDS incidence spreading across space and over time have convinced young people that AIDS is a serious threat. For an analyst trying to use visualization tools to uncover a locational structure in the AIDS phenomenon, however, greater success might be achieved by making direct use of both indispensable variables (space and time). Space could be used to represent relative position in AIDS interaction space (a multidimensional space analogous to that of Monmonier's geographic biplots discussed above) while time could be used to represent moments separated in chronological time (Gould et al., 1991). In this way, changes in the linkage structure over time would be most noticeable because they would be depicted by change in location (e.g., a new corridor or transmission opening up directly between two cities that had been previously linked only through secondary connection).

Human visual information processing clearly has capacity constraints. Based on viewing what now runs into the hundreds of map animations, I have developed a hypothesis that making use of time as a cartographic variable is not a cognitively zero-cost addition to GVIS. There is a tremendous amount of information presented by a temporal display (just consider Dorling and Openshaw's 25 space–time filtered leukemia time slice maps per second). When we push vision to capacity (as we certainly do with some map animations), we should expect a decrease in the ability of vision to notice contrast in nondynamic cartographic variables (e.g., color, orientation, etc.). If the map is dynamic, we will need to reduce the amount of aspatial and attribute information provided to allow vision to cope. Dorling and Openshaw (1992, p. 649) seem to disagree when, in summarizing the results of their animation research, they state: "There is no need to amalgamate observations, or (worst still) dimensions." Interestingly, in another article in the same journal issue, Dorling (1992c, p. 630) points out (in relation to space–time analysis of voting patterns) that "the most enlightening animations were ones in which the spatial relations were dispensed with and the distributions of constituency votes (by party) were shown as bouncing points inside a triangle." I have seen videos of some of Dorling and Openshaw's efforts and agree. GVIS tools that allow an analyst to turn the temporal or spatial dimensions on and off, to isolate patterns individually in time and space, before trying to understand patterns in space–time may be particularly useful. A similar strategy has been used for analysis of n-dimensional space for which three dimensions at a time are sequentially represented in a perspective view (Beshers and Feiner, 1992).

Using display space (one of our indispensable variables) to represent attributes (rather than geographic space) directs vision to changes in proximity of places in attribute space over time. Similarly, using display time to represent attributes (rather than chronological time) can highlight temporal patterns for attributes of places. The procedure labeled "re-expression" involves matching attribute order to display order (DiBiase et al., 1992). One example is to display a sequence of maps representing deaths from earthquakes around the world in the order from fewest to most deaths. The date that each map represents is signified by a pointer on a time line. Since vision is particularly adept at recognizing movement, apparent movement should be noticed. Temporal patterns will be signified by the kind of spatial movement that the pointer exhibits, with steady progression toward the left indicating a negative correlation between time and earthquake fatality (which would suggest that risk amelioration and communication measures were working), a steady progression in the opposite direction indicating a positive correlation (with increasing population pressures forcing people into high-risk environments), a random pattern indicating that likelihood of earthquakes in unprepared populated places was temporally random, and a regular cycle back and forth on the time line indicating some temporally cyclic process at work.

Geographers and researchers in other disciplines have become used to information graphics in which space is used to represent things other than space. Most analysts, then, can be expected to have accessible schemata that allow them to deal with this rather severe metaphorical transfer to nonspatial domains where spatial proximity has come to mean "associated." The projection of attributes onto time, however, is less common. It is quite likely that analysts will apply the wrong schema to a dynamic display in which temporal order is used to depict attribute order, unless a cue is provided that prompts them toward the correct more abstract schema. Krygier (1994) has suggested use of sonic cues that employ a nonvisual information channel, thus reducing the potential interference with visual interpretation. Krygier describes a re-expression in which a time series of presidential landslide elections is analyzed. On the earthquake example above, the changing location of the pointer on a time line competes visually with the changing map representation. Since motion is probably more salient in vision than change in place, attention is likely to favor the time line over the map. Krygier tried to overcome this visual competition by replacing the time line and pointer with sound in his animated map series. Pitch (an ordered sonic variable) is matched to chronological order. Viewers can begin by examining maps in temporal order looking for changing spatial patterns over time. As they do, they hear a steadily increasing pitch. The data can then be re-expressed using display order to depict magnitude of the landslides. As the analyst sees the spatial distribution of larger and larger landslides, she can tell by the order and regularity of pitch whether or not a temporal pattern exists.

Process Steering

Most attempts at process steering have dealt with models. One example of real-world process steering using what might be considered a kind of GVIS environment is air traffic control. Here, a controller uses visual (and auditory) input to steer a process by routing planes, putting them in holding patterns, and so on. Recent research into design of air traffic control displays bring them closer to the concept of GVIS presented here, as an interactive spatial information analysis tool. Kaiser and Proffitt (1992), for example, describe a system in which the SKE depth effect is used (at controller discretion) to simulate depth in the display. Their display, when switched to 3-D mode, provides a ladderlike sign-vehicle that uses perspective and size cues together with SKE to allow the controller to view aircraft positions in three dimensions rather than two. The difference between real-world and simulation process steering seems to be that (at least for the present) real-world steering has fewer controllable parameters and feedback is not usually provided through the display.

A variety of examples exist of simulation models that allow some level of interaction or steering by a user as the model runs, but most are nongeographic. Marshall et al. (1990), for example, mention driving and flight simulators in which a numerical model defines what a user can see from various points and how the virtual vehicle behaves in response to user input. They describe their own work on a process steering system in which maps form a major component of the visual information provided to the users. The system described deals with a three-dimensional hydrologic forecast model for Lake Erie. The model forecasts changes in lake water level spatially and temporally. Visual display options include representing water levels with a shaded perspective polygon mesh or color layer tint isoline map, and adding overlays of isometric lines representing temperature and heat flux. Users can control several parameters of the model, including scheduling of storm events, setting strength and direction of wind, and adjusting the heat flux associated with storms. Although the authors demonstrate that the system does allow real-time model steering, no evaluation of system use is provided and no speculation is made concerning implications for spatial thinking (or the hypotheses derived from dynamic exploration).

In addition to scientific models designed to allow researchers to explore the characteristics of models they are building, interactive model steering has found its way into some software intended for entertainment or education. In particular, work in what has come to be termed "virtual reality" generally involves model steering of some sort. Perhaps the most widely known examples of what might (loosely) be categorized as virtual realities using geographic models driven by user steering are the computer game/learning packages known as *SimCity* and *SimEarth*. Both can be described as a GVIS environment that acts as an interface to a model which reacts to user input. At this point, I know of no research concerning how users process the visual feedback provided by these packages, how domain expertise (schemata) influences strategies for interaction, and so on. This seems to be a fruitful ground for research.[7]

DISCUSSION

This chapter has considered various ways in which GVIS might facilitate analysis of geographic relationships among variables in space and across time. The common thread throughout has been an emphasis on taking multiple perspectives on data rather than trying to find one optimal view. Taken to extremes, however, this approach has two flaws. First, there are too many possible options to be considered, so the chance of stumbling upon a useful one may be low, unless prior constraints are imposed on the analysis. Second, once a feature or relationship among features is identified from among the many perspectives, how will an analyst know whether it is real?

The first problem is being addressed by attempts to merge expert systems and human interaction to facilitate knowledge discovery. These attempts are closely related to using GVIS as a process model steering tool. In this case, we can think of a model steering the human analyst through the visualization tools. Grinstein et al. (1992) have speculated on the linking of AI tools with automated knowledge discovery approaches based on AI. For geographic information, Monmonier (1989b) has described a research agenda directed to automated production of graphic scripts. He envisions these rule-generated scripts as a "pump-primer" for database exploration. This pump-primer would be used to sift out the most promising features for visual examination, at which point the analyst would have an opportunity to extend from this base in a less structured way.

The second question, how to judge the verity of GVIS, brings up the issue of meaning at multiple levels, both denotative and connotative (topics of Chapter 7). How to judge verity (or truth) in visualization and just what truth in visualization might mean are taken up in the final chapter.

NOTES

1. See Liu and Wickens (1992) for related ideas on cluster analysis and graphic depiction of similarity matrices.

2. See Chapter 10 for a discussion of symbol stability.

3. For a review of some of the evidence, see Chapter 3.

4. The concept of linking a map and scatterplot, with both "brushable," was also proposed (more recently) by Buja et al. (1991) in what seems to be a case of independent invention.

5. Some versions of Cyberspace might be considered one-to-one representations of time and space because they (will eventually) simulate a real world that the user feels they are behaving in. Whether Cyberspace is a spatially and temporally one-to-one representation depends, of course, on whether anything of the world is being represented.

6. Their highlighting method might also be less irritating than Monmonier and Gluck's blinking because it is applied to only one location at a time.

7. Exploring human–map–environment interaction through these games was first suggested to me by Christopher Board.

CHAPTER TEN

GVIS

SHOULD WE BELIEVE WHAT WE SEE?

In the previous two chapters, the multiple approaches to map representation and design delineated in Parts I and II were applied to design of GVIS environments. The perspective taken was that visual thinking can be facilitated by tools that integrate what we know about vision, visual cognition, and cognitive representation with an understanding of the semiotic structure of cartographic representation. In this final chapter, I address issues of private and public *truth* in seeing. At the private level of the analyst, the questions are how to judge the quality (i.e., truth) of GVIS displays and how an analyst recognizes truth in the mental visualizations they prompt. At the public level, the issue is how to deal with the multiplicity of representation levels with which our visual displays are infused when those displays "go public." This discussion will be divided into two primary questions: *How can "truth" be judged in the displays that GVIS provides?* and *What is "truth" in the context of GVIS?*

HOW TO JUDGE "TRUTH" IN GVIS

The truth of a GVIS representation can be assessed at two levels. First, we can consider the veracity of individual signs as they serve to define the sign relationships between a phenomenon and data element and between a data element and a map mark. If a triadic model of the sign as relation is adopted (as it was in Part II), then the truth of the sign depends upon the consistency with which specific categories of referent are matched with specific categories of sign-vehicle through their common interpretant. It also will depend on the extent to which the particular referent matches the prototype of the category to which it is assigned. In most cases we cannot determine the accuracy of the match for individual signs, but can often estimate the *probability* that the match is a "true" one. The second level at which truth can be assessed is that of the map itself as a complex sign. The truth of a map will be a function not only of a map–world sign relation, but of the inherent logic of map syntactics as well as the viewer's processes of vision, visual cognition, and knowledge structuring via general and domain-specific schemata. The result of visual–cognitive processing is considered here to be analogous to hypothesis formulation and testing. As with statistical analysis, errors can be of two types: failure to see geographic features that are there in the world (i.e., not seeing, or Type II errors) and identification of illusory geographic features (i.e., seeing wrong, or Type I errors) (MacEachren and Ganter, 1990).

Both truth of signs and truth of maps as signs are intimately linked to processes of conceptual categorization. We can only know the "true" world in the context of conceptual categories that we apply to organize that world. Different ways of categorizing are likely to yield different truths, as Lakoff (1987) points out in the context of biological classification by those adopting "phenetic" versus "cladistic" criteria.

Truth of Signs in the Display

The truth of a map sign in a GVIS display will be a function of the accuracy with which sign-vehicles are matched to referents, the degree to which the referent is representative of the category prototype (i.e., the variance within categories), and the level of categorization hierarchy at which distinctions are made (MacEachren, 1992b). In practice, viewers will evaluate truth on the basis of whether an element in the real world fits in the category signified by the sign-vehicle representing it and whether things signified as similar on the map are similar in reality (and things signified as different are different). Truth is thus linked to the standard container schema for understanding categories, which imposes an assumption that category members are similar. On a grid cell map of soils, for example, a cell represented as inceptisol that contains 50% inceptisol, 25% histosol, and 25% alfisol soil is accurately categorized. Most people would, however, consider it a less "true" representation than that provided for a cell using the same sign-vehicle to represent an area that is 100% inceptisol. The truth of a judgment about similarity of entities represented by two map sign-vehicles, then, will depend on how much internal category variation there is (one aspect of category precision). Accuracy in GVIS will never be known, but can be estimated. Precision of categories can usually be determined.

I have elsewhere proposed that certain static graphic variables are particularly well suited to representing reliability (or potential truth) (MacEachren, 1992b).[1] Color saturation and clarity seem particularly amenable to reliability representation because these variables have a more iconic relationship to reliability than do the other graphic variables. Both denote a range of specificity. Low-to-high saturation constitutes a range from an uncertain hue mixture to a single definite hue. In Chapter 6, I partitioned clarity into three distinct subvariables: crispness, resolution, and transparency. Crispness refers to a range from highly detailed, precisely defined sign-vehicles to sign-vehicles with limited detail and/or fuzzy edges.[2] Resolution refers to the spatial precision of the map's geographic base, with a coarse base (possibly) suggesting lack of certainty about data depicted on that base. While crispness and resolution are embedded in the sign-vehicles used to represent primary information, transparency is most logically linked directly to metadata (i.e., reliability measures for the data). Transparency can be used to block user access to unreliable data.

Traditional static paper maps, whether used for communication or visualization, have limited capacity to provide users with clues to their truth. All of the uses of color saturation or clarity mentioned above, for example, are likely to make the map harder to read. With GVIS, the addition of interactivity, animation, and sound opens up several possibilities for providing such information without interfering with the analyst's ability to see any features that are present in the display. There are several ongoing efforts to develop techniques for integrating quality, reliability, or potential truth information into GVIS.[3] Side-by-side comparison of data and metadata, overlay, merging into a conjunctive sign-vehicle, alternating syntagms, and several other strategies have all been tried. Although the issue of measuring the reliability of sign relations has received more attention than that of representing that reliability, it is the representational issues that concern us here. A few examples from this growing literature will illustrate ways in which principles outlined in this book might be applied to tools that help analysts judge the truth of individual map signs.

I will begin with the first effort by our Pennsylvania State University visualization group to depict reliability. The project goal was comparison of different global climate model (GCM) predictions for temperature and precipitation in selected cities in Mexico (DiBiase et al., 1992). Data examined were provided by colleagues whose research was focused on the implications of global change for agriculture in marginal areas (Liverman and O'Brian, 1991). The particular interest of these colleagues was with differences among several GCMs as an indication of reliability (or truth) of the estimates generated. We tried several representation methods to look at spatial, temporal, and space–time differences among the GCMs.

In our initial GCM comparison, measured temperature was matched to the fill of a "frame rectangle," predicted temperature was matched to a horizontal line across the frame rectangle, and variance among models was matched to a pair of triangular pointers bounding the predicted temperature line (Figure 10.1). The analyst, therefore, had to attend to movement up and down the scale of actual temperature (shown with a color fill) and predicted temperature (shown with a line). At the same time, the tolerance band around the predicted temperature line moves (with the line) and changes size. It is not surprising that viewers could not accomplish the difficult task of visually integrating data for 12 cities at once. Although the GVIS tools applied to the GCM data resulted in some unanticipated relations being noticed, they did not allow space and time to be explored together.

As occurred for Dorling (1992b), our greatest success came when we ignored space and examined attributes over time for single places. A display was produced in which data for each city could be visually isolated by zooming in on a single dynamic graph (Figure 10.2). When played in chronological order with sign-vehicle size linked to among-model variance, all that was initially evident was an impression of change over time.

FIGURE 10.1. A single frame pair from an early version of the GCM animation. Similar depictions were used for each of the 12 Mexican cities being considered. Temperature bars (on the left) were paired with precipitation bars (on the right). Temperature was indicated in shades of red and purple and preceipitation in shades of blue. *After DiBiase et al. (1992, Fig. 9, p. 265). Adapted by permission of the American Congress on Surveying and Mapping.*

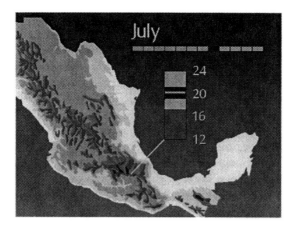

FIGURE 10.2. A single frame from the redesigned animation of global climate model variance. In this animation, only temperature (or precipitation) is depicted at one time. In addition, the tolerance pointers at the sides of the bar are replaced by tolerance bands inside the bar. *After DiBiase et al. (1992, Fig. 9, p. 265). Adapted by permission of the American Congress on Surveying and Mapping.*

Re-expression was then tried in which the graphic displays were played in order of increasing model discrepancy (and the additional temporal variable of duration was used redundantly to put emphasis on months with highest discrepancy). A pattern of high model variation in spring (planting season) months became immediately apparent (see Figure 6.36).

The goal for the analyst using the tools described is to understand spatial, temporal, and space–time relations for temperature across Mexico, predicted temperature across Mexico, and correlation between actual and predicted temperature. This task will certainly require several different views of the data, but would be facilitated by sign-vehicles that allow both individual variables and their conjunctions to be easily judged and compared. The thermometer-like "frame rectangle" symbol was demonstrated by Cleveland and McGill (1984) to be easy to compare in spite of varied position of the base. Our experience suggests that a trivariate frame rectangle (showing actual and predicted temperature along with variance in predictions) does not work well, particularly if temperature–precipitation pairs are to be examined spatially and temporally. Here I offer two (as yet untested) alternatives to the visually complex frame-rectangle approach. They build upon concepts of perceptual organization and research on feature identification discussed above.

One possibility is to separate the data from its reliability (i.e., variance) by "mapping" them to alternative perceptual channels, with data to vision and metadata to sound. Even by expanding the available information channels, however, it will probably be impossible to simultaneously deal with 12 (or more) pairs of changing positional symbols. A second alternative, one that takes advantage of the ability of vision to group by orientation, is bivariate glyphs similar to those proposed by Carr et al. (1992) (see Figure 3. 39). For the temperature problem, one possibility is to use two lines oriented to the same side of the symbol, with actual temperature black (as a longer and wider line in the background) and mean predicted temperature red (as a narrow, short line in the foreground). To facilitate assignment of an up–down schema, scaling should be set so that the mean annual (actual) temperature is represented by a horizontal line and straight up or down is set at the maximum deviation of values from the mean (either actual or predicted). For variance around the mean predicted temperatures, an error fan (like that used in Carr et al., 1992, p. 233, Figure 5), seems appropriate. Individually, bivariate ray glyphs may be less accurately judged than frame rectangles, but vision is capable of noticing similarity or difference in orientation almost immediately (a requirement for a dynamically changing map).

Both the frame rectangles and bivariate glyphs described combine data and reliability through superposition. This allows the data to be visually isolated from the metadata, but can impede visual integration. An alternative is to merge data–metadata into compound symbols. The multidimensionality of color together with the range of discriminable colors has led to several attempts in which color has been applied to merging of data and potential truth of that data. Value and saturation can be combined on bivariate quantitative maps that depict both data and their reliability, while a hue–saturation combination can be used to represent nominal attributes and reliability of categorization (MacEachren, 1991b, 1992b). For the latter application, Brown and van Elzakker (1993) conducted a detailed analysis of technical issues involved in generating hue–saturation color tables for computer display and color hardcopy. They conclude that the limited human visual capacity for differences in saturation combined with the variation in number of discriminable saturation steps for different hues (and different hardware) limits practical use of saturation to three steps. They propose that up to 12 data categories can be represented by variations in hue with three data reliability levels, represented by variations in saturation.

Schweizer and Goodchild (1992) conducted what was possibly the first empirical evaluation of color bivariate maps as a representation of quantitative data and reliability of those data. They attempted to represent 15 categories of data and 15 categories of "data quality" (reliability) using color value and saturation combinations. Not surprisingly, they found that subjects were not able to make much sense out of the color set in the absence of a legend. Olson (1981), for example, in the most comprehensive analysis of bivariate color maps to date, demonstrated that even a four-by-four bivariate map was not intuitively obvious (without prior training in how to interpret it). Schweizer and Goodchild drew on the metaphor of "fog" or "foggy" as a suggestion of uncertainty, then made a connection between fog and "gray" to arrive at the same iconic relationship I have suggested between reliability and saturation (more "gray" equals more uncertain). Schweizer and Goodchild, however, mistakenly equate "grayness" with color value (perhaps because color value results in a "gray scale" when hue is absent—when saturation is zero). This confusion of terminology leads them to hypothesize that color value rather than color saturation has an iconic relationship to uncertainty. Their experiment demonstrates the opposite, that color value to represent data and color saturation to represent certainty works better then the reverse, thus supporting the hypothesis that I originally proposed (MacEachren, 1992b). Overall, results indicate that subjects had a difficult time trying to judge maps with 15 value and 15 saturation steps, and in many cases ignored the value–saturation distinction entirely (noticing only value differences regardless of whether the question being responded to was about data or metadata). Analysts who know what they are looking at, understand the symbolization, and are motivated to find patterns probably will not make this latter mistake (although they will still find 15 saturation steps on a 225-color map perceptually impossible).

Although 15 categories of reliability is clearly beyond the discriminable range for saturation, and there is evidence that three saturation categories may be "ideal," a GVIS environment eliminates the need to select a fixed number of categories. Instead, we can build dynamic tools that allow analysts to explore the uncertainty information using an extension of the "focusing" or "brushing" concepts discussed in the previous chapter. One possibility is to create a map in which only two saturation categories are portrayed (fully saturated for more reliable and gray for less reliable). The analyst would have a slider bar allowing her to move the class break point back and forth, first highlighting those areas that are least reliable (by having the certainty threshold quite low), then gradually shifting the threshold until most of the map is in shades of gray with only the most certain regions highlighted by remaining in a nongray hue. My graduate students and I have implemented the general principle of a sliding reliability threshold with side-by-side data and metadata maps (MacEachren et al., 1993). The approach to reliability thresholds implemented was adapted from a similar representation system proposed by van der Wel (1993). In a GVIS environment called METAGIS, van der Wel (and a group of colleagues) implemented a dynamic link between map category boundaries and a reliability threshold. With the threshold set at one extreme, the boundaries are treated as reliable and represented with typical crisp (unblurred) lines. As the threshold is changed, reliability in the boundaries is represented through a change in crispness, via blurring of lines (Figure 10.3).

A dynamic environment that allows change in visual display via interaction also makes sonic representation possible. A manipulable sonic probe for certainty information was developed by Fisher (1993) for soils maps. Fisher treated the classification of grid cells of a map and known variability for soil within those cells as a bivariate problem. Soil classification was represented visually via color assignments. Because soil has a complex system of categorization, soil maps are visually complex. Although it might be possible to add reliability information to the maps with a pattern overlay (using EXVIS-like sign-vehicles) or to embed reliability information into the color by using saturation differences (as discussed above), a map with as many categories as a typical soils map would certainly be difficult to read. As an alternative to these techniques, Fisher linked reliability information for each grid cell to sounds produced when the cell was probed. He experimented with loudness, pitch, and rate of change as simple, easily produced sounds with an intuitive order that could be matched to reliability magnitudes. As in the EXVIS implementation described above, sound changes dynamically as the user drags the sound probe around the map. Because geographic data typically exhibit relatively high spatial autocorrelation, adjacent sounds tend to be closely related, resulting in a fairly coherent sonic "picture" if the user scans in a systematic way across the map. An important feature of Fisher's implementation, particularly in a GVIS context, is a tool for exploring scale dependence of the data–reliability relationship. Fisher's sonic probe is scalable, allowing both individual cells and rectangular patches of cells to be probed.

Fisher's sonic probe could be adapted to the GCM comparisons discussed above. Using sound redundantly with a visual depiction seems like the most viable approach in this case. If variance is treated as a single value (as it was in our earlier analysis of this data), pitch (together with the angle formed by error fans) could be matched to magnitude of variance. Alternatively, five pitches might be sounded simultaneously with each matched to the difference between one model and the mean. The combined sound would be a single pitch if the models agreed, and a discordant combination of pitches if models disagreed. An advantage of this technique is that the data could be "played" repeatedly turning on (or off) the variance for single models in an effort to determine relative contributions of each model to the overall variance. Also, unlike the visual error fans, variance due to a single aberrant model will "sound" different than that due to a general discrepancy among models, even if magnitude of variance is the same. The sonic depiction (like several of the sonic probes

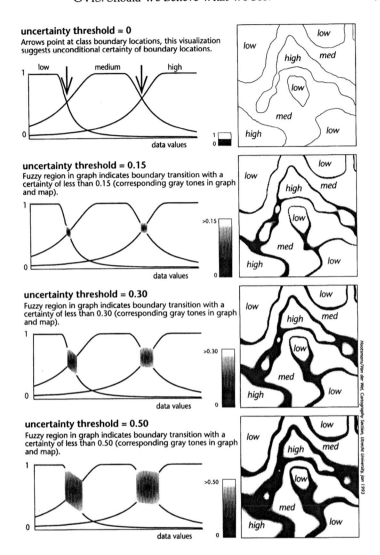

FIGURE 10.3. Example of four reliability thresholds for map boundaries (called "uncertainty thresholds" by these authors). As the threshold changes, boundaries are more or less fuzzy (i.e., line crispness varies). *Reproduced from van der Wel (1993, Plate 4, p. 897); figure supplied by Rob Hootsmans and Frans van der Wel, with slight modifications from the original published figure.*

discussed above) is viable only for examining variance of one city at a time.

The above approaches to depicting truth of signs are based on distilling truth to a single measure of probability or reliability. Fisher (1994) has provided an alternative that relies on the ability of vision to notice change (and ignore stability). The technique is applied to a classified land-use image derived from remotely sensed information. It involves generating a dynamic display (grid cell based) in which the probability of various land uses for each cell is matched to the temporal variable of duration. For example, if a cell is most likely to be crop land, but could possibly be grazing land, the cell would be depicted as crops most of the time but as grazing part of the time. The exact display "date" matched to the crop versus the grazing depictions is randomized. The result is a map in which uncertain areas (with multiple possible classifications) appear in continual flux and certain areas are stable. Some viewers of this map have criticized the technique because it might imply that the land use is changing over time. No analyst who understands land use change would make this mistake because their schemata for dealing with land use patterns would not allow for continuous alternation in land use. The more serious problem with Fisher's dynamic depiction of reliability is that the changing areas are almost impossible to ignore. It may therefore be difficult to pay attention to areas of stability. The solution to this problem seems to be (as it is with most GVIS representation of complex information) to allow analysts to decide when the reliability representation is used, as well as to control various display parameters such as rate of change, threshold at which alternative classifications are first depicted, and the like.

Truth of the Display as Sign: Seeing Wrong versus Not Seeing

Gregory (1990) points out that when perceptual and conceptual processes are in conflict,[4] the conceptual ones will not always win out. It is possible, for example, to experience, and be convinced by, a visual illusion, even when we "know" it is illusory. Vision, linked to cognitive representations through schemata, has evolved to yield plausible interpretations of the visual scene. It is a process with powerful mechanisms for finding order in potentially chaotic input. When we see a feature or spatial pattern that "makes sense," we are sometimes too easily convinced. In contrast to this tendency to find patterns where none exist, the extremes of data abstraction involved in collecting and processing data for GVIS can easily disguise or hide a pattern that is really there. With statistical analysis we have devised formal tests to limit the likelihood of making these Type I or Type II errors. For GVIS (or scientific visualization in general) we have no such tests. The best that we can do is to adapt the same GVIS tools that created the displays (in which we think we see, or do not see, geographic features) to the task of assessing the truth of the displays as signs.

If we are depicting a visible geographic feature such as terrain, we can judge the truth of any particular map against the real terrain (if we have access to that terrain). For most GVIS applications, however, we do not have direct access to the reality being represented (either due to geographic scale, distance in space or time, or abstractness of the information being represented). One perspective on the results of any particular GVIS operation, then, is that it is analogous to a sample from a large set of possible ways to parse the available data. Treating any single display as true is like treating a single sample from a large population as representative of that population. It might be, particularly if we select the sample using various criteria that we think characterize the prototypic population member, but with a single sample there is no way to know for sure. As with statistical confidence tests, the more "sample" representations we have to compare, the more certain we can be about the truth of the picture they provide. The only practical method for assessing confidence in what we see on maps, then, is to compare multiple views.

Muehrcke (1990) introduced the concept of map *stability*, a concept that serves as a basis for the approach to confidence assessment suggested above. Stability describes the extent to which the map representation remains consistent with small changes in the method used to create it. A map feature that is robust enough to remain through several representations is more likely to be true—that is, we can have more confidence in it. A map feature that only appears fleetingly with an extreme manipulation of map parameters should be accepted only with caution.

To use map stability as a clue to the truth of GVIS, we must consider the probability that certain map symbolization techniques are inherently more or less stable than others. For example, elsewhere I have provided some evidence that perspective terrain views are more stable than isometric views (MacEachren and Ganter, 1990). As the density of input data changes, isometric map representations can change dramatically while perspective views are less drastically affected. This is probably because perspective views put more emphasis on global features while isometric maps draw attention to local details.

A study of viewer ability to judge correlation between map pairs represented with several different symbolization methods provides a further piece of evidence concerning the instability of isometric symbolization. In this study, Muehrcke (1973) compared judgments for choropleth, isarithm, shaded isarithm, and perspective maps on a rank-ordering task (in which 10 maps were arranged in order of strongest association to weakest association with an eleventh standard map). At issue was both how accurate each specific judgment was and how accuracy varied for a range of different kinds of distribution. Errors in correlation judgments (as judged by rank-order mistakes) exhibited the greatest variation for isarithmic maps, with both the lowest and highest error rate associated with this representation method (Figure 10.4). At the other extreme, dot representation seems to result in maps that are less susceptible to the vagaries of a particular data set. Although dot maps did not always lead to the most accurate correlation rankings, correlation accuracy was consistent across all surfaces (except for a linear trend surface—the least likely pattern for real data) and generally toward the high end for symbolization methods compared. Based on principles of visual organization and the idea of indispensable variables, this result makes sense. On dot maps, the sole visual variable used is location in space. An analyst can rely primarily on low-level visual processes to judge similarity rather than needing to compare conjunctions of location and color value (in the choropleth map) or activate a domain-specific schema required to interpret abstract symbolization (in the case of isarithmic maps). While the perspective maps used in the study also rely on space, the key third dimension is only simulated (via a limited set of depth cues) and portions of the map are always obscured.

Muehrcke's (1973) research and that of others (e.g., Steinke and Lloyd, 1983b) indicate that correlation judgment for maps is often inaccurate. Simply relying on human visual ability to compare a set of side-

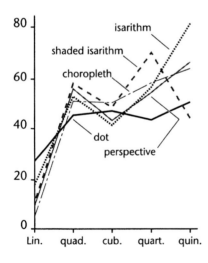

FIGURE 10.4. Comparison of map types for different levels of data complexity on Muehrcke's (1973) similarity judgment task. Maps used in the experiment were derived mathematically using linear, quadratic, cubic, quartic, and quintic trend surfaces (represented on the X-axis of the graph). *After Muehrcke (1973, Fig. 1, p. 193). Adapted by permission of the Association of American Geographers.*

by-side maps to assess map stability, therefore, may not work. An alternative is to treat various versions of a map as if they were maps of different variables and apply the feature comparison techniques discussed in Chapter 9. In our work on visualizing the health of Chesapeake Bay, my students and I have proposed using the alternating syntagm (or flickering) technique for this purpose (MacEachren et al., 1993). Among other uses, we have suggested this method for comparison of dissolved nitrogen surfaces generated from sample points using different interpolation techniques. For each month over the six-year period for which data are available, two image maps generated by alternative interpolation algorithms can be "flickered." When these flickering images are run in sequence as an animation, they should provide both a reliability assessment of maps at each time period, and a way to assess changes in pattern stability over time.

WHAT "TRUTH" IS IN GVIS

The truth of GVIS is, of course, a function of the questions one tries to answer. At the private level of the scientist or the analyst exploring a problem context, we cannot go too far wrong if we simply define truth as what we believe it to be. GVIS extends the ability of analysts for apparently direct observation. Associated with this extension of human observation ability are a variety of implications due to categories and schemata that provide the interface between knowledge and visual depiction. There is considerable evidence to suggest that the "mind-set" embodied in a domain-specific schemata complex brought to bear on a problem will exert a profound bias on what the scientist will see and what is disregarded—on what is accepted as true.

Giere (1988) contends that, for scientists, theoretical models are intended to be "at least partial representations of systems in the real world," but that these models are not expected to have truth relationships with the world. They are instead matched to the world in "respects of similarity" (not degrees of similarity). Models can be considered iconic signs (hypoicons) that have some specified link to a referent through their shared interpretant. As with other hypoicons, similarity may be in respect to visual qualities, relations of parts, or parallelism to another relation (achieved through metaphor). Maps and other graphics displayed by GVIS are models (i.e., signs) representing some aspect of the world. Like other scientific models, they are matched to the world in respects of similarity. From this perspective, GVIS displays are not true or false, only similar or dissimilar, in some respects, to what is represented. Whether GVIS works, then, might be based upon whether those presented with the

GVIS environment recognize the ways in which the GVIS displays are intended to be similar to the phenomenon represented and also recognize that some apparent similarities are in fact coincidental.

As GVIS displays interact with the social context of which they are a part (particularly as they are made public), they take on multiple levels of representation with multiple potential meanings (both denotative and connotative). Some of the connotations suggested will be "true," from several perspectives, and some will be true only from a limited perspective, or not at all. As with any form of information representation, the author of the representation can exploit the likelihood that viewers will take connotations for granted (i.e., accept them as true).

This section is divided (as much of the book has been) on a private–public basis. How truth is defined at the level of individual scientists is considered initially. The section then concludes with a brief look at the way truth can be approached when GVIS provides an interface between science and society.

Visual Thinking and Cognitive Gravity

In outlining a cognitive approach to how science works, De Mey (1992) introduces the concept of *cognitive gravity*. Although De Mey probably does not intend that the gravity metaphor be taken too seriously, he does contend (and this echoes Kuhn, 1970) that many cases of significant discovery in science have required an "extraordinary effort" to overcome the "momentous strength of established conceptual systems." Researchers face the "gravitational pull of established doctrine" (p. xiv), which guides ordinary research but must be overcome to achieve profound insight. Cognitive gravity, then, is the pull exerted by the large volume of accumulated facts and rules.

De Mey's cognitive gravity is instantiated in the form of "global cognitive structures" (i.e., schemata). It might be described as the net result of a shared general and domain-specific schemata complex. GVIS, for the most part, will build upon this schemata complex, making use of category systems and expectations that allow vision to sift and filter vast amounts of information in the search for unique features, spatial patterns, anomalous situations, and so on. The very reasons why GVIS works, however, are why we must be cautions about accepting what it suggests to an analyst. It is hard to see the unexpected (or at least hard to interpret it correctly) because it does not match neatly with existing schemata. In a complementary fashion, if we see something that does "fit" our schemata, it is hard to judge what we see critically.

Schustack (1988) reviews empirical evidence on tasks of rule determination that suggests it is particularly hard to avoid adopting a hypothesis that is too specific. Typical experiments discussed involve tasks in which stimuli that differ in color, shape, and size are organized into categories and the subject must determine the rules for category membership. If a subject, due to initial evidence, develops a hypothesis that one category consists of "red squares," further exploration is typically directed toward testing this hypothesis. If the true rule is more general (e.g., all squares), it may not be found. Schustack (1988, p. 102) contends that illusory correlations and relationships often result from "*confirmatory bias.*" He describes confirmatory bias as a pervasive tendency of human reasoners to perceive data (often erroneously) as confirmatory of their hypotheses, to seek out data selectively that are confirmatory, and to undervalue evidence that is contradictory.

Darius (1990) suggests that scientists may be more susceptible to the "quirks of visual perception" than others. He contends that there is a strong tendency among scientists to believe what they see, particularly if it fits their theoretical perspective (the schemata complex they have developed). Among the many examples he provides, an astronomical one comes closest to a typical GVIS problem. The example involves astronomers using two-dimensional "maps" of star location to identify star "chains" and "circles." Darius (1990, p. 341) quotes the physicist Kastner, who (in the 1950s) contended that "it is obvious that this lining-up could not have occurred by chance alone." The potential chimerical nature of these chains and circles, however, was demonstrated by Lindermann and Burki (cited in Darius, 1990), who generated a series of simulated random star fields. Within these random fields were chains and rings similar to those "identified" for real star fields. As Lakoff (1987) notes for chemists analyzing NMR-produced representations, metaphorical mapping of image schemata to concrete representations can result in the representations being treated as real (particularly through repeated use of a category of representation). One reason that illusory astronomic features were seen as real is that Gestalt grouping principles acted to identify groups according to apparent proximity. Positions seen as proximate in two spatial dimensions were taken (mistakenly) to be proximate in three dimensions.

In relation to the dangers of scientists putting unwarranted faith in what they can observe, Judson (1987) quotes a personal conversation with Francis Crick (one of the discoverers of DNA structure) about research into protein structure by some colleagues. Crick stated that the research had gone "embarrassingly wrong" because the researchers were "mislead" by their observational data. Crick claimed that part of the success he and Watson achieved was due to knowing about this experience and as a result being prepared to "throw away" any data. Crick is quoted as stating that "the point is that evidence can be unreliable, and therefore you should use as little of it as you can." This statement reinforces the potential role of cognitive gravity in truth of visualization. When scientists have a firmly held set of views they are more likely to judge visual evidence that does not support those views as wrong and supporting evidence as true. Judson (1987, p. 208), in reference to this process, calls it "coaxing the evidence into view, creating the phenomenon." From this perspective, there is not truth, the phenomenon does not even exist, until the scientist creates it. In a more geographic context, but arguing from a similar perspective, Penn (1993, p. 54) contends that "useful urban theory will only come about if we first create some phenomena about which to theorize."

Considerable evidence has been amassed that scientists make use of schemata (or patterns) as a means to link observation with knowledge (e.g., Larkin et al., 1980; Margolis, 1987; Giere, 1988). This evidence ranges from philosophical speculation through sociological analysis, to results of empirical cognitive experiments. A study by Larkin et al. (1980) exemplifies the latter. Using protocol analysis of physicists solving real problems in classical mechanics, they provided evidence for "rich schemata" that guide interpretation and at the same time fill in missing details. They contend that "this capacity to use pattern-indexed schemata is probably a large part of what we call physical intuition" (Larkin et al., 1980, p. 1342).

In the case of the Watson–Crick discovery of DNA structure, their rich schemata (that included an understanding of helical structures in amino acids identified by Pauling), willingness to disregard contrary evidence, and ability to build upon supporting evidence led to important breakthroughs. These breakthroughs were prompted, in part, by an ability to recognize the "true" patterns in visual evidence. Helical molecules produce a unique pattern on X-rays. Crick had worked out the mathematics of the pattern and taught Watson how to recognize it. Watson, examining an X-ray of DNA produced in another laboratory, "saw" the telltale pattern. Of course, not all applications of domain-specific schemata have the successful results of Watson and Crick's work on DNA. Holyoak and Nisbett (1988) cite a study of clinical psychologist's reports on differences between male homosexuals and heterosexuals responding to Rorschach inkblot tests. The clinicians reported that the homosexual men saw more genitals in the inkblots. Subsequent analysis determined that this apparent association did not exist, but instead was the result of the clinicians' preparedness to find it.

In an example closer to GVIS, Giere (1988) recounts the sequence of events associated with acceptance of the theory of continental drift. A key component in "proving" the theory was pattern noticing. Specifically, bilateral symmetry was almost immediately noticed when a set of magnetic anomaly profile data spanning the Pacific–Antarctic Ridge was visually examined. Although the symmetry seems quite conclusive today, when continental drift is taken for granted, there were distinct differences of opinion at the time among researchers working at the lab where the pattern was first noticed. According to Giere, Neil Opdyke immediately accepted the pattern as proof of seafloor spreading. James Heirtzler, in contrast, contended that the profile results were too perfect and must be due to electrical currents in the earth's upper mantle. This argument discounted the profiles as signifying anything that happened in the past. Instead they were treated as an interaction between present day electrical fields and the measurement technology being used. It was more than a month before Heirtzler was convinced that the pattern was real. Giere's explanation for the difference in reaction of Opdyke and Heirtzler emphasizes the difference in scientific background that the two researchers brought to the situation. Opdyke was trained in geology and had a thorough understanding of the theory of continental drift (although he previously considered the theory wrong). Heirtzler was trained in physics and had only recently encountered the theory, in the form of the Vine–Mathews hypothesis published two years before. Giere contends that these two scientists brought a different set of "cognitive resources" to bear on interpretation of the profile. I would call this a difference in available schemata to which observations could be matched.

In introducing the expanded model of GVIS feature identification above, I proposed a distinction between spatial and domain-specific schemata. In the continental drift example, both researchers applied the same spatial schema and "saw" bilateral symmetry in the magnetic anomaly profiles. They differed, however, in the domain-specific schema applied, with one matching observations to a temporal geologic model (at a scale of millions of years) while the other applied a model based on physical processes in the earth's crust (operating at human time scales).

The above discussion emphasizes the role of knowledge schemata in determining how truth is to be assessed. The issues of seeing wrong and not seeing are related to how mismatches between visual input and domain-specific schemata are dealt with. To some extent the issue involves the ways in which general (i.e., spatial) schemata interact with domain-specific schemata. A number of researchers in GVIS (and even more in scientific visualization in general) advocate making displays as realistic as possible (e.g., Friedhoff and Benzon, 1989; Pittman, 1992). Increasingly iconic representation, however, will change the balance between general and domain-specific schemata toward general schemata, and will have implications that are hard to predict. Mark and Gold (1991, p. 1428) propose that "when the user sits at a workstation and uses a GIS, he or she should be thinking about real-world phenomena, and not about computers or peripherals, commands or syntax, layers or pixels." They do not, however, discuss the implications, for individual cognitive interaction or for public sharing of insights, of treating the simulated world as "real"—something that is almost certain to happen as we make interfaces more and more transparent.

Public Presentation and Implicit Connotation

Friedhoff and Benzon (1989) argue that the goal of visualization should be to make representations of phenomena as pictorial as possible. Their argument is based on an assumption that human vision (and hearing and touch) has evolved to deal with the real world. If we can make our scientific representations more like the real world, they reason, we can make better use of the full power of human sensory processing systems. That this argument has practical flaws is demonstrated by the centuries of abstract maps that have proven to be more useful than detailed drawings for many kinds of knowledge acquisition and problem solving. What may not be as obvious is that the move toward realism in some aspects of GVIS, and in scientific visualization in general, has implications well beyond those for the nature of visual and cognitive interaction by individual analysts.

As the results of GVIS manipulation of information become public, realism will have a substantial impact on the range of connotations generated. These connotative implications begin with the immediate community of other scientists/analysts with whom the individual scientist/analyst interacts. They extend to the broader community of scientists/analysts to whom visualization results are presented in support of a particular theory or proposal for action. Ultimately, at the societal level, visual displays become a substantial component of the interaction between science and society. They are signs that can be imbued with multiple (perhaps contradictory) meanings. Adding realistic cues to what are otherwise abstract illustrations can have an impact on connotations attached to these illustrations. This potential has even been demonstrated for simple graphs. Carswell et al. (1991), found that people judged a university to be more modern and distinctive if information about the university was presented on 3-D graphs versus 2-D graphs.

We should not assume that it is only naive college freshmen for whom information graphics connote more than they denote. If we judge by how scientists present their work to others in their field, we can uncover considerable evidence that connotation is as important as denotation in illustrations used to "sell" a scientific theory. Meyers's (1988) detailed analysis of illustrations in E. O. Wilson's *Sociobiology* is a case in point. He focuses specifically on the implications of making information graphics look more like objects in the world and less like contrived representations. He presents a continuum of display forms from photographs through drawings, maps, and graphs to imaginary figures (Figure 10.5). This continuum corresponds to ideas about sign iconicity presented in Chapter 6 and matches one axis of my "map as a radial category" concept (introduced in Chapter 4). In the context of scientific visualization, John Ganter and I (MacEachren and Ganter, 1990) introduced a similar continuum based on the "degree of abstraction" (Figure 10.6) that was subsequently adapted to the more specific case of GVIS (MacEachren et al., 1992).

Although Meyers's continuum arranges kinds of illustrations in the same order as I have, he focuses not on abstractness or iconicity of signs, but on the degree to which illustrations include "gratuitous detail." Implicit in his approach is a view that the continuum deals with generality, from "particularity of one observation to the generality of a scientific claim" (i.e., a range from singular to general signs) (Meyers, 1988, p. 235). Meyers contends that "the squiggles and splotches that do not seem relevant to the claim the picture illustrates have their own significance, as part of what makes the picture seem continuous with our own world" (p. 235). In a footnote he cites Bastide (1985), who developed a complementary continuum of illustration with an emphasis on an author's increasing "intervention" in the image. Photographs are accepted as being unbiased because they contain less evidence of author intervention. In

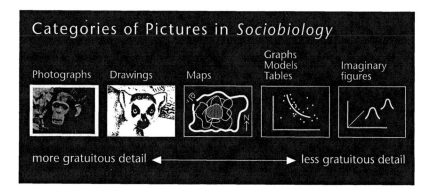

FIGURE 10.5. Meyers's continuum of "pictures" in *Sociobiology*. After Meyers (1988, Fig. 1, p. 234). In M. Lynch and S. Woolgar (Eds.), Representation in Scientific Practice. Cambridge, MA: MIT Press. *Adapted by permission of MIT Press.*

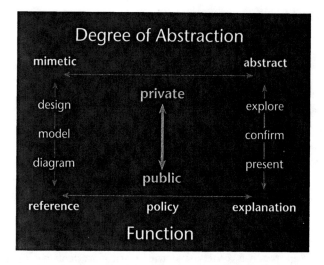

FIGURE 10.6. The range of visualization tool application in relation to abstractness and function. The diagram is intended to suggest that visualization tools incorporating abstract representations are most appropriate for scientific exploration and that as the problem context moves toward applied science, environmental planning, environmental design, and advertising design, mimetic (or image-related) tools become more appropriate.

Meyers's view, it is removal (or inclusion) of gratuitous detail that provides the cue for author intervention (or lack of it).

Meyers, in discussing the extreme of photographs, argues that while giving the appearance of naturalness and lack of observer intervention, photographs are really carefully constructed images that work because they are taken from a particular perspective, appear to be from a certain distance, allow vision to distinguish figure from ground through value and texture contrast, and so on. Because we are used to looking at natural scenes, we immediately assume that the world depicted extends in space beyond the edges of the picture and in time before and after. Much of the detail in Wilson's photograph of chimpanzees is irrelevant to the explicit purpose it seems to serve (to point out various kinds of behavior). Meyers (1988, p. 238) contends that while the detail carries no "relevant" information, it has an important function in making the illustration "seem to be a document recording an unmediated perception of a particular piece of nature." In discussing a map, Meyers points to what he considers unnecessary detail included, like that in the photo, to convince the viewer that the map depicts something real rather than some scientist's abstract theory. Meyers draws particular attention to Wilson's use of maps in arguments about animal territories. For example, a detailed plot of the travels of a band of coati on Barro Colorado Island is interpreted as gratuitous detail that provides "testimony" to the reality of territory. Meyers demonstrates how photographs can be linked to maps, producing an even more compelling "argument" for the reality of a map depiction. He (Meyers, 1988, p. 254) contends that a photo of fish "in their own territories" acts to show "a map of territories that is an inscription of the animals themselves, suggesting that 'territory' is not a theoretical concept created by biologists in interpreting their observations, but is a fact that can be directly observed."

Abstract images can take on characteristics of real objects, and eventually come to be viewed as uncontested facts. Meyers, in the context of the science–society interface, cites the "master images" of the double helix and the Bohr atom. The power of such master images to popularize theory is clear. Part of their power probably stems from their three-dimensionality, a feature that brings the images closer to real objects, that "objectifies" them. Holton (1978; cited in Meyers, 1988) even suggested that sociobiology would prove to be a theory with limited popular appeal due to a lack of a similar single master image with which it can be identified. This view seems to match that of Lakoff (1987), who contends that it is the role of concrete representations in extending basic-level perception and manipulation that gives us confidence that science provides real knowledge (see Chapter 4).

Many of Meyers's arguments about detail in information graphics and evidence of author manipulation apply to maps and other illustrations generated with GVIS. In linking his own analysis of Wilson's book to television portrayals of evolution via natural selection, Meyers notes that television offers a variety of additional cues to realism or naturalness not available to Wilson in the printed form. Color, motion, and change can be used to objectify the portrayal. In addition, sound effects can be added to enhance realism. Just as with the example of 3-D versus 2-D graphs above, the simple addition of these natural cues is likely to make maps included in visualization activities seem more real than maps on a printed page.

Several years ago, Eastman (1985a) suggested that we should consider maps as facilitators rather than communicators of information. This suggestion is complementary with Pinker's (1990) view that displays leading to clear perceptual organization will be most effective. In both cases, it is anticipated that characteristics of the display will influence how easily existing schemata are brought to bear on the problem of deriving information through viewing the display. An overriding issue in the analysis of Wilson's *Sociobiology* is that what the viewer interprets an illustration to be will in large part determine her interpretation of it. The viewer's interpretation will be a function of the entire sign relationship established at both a denotative and a connotative level. In the case of maps, their position along the abstractness (gratuitous-detail) continuum discussed above will play a major role. For GVIS, what viewers interpret spatial displays to be will be related to the concept of "map" as a fuzzy, radial category. Realism in cartographic representation may lead to interpretation of displays as something other than a map. For example, photo-realism in terrain depiction may prompt a landscape photo rather than a map schema. The schema choice may, in turn, lead to unwarranted expectations of veracity.

Wood's (1992) analysis of the Van Sant global image map (see Chapter 7) suggests an interesting mechanism by which subtle connotations can result from the category of illustration we interpret a display to be. If the Van Sant "map" is taken to be a photograph, then the absence of things usually in photographs take on connotations. Wood's argument about the suppression of clouds connoting a fragile earth waiting for something to happen could not be valid for a viewer who applies a map schema. For a viewer relying on a photo schema, however, these connotations may be impossible to ignore.

An interesting piece of anecdotal evidence concerning the power of realism is provided by a comment made in the question period following delivery of a paper at RT '92 (Resource Technology '92, held in Washington, DC, August 1992). The comment is directly relevant to the issue of choosing between photo and map schemata when a landscape representation is highly realistic. Brian Orland presented a talk on visualization of plant growth and pest models. After the presentation, he was asked about his lab's use of rather abstract trees in visualizations supporting a forest-simulation model. He responded that a main reason for choosing these schematic depictions was that less abstract depiction did not allow them to achieve real-time animation/interaction using their current hardware. A move to more realistic depictions was anticipated as faster hardware became available. Someone from the audience then followed with an observation based on experiences in presenting landscape/environmental management plans to the public. This person indicated that in his experience, the degree of realism in depictions will determine the level of public reaction. If a realistic landscape representation is used, the audience tends to treat the plan as final—as not subject to question. In contrast, if maps and/or drawing are used, the plan is treated as a proposal open for comment.

DISCUSSION

This final chapter can only hint at a variety of complex issues related to truth in visual displays. The overall perspective presented is closely linked to the model of GVIS detailed in Chapter 8. This feature ID model presents GVIS as a pattern-matching process (see Figure 8.1). The main contention is that scientists (and humans in general) make decisions by matching present situations against a collection of patterns representing past experience and "knowledge." These patterns will be based on prototypic categories of things, arrangements, and processes. As a consequence, schemata complexes containing propositional, image, and event schemata will provide the mechanism for pattern matching. The schemata applied will, in large part, determine the parameters for judging truth.

An analogy can be drawn between the feature-matching model of GVIS proposed here and Kuhn's theory of paradigm shift in science. As Giere (1988) points out, Kuhn emphasized a sequence in which scientists cope with anomalies, adjust theories, and ultimately face a crisis in which current theory no longer adequately accounts for observations. A paradigm shift may then take place in which one set of exemplars (interpreted here as schemata) is abandoned in favor of another. With the feature-matching model, a schema shift for an individual is analogous to a paradigm shift for the discipline. If the results of GVIS are different enough from expectations, and an appropriate schema is found to structure features seen, the schema shift by an individual may be the catalyst for a subsequent paradigm shift by the discipline as a whole (as it seems to have been for seafloor spreading).

As the results of GVIS move to the public realm, either in the context of interaction among scientists or that of broader communication of science to the public, issues of truth become embroiled in multiple levels of denotation and connotation. A particularly important issue in GVIS, as it is for scientific visualization in general, is the implications of a trend toward realism in visualization tools. Knowlton (1966), as described in Chapter 6, depicts the sign relations for "illustrations" as ranging from no overlap of sign-vehicle and referent at the arbitrary extreme to virtually complete overlap at the perceptual extreme (at which point the sign-vehicle and the referent become confused and the sign relationship no longer exists). There is likely to be a growing tension between an attempt to maintain the impression (if not the reality) that science is objective, and the trend toward elaborate 3-D, dynamic, realistic imagery of scientific visualization where the image becomes reality. It is not certain that scientists who make extensive use of these highly realistic images will be able to hang on to the claim of objectivity and lack of bias that is more readily accepted when their communication tools are abstract and devoid of apparent individuality.

Whatever else it does, realism in scientific representation leads to ambiguous and vague signs and conflicting measures of truth. Whether this is good or bad depends upon your view of science.

NOTES

1. In the paper cited and elsewhere, I used the term "uncertainty" rather than "reliability." Uncertainty was originally selected over "data quality" because the former has broader connotations than the latter. I have decided to opt for "reliability" here as preferable to either of the other terms. Like "data quality," "reliability" is a positive term and thus is preferable to "*un*certainty." In addition, reliability has become a common term in various statistical contexts in which maps and statistical analysis are used in conjunction.

2. With crispness (and to some extent resolution), however, evidence presented in Chapter 8 suggests that blurring, or reducing the crispness, of an entire map rather than of individual sign-vehicles can actually enhance clarity (or certainty) at a global level. While crispness may be a logical way to explicitly signify uncertainty for individual signs, therefore, we must remain cognizant of the implications it might have for apparent truth of the map as sign.

3. The National Center for Geographic Information and Analysis (NCGIA) has played a major role in stimulating much of this research. See Buttenfield (1993) for a presentation of results from the initial NCGIA research initiative and Buttenfield and Beard (1994) for a conceptual framework for further research.

4. The conflict is an example of what Margolis (1987) refers to as "paradox."

Postscript

The foregoing chapters argue for a multilevel, multiperspective, representational approach to maps and mapping. My principal thesis is that to more fully understand how maps work, we need to investigate mechanisms by which maps both represent and prompt representations. The communication paradigm took us a step in this direction but floundered due to a fundamental assumption that matched only a small proportion of mapping situations: maps as primarily a "vehicle" for transfer of information. A representational perspective, in contrast, begins with an assumption that the process of representation results in knowledge that did not exist prior to that representation; thus mapping and map use are processes of knowledge construction rather than transfer. To more fully understand how maps work, then, we must consider the ways in which mapmakers structure knowledge and the ways in which cognitive and social processes applied to the resulting cartographic representations restructure that knowledge and yield multiple alternative representations.

The view of mapping as a representational rather than a communication problem, along with perspectives gleaned from approaches to representation taken in a diverse range of disciplines, led me to the two-pronged attack on representation presented in Parts I and II. When I began this book, I had the general goal of linking cognitive research with cartographic decision making. As I explored an increasing array of literatures, I became convinced that a semiotic approach to cartographic decision-making issues could be a productive one. At the same time, a number of writers were drawing attention to the links between cartography and society. A semiotic approach, again, seemed to provide a useful structure through which these ideas might be further explicated, and a way to link cartographic decisions with their societal consequences. What evolved during the course of writing was an integrated cognitive–semiotic approach to representation. I contend that this approach will result in more robust theory than can either approach in isolation or the sum of the two approaches if pursued independently. The contention remains largely untested, and I hope that this book will encourage people to probe and challenge it.

As a first step toward demonstrating the viability of a cognitive–semiotic approach to cartographic–geographic representation, I sketched (in Part III) a framework through with we might investigate how maps work in the context of geographic visualization. GVIS was selected here as a "case study" because *I view GVIS as the most important development in cartography since the thematic mapping "revolution" of the early 19th century.* For map users, GVIS represents nothing less than a new way to think spatially. For cartographers, GVIS represents a substantial change in emphasis from maps as a presentation tool to maps as part of a thinking–knowledge construction process. Whether that process is one directed toward generating scientific hypotheses or toward making policy decisions, issues of interactivity, multiple views of information, integration of maps with other representation forms, and so on, present cartographers with a plethora of unexplored questions. The extent to which a cognitive–semiotic representational approach can provide useful answers to some of these questions will determine the viability of the approach.

Although I chose to emphasize applications in GVIS, I believe the representational approach delineated in Parts I and II will be applicable to all aspects of cartographic representation. In particular, I think that the approach is well suited to grappling with the many rapidly developing map application areas in which we are faced with a need for new representational forms and new ways to link maps with other information tools. These application areas include wayfinding/travel aids, interactive learning systems, and expert systems for map design.

The representational perspective at the core of this book (and the literature drawn upon to support it) suggests a need for stronger links between cartographic research efforts and other research directed toward spatial representation. These quite obviously include efforts in GIS to develop a theoretical framework for spatial representation and those in remote sensing related to image analysis. In addition, however, we need to forge links to behavioral geography and environmental psychology, with research directed toward understanding spatial knowledge acquisition by humans and the impact of spatial representations on behavior in space; to cognitive psychology, with its concern for understanding the processes by which cognitive spatial representations are developed; to cognitive science, with its concern for modeling thought processes and decision making; to sociology, where there is a growing concern with the role of explicit representation in the functioning of science; to semiotics, where issues of representation, sign meaning, and sign systems are systematically addressed; and to philosophy, where questions of space and representation are pursued as fundamental issues underlying the human condition.

Those of us who study maps and mapping, along with those of us who produce maps, face substantial challenges in the years ahead. Ideas about how maps represent and what it means to represent are in flux. As technologies for exploratory visualization, hypermedia, interactive wayfinding aids, and so on, continue to emerge, a view of maps as a stable product of some cartographic process is being replaced by one in which maps are an element in a larger process of spatial information access and knowledge construction. As mapping moves from the cartographer's darkroom into the light of day—where anyone with access to the Internet, a computer with CD-ROM, or a public library can generate her own map representations—understanding how maps work will become an increasingly important, while increasingly interdisciplinary, endeavor.

References

Ahmad-Taylor, T., and Montesino, D. (1992, December 6). In my backyard? (accompanying map). *New York Times*, p. 54.

Allis, S. (1979, August 5). Underground artistry in New York: When the city changes its subway map, art critics and cartographers crashed head-on. *Washington Post*, p. G3.

Anderson, J. R. (1982). Acquisition of cognitive skill. *Psychological Review, 89*, 369–406.

Andrews, S. K., Otis-Wilborn, A., and Messenheimer-Young, T. (1991). *Beyond Seeing and Hearing: Teaching Geography to Sensory Impaired Children. An Integrated Base Curriculum Approach*. Indiana, PA: NCGE.

Antes, J. R., and Mann, S. W. (1984). Global–local precedence in picture processing. *Psychological Research, 46*, 247–259.

Antonietti, A. (1991). Why does mental visualization facilitate problem-solving? In R. H. Logie and M. Denis (Eds.), *Mental Images in Human Cognition* (pp. 211–227). Amsterdam: Elsevier.

Armstrong, S. L., Gleitman, L. R., and Gleitman, H. (1983). What some concepts might not be. *Cognition, 13*, 263–308.

Arnheim, R. (1974). *Art and Visual Perception*. Berkeley and Los Angeles: University of California Press.

Arnheim, R. (1985). The double-edged mind: Intuition and the intellect. In E. Eisner (Ed.), *Learning and Teaching the Ways of Knowing: Eighty-fifth Yearbook of the National Society for the Study of Education* (pp. 77–96). Chicago: National Society for the Study of Education.

Axelsen, B., and Jones, M. (1987). Are all maps mental maps? *GeoJournal, 14*, 447–464.

Baddeley, A. (1988). Imagery and working memory. In M. Denis, J. Engelkamp, and J. T. E. Richardson (Eds.), *Cognitive and Neuropsychological Approaches to Mental Imagery* (pp. 169–180). Dordrecht: Martinus Nijhoff.

Baddeley, A. D., and Hitch, G. (1974). Working memory. In G. Bower (Ed.), *The*

Psychology of Learning and Motivation, Volume VIII (pp. 47–89). New York: Academic Press.

Balchin, W. G. V. (1988). The media watch in the United Kingdom. In M. Gauthier (Ed.), *Cartographie dans les Médias* (pp. 33–48). Quebec: Presses de l'Université Québec.

Barlow, H. (1990). What does the brain see? How does it understand? In C. B. H. Barlow and M. Weston-Smith (Eds.), *Images and Understanding* (pp. 5–25). Cambridge: Cambridge University Press.

Barsalou, L. W. (1983). Ad hoc categories. *Memory and Cognition, 11*, 211–227.

Barsalou, L. W. (1985). Ideals, central tendency, and frequency of instantiation as determinants of graded structure in categories. *Journal of Experimental Psychology: Learning, Memory, and Cognition, 11*, 629–654.

Barthes, R. (1967). *Elements of Semiology*. London: Cape.

Barthes, R. (1977). *Image—Music—Text*. New York: Hill & Wang.

Bastide, F. (1985, June). Iconographie des textes scientifiques: Principes d'analyse. In B. Latour and J. d. Noblet (Eds.), *Les "Vues" de l'Esprit. Culture Technique, 14* [Special Issue].

Beck, J. (1966). Perceptual grouping produced by changes in orientation and shape. *Science, 154*, 538–540.

Becker, R. A., and Cleveland, W. S. (1987). Brushing scatterplots. *Technometrics, 29*, 127–142.

Bemis, D., and Bates, K. (1989). *Color on Temperature Maps: A Second Look*. Unpublished manuscript, Department of Geography, Pennsylvania State University.

Berlin, B. (1972). Speculations on the growth of ethno-botanical nomenclature. *Language in Society, 1*, 51–86.

Berlin, B., and Kay, P. (1969). *Basic Color Terms: Their Universality and Evolution*. Berkeley and Los Angeles: University of California Press.

Bertin, J. (1981). *Graphics and Graphic Information Processing*. Berlin: Walter de Gruyter. (French edition, 1977)

Bertin, J. (1983). *Semiology of Graphics: Diagrams, Networks, Maps*. Madison: University of Wisconsin Press. (French edition, 1967)

Berton, J. A. J. (1990). Strategies for scientific visualization: Analysis and comparison of current techniques. *Proceedings, Extracting Meaning from Complex Data: Processing, Display, Interaction*, February 14–16, Santa Clara, CA, SPIE—The International Society for Optical Engineering, pp. 110–121.

Beshers, C., and Feiner, S. (1992). Automated design of virtual worlds for visualizing multivariate relations. *Proceedings, Visualization '92*, October 19–23, Boston, IEEE Computer Society Technical Committee on Computer Graphics, pp. 283–290.

Beveridge, M., and Parkins, E. (1987). Visual representation in analogical problem solving. *Memory and Cognition, 15*(3), 230–237.

Beyls, P. (1991). Discovery through interaction: A cognitive approach to computer media in visual arts. *Leonardo, 24*(3), 311–315.

Biederman, I. (1987). Recognition-by-components: A theory of human image understanding. *Psychological Review, 94*(2), 115–147.

Bishop, I. D. (1992). *The Role of Visual Realism in Communicating and Understand-*

ing *Spatial Change and Process*. Working paper for the AGI Visualization Workshop, July 12–14, Loughborough University of Technology.

Bjorklund, E. M. (1991). Culture as input and output of the cognitive–linguistic processes. In D. M. Mark and A. U. Frank (Eds.), *Cognitive and Linguistic Aspects of Geographic Space* (pp. 65–70). Amsterdam: Kluwer Academic.

Board, C. (1967). Maps as models. In R. J. Chorley and P. Haggett (Eds.), *Models in Geography* (pp. 671–725). London: Methuen.

Board, C. (1973). Cartographic communication and standardization. *International Yearbook of Cartography*, 13, 229–236.

Bowman, W. J. (1968). *Graphic Communication*. New York: Wiley.

Brachman, R. J. (1985). On the epistemological status of semantic networks. In R. Brachman and H. Levesque (Eds.), *Readings in Knowledge Representation* (pp. 191–215). San Mateo, CA: Morgan Kaufmann.

Brewer, C. A. (1989). The development of process-printed Munsell charts for selecting map colors. *American Cartographer*, 16(4), 269–278.

Brewer, C. A. (1991). *The Prediction of Surround-Induced Changes in Map Color Appearance*. Unpublished Ph.D. dissertation, Michigan State University.

Brewer, C. A. (1992). *Color Selection for Geographic Data Analysis and Visualization*. Paper presented at GIS/LIS, November, 12, San Jose, CA.

Brewer, C. A. (1993). *Color Selection for Maps: Interactive Tutorial*. Unpublished hypermedia document, Department of Geography, San Diego State University.

Brewer, C. A. (1994). Color use guidelines for mapping and visualization. In A. M. MacEachren and D. R. F. Taylor (Eds.), *Visualization in Modern Cartography* (pp. 123–147). Oxford, UK: Elsevier.

Brewer, C. A., and Marlow, K. A. (1993). Color representation of aspect *and* slope simultaneously. *Proceedings, Auto-Carto 11*, October 30–November 1, Minneapolis, MN, ASPRS & ACSM, pp. 328–337.

Brown, A., and van Elzakker, C. P. J. M. (1993). The use of colour in the cartographic representation of information quality generated by a GIS. *Proceedings, 16th Conference of the International Cartographic Association*, May 3–9, Cologne, Germany, pp. 707–720.

Brown, M. H., and Hershberger, J. (1992, December). Color and sound in algorithm animation. *Computer*, pp. 52–63.

Brown, R. (1958). How shall a thing be called? *Psychological Review*, 65, 14–21.

Bruce, V., and Green, P. R. (1990). *Visual Perception: Physiology, Psychology, and Ecology* (2nd ed.). Hove, UK: Erlbaum.

Buja, A., McDonald, J. A., Michalak, J., and Stuetzle, W. (1991). Interactive data visualization using focusing and linking. *Proceedings, Visualization '91, IEEE Conference on Visualization*, October 22–25, San Diego, CA, pp. 156–163.

Bunn, J. H. (1981). *The Dimensionality of Signs, Tools, and Models*. Bloomington: Indiana University Press.

Buttenfield, B. (Ed.). (1993). *Mapping Data Quality*. *Cartographica*, 30(2 & 3) [Special Content Issue].

Buttenfield, B., and Beard, K. (1994). Graphical and geographical components of data quality. In H. Hearnshaw and D. Unwin (Eds.), *Visualization in Geographical Information Systems* (pp. 150–157). London: Wiley.

Buxton, B. (1990). Using our ears: An introduction to the use of nonspeech audio cues. *Proceedings, Extracting Meaning from Complex Data: Processing, Display, Interaction*, February 14–16, Santa Clara, CA, SPIE—The International Society for Optical Engineering, pp. 124–127.

Caivano, J. L. (1990). Visual texture as a semiotic system. *Semiotica*, 80(3/4), 239–252).

Campbell, J., and Davis, J. (1979). *The Northern Part of the Wakarusa Quadrangle, Kansas Atlas*. Kansas Geological Survey, University of Kansas, Lawrence, KS, and the experimental Cartography Unit, Natural Environmental Research Council, London, UK.

Carr, D. B., Olsen, A. R., and White, D. (1992). Hexagon mosaic maps for display of univariate and bivariate geographical data. *Cartography and Geographic Information Systems*, 19(4), 228–236.

Carswell, C. M., and Wickens, C. D. (1990). The perceptual interaction of graphical attributes: Configurality, stimulus homogeneity, and object integration. *Perception and Psychophysics*, 47, 157–168.

Carswell, C. M., Frankenberger, S., and Bernhard, D. (1991). Graphing in depth: Perspectives on the use of three-dimensional graphs to represent lower-dimensional data. *Behavior and Information Technology*, 10(6), 459–474.

Castner, H. W. (1983). Tactual maps and graphics: Some implications for our study of visual cartographic communication. *Cartographica*, 20(3), 1–16.

Castner, H. W. (1990). *Seeking New Horisons: A Perceptual Approach to Geographic Education*. Montreal: McGill–Queen's University Press.

Castner, H. W., and Eastman, J. R. (1985). Eye-movement parameters and perceived map complexity—II. *American Cartographer*, 12(1), 29–40.

Castner, H. W., and Robinson, A. (1969). *Dot Area Symbols in Cartography: The Influence of Pattern on Their Perception*. Washington, DC: American Congress on Surveying and Mapping.

Cave, K. R., and Wolfe, J. M. (1990). Modeling the role of parallel processing in visual search. *Cognitive Psychology*, 22, 225–271.

Cavenagh, P. (1987). Reconstructing the third dimension: Interactions between color, texture, motion, binocular disparity and shape. *Computer Vision, Graphics and Image Processing*, 37, 171–195.

Cavenagh, P. (1988). Pathways in early vision. In Z. Pylyshyn (Ed.), *Computational Processes in Human Vision* (pp. 239–261). Norwood, NJ: Ablex.

Chang, K.-T., Antes, J., and Lenzen, T. (1985). The effect of experience on reading topographic relief information: Analysis of performance and eye movements. *Cartographic Journal*, 22(2), 88–94.

Chase, W. G., and Simon, H. A. (1973). The mind's eye in chess. In W. Chase (Ed.), *Visual Information Processing* (pp. 215–281). New York: Academic Press.

Cheal, M., and Lyon, D. R. (1992). Attention in visual search: Multiple search classes. *Perception and Psychophysics*, 52(2), 113–138.

Cheng, P. W., and Holyoak, K. J. (1985). Pragmatic reasoning schemas. *Cognitive Psychology*, 17, 391–416.

Cheng, P. W., and Pachella, R. G. (1984). A psychological approach to dimensional separability. *Cognitive Psychology, 16,* 279–304.
Cleveland, W. S. (1991). *A Model for Studying Display Methods of Statistical Graphics.* Unpublished manuscript, AT&T Bell Laboratories.
Cleveland, W. S. (1993). A model for studying display methods of statistical graphics. *Journal of Computational and Graphical Statistics, 2*(4), 323–343.
Cleveland, W. S., and McGill, R. (1984). Graphical perception: Theory, experimentation, and application to the development of graphical methods. *Journal of the American Statistical Association, 79,* 531–554.
Coulson, M. R. C. (1987). In the matter of class intervals for choropleth maps: With particular attention to the work of George F. Jenks. *Studies in Cartography, Monograph 37, Cartographica, 24*(2), 16–39.
Coulson, M. R. C. (1991). Progress in creating tactile maps from geographic information systems (G.I.S.) output. *Proceedings, 15th Conference of the International Cartographic Association,* September 23–October 1, Bournemouth, UK, pp. 167–174.
Cox, C. W. (1976). Anchor effects and the estimation of graduated circles and squares. *American Cartographer, 3,* 65–74.
Crampton, J. (1992a). A cognitive analysis of wayfinding expertise. *Cartographica, 29*(3/4), 46–65.
Crampton, J. (1992b). New directions in the information era—a reply to Taylor (1991). *Cartographic Journal, 29*(2), 145–150.
Crampton, J. (1994). *Alternative Cartographies: New Frontiers of Human Issues.* Unpublished Ph.D. dissertation, Department of Geography, Pennsylvania State University.
Crawford, P. V., and Marks, R. A. (1973). The visual effects of geometric relationships on three-dimensional maps. *Professional Geographer, 25*(3), 233–238.
Cuff, D. J. (1973). Colour on temperature maps. *Cartographic Journal, 10,* 17–21.
Dacey, M. (1970). Linguistic aspects of maps and geographic information. *Ontario Geography, 5,* 71–80.
Darius, J. (1990). Scientific images: Perception and deception. In C. B. H. Barlow and M. Weston-Smith (Eds.), *Images and Understanding* (pp. 333–357). Cambridge: Cambridge University Press.
DeLucia, A. A., and Hiller, D. W. (1982). Natural legend design for thematic maps. *Cartographic Journal, 19,* 46–52.
DeMey, M. (1992). *The Cognitive Paradigm.* Chicago: University of Chicago Press.
Dennett, D. (1990). Thinking with a computer. In C. B. H. Barlow and M. Weston-Smith (Eds.), *Images and Understanding* (pp. 297–309). Cambridge: Cambridge University Press.
Dent, B. D. (1970). *Perceptual Organization and Thematic Map Communication: Some Principles for Effective Map Design with Special Emphasis on the Figure–Ground Relationship.* Worcester, MA: Department of Geography, Clark University.

Dent, B. D. (1972). Visual organization and thematic map communication. *Annals of the Association of American Geographers, 62*(1), 79–93.
Dent, B. D. (1993). *Cartography: Thematic Map Design* (3rd ed.). Dubuque, IA: Wm. C. Brown.
DeValois, R. L., and Jacobs, G. H. (1968). Primate color vision. *Science, 162,* 533–540.
DeYoe, E., Knierem, J., Sagi, S., Julesa, B., and van Essen, D. (1986). Single unit responses to static and dynamic texture patterns in macaque V2 and V1 cortex. *Investigative Ophthalmology and Visual Science, 27,* 18.
DiBiase, D. (1990). Visualization in the earth sciences. *Earth and Mineral Sciences, Bulletin of the College of Earth and Mineral Sciences, Pennsylvania State University, 59*(2), 13–18.
DiBiase, D., Krygier, J., Reeves, C., MacEachren, A. M., and Brenner, A. (1991a). Animated cartographic visualization in earth system science. *Proceedings, 15th Conference of the International Cartographic Association,* Bournemouth, UK, September 23–October 1, pp. 223–232.
DiBiase, D., Krygier, J., Reeves, C., MacEachren, A. M., and Brenner, A. (1991b). *An Elementary Approach to Cartographic Animation.* University Park, PA: Deasy GeoGraphics Laboratory, Department of Geography, Pennsylvania State University [video].
DiBiase, D., MacEachren, A. M., Krygier, J. B., and Reeves, C. (1992). Animation and the role of map design in scientific visualization. *Cartography and Geographic Information Systems, 19*(4), 201–214.
DiBiase, D., Paradis, T., and Sloan, J. L. I. (1994a). Weighted isolines: An alternative method of isoline symbolization. *Professional Geographer, 46*(2), 218–228.
DiBiase, D., Reeves, C., Krygier, J., MacEachren, A. M., von Wyss, M., Sloan, J., et al. (1994b). In A. M. MacEachren and D. R. F. Taylor (Eds.), *Visualization in Modern Cartography* (pp. 287–312). Oxford, UK: Elsevier.
DiBiase, D., Reeves, C., MacEachren, A. M., Krygier, J., von Wyss, M., Sloan, J., et al. (1993). A map interface for exploring multivariate paleoclimate data. *Proceedings, Auto-Carto 11,* Minneapolis, MN, October 30–November 1, ASPRS & ACSM, pp. 43–52.
Dobson, M. W. (1973). Choropleth maps without class intervals? A comment. *Geographical Analysis, 5,* 262–265.
Dobson, M. W. (1979a). The influence of map information on fixation localization. *American Cartographer, 6,* 51–65.
Dobson, M. W. (1979b). Visual information processing during cartographic communication. *Cartographic Journal, 16,* 14–20.
Dobson, M. W. (1983). Visual information processing and cartographic communication: The utility of redundant stimulus dimensions. In D. R. F. Taylor (Ed.), *Graphic Communication and Design in Contemporary Cartography, Volume 2, Progress in Contemporary Cartography* (pp. 149–175). New York: Wiley.
Dobson, M. W. (1985). The future of perceptual cartography. *Cartographica, 22*(2), 27–43.
Dorling, D. (1992a). *Persuading Older Research Workers of the Wonders of Car-*

tograms. Paper presented at the IBG Annual Conference, January 6, Royal Holloway, London.
Dorling, D. (1992b). Stretching space and splicing time: From cartographic animation to interactive visualization. *Cartography and Geographic Information Systems, 19*(4), 215–227.
Dorling, D. (1992c). Visualizing people in time and space. *Environment and Planning B: Planning and Design, 19*, 613–637.
Dorling, D., and Openshaw, S. (1992). Using computer animation to visualize space–time patterns. *Environment and Planning B: Planning and Design, 19*, 639–650.
Downing, C. J., and Pinker, S. (1985). The spatial structure of visual attention. In M. I. Posner and O. S. M. Martin (Eds.), *Attention and Performance, Volume XI* (pp. 171–188). Hillsdale, NJ: Erlbaum.
Downs, R. M., and Liben, L. S. (1987). Children's understanding of maps. In P. Ellen and C. Thinus-Blanc (Eds.), *Cognitive Processes and Spatial Orientation in Animal and Man, Volume II, Neurophysiology and Development Aspects* (pp. 202–219). Dordrecht: Martinus Nijhoff.
Downs, R. M., Liben, L. S., and Daggs, D. G. (1988). On education and geographers: The role of cognitive development theory in geographic education. *Annals of the Association of American Geographers, 78*, 680–700.
Duncan, J. (1984). Selective attention and the organization of visual information. In M. I. Posner and O. S. M. Marin (Eds.), *Attention and Performance, Volume XI* (pp. 85–106). Hillsdale, NJ: Erlbaum.
Duncan, J., and Humphreys, G. W. (1987). Visual search and stimulus similarity. *Psychological Review, 96*, 433–448.
Eastman, J. R. (1981). The perception of scale change in small-scale map series. *American Cartographer, 8*(1), 5–21.
Eastman, J. R. (1985a). Cognitive models and cartographic design research. *Cartographic Journal, 22*(2), 95–101.
Eastman, J. R. (1985b). Graphic organization and memory structures for map learning. *American Cartographer, 22*, 1–20.
Eastman, J. R. (1986). Opponent process theory and syntax for qualitative relationships in quantitative series. *American Cartographer, 13*(4), 324–333.
Eastman, J. R. (1987). Graphic syntax and expert systems for map design. *Technical Papers, ACSM & ASPRS, 4*, 87–96.
Eastman, J. R., and Castner, H. W. (1983). The meaning of experience in task-specific map reading. In D. R. F. Taylor (Ed.), *Graphic Communication and Design in Contemporary Cartography, Volume 2, Progress in Contemporary Cartography* (pp. 115–148). New York: Wiley.
Eco, U. (1972). *Einführung in die Semiotik*. Munich: Fink.
Eco, U. (1976). *A Theory of Semiotics*. Bloomington: Indiana University Press.
Eco, U. (1977). *Zeichen: Einführung in einen Begriff und seine Geschichte*. Frankfurt: Suhrkamp. (Italian edition, 1973)
Eco, U. (1985a). How culture conditions the colors we see. In M. Blonsky (Ed.), *On Signs* (pp. 157–175). Baltimore: Johns Hopkins University Press.
Eco, U. (1985b). Producing signs. In M. Blonsky (Ed.), *On Signs* (pp. 176–183). Baltimore: Johns Hopkins University Press.

Edney, M. H. (1993). Cartography without progress: Reinterpreting the nature of historical development of mapmaking. *Cartographica, 30*(1), 54–68.

Edwards, G. (1991). Spatial knowledge for image understanding. In D. M. Mark and A. U. Frank (Eds.), *Cognitive and Linguistic Aspects of Geographic Space* (pp. 295–307). Amsterdam: Kluwer Academic.

Ellis, W. (1955). *Sourcebook of Gestalt Psychology*. London: Routledge & Kegan Paul.

Engel, F. (1977). Visual conspicuousness, visual search and fixation tendencies of the eye. *Vision Research, 17*, 95–108.

Eriksen, C. W. (1952). Location of objects in a visual display as a function of the number of dimensions on which the objects differ. *Journal of Experimental Psychology, 44*, 126–132.

Eriksen, C. W., and Murphy, T. D. (1987). Movement of attentional focus across the visual field: A critical look at the evidence. *Perception and Psychophysics, 42*, 299–305.

Eyton, J. R. (1990). Color stereoscopic effect cartography. *Cartographica, 27*(1), 20–29.

Familant, M. E., and Detweiler, M. C. (1993). Iconic reference: Evolving perspectives and an organizing framework. *International Journal of Man–Machine Studies, 39*, 705–728.

Ferreia, J., and Wiggins, L. (1990). The density dial: A visualization tool for thematic mapping. *GeoInfo Systems, 1*, 69–71.

Finke, R. (1980). Levels of equivalence in imagery and perception. *Psychological Review, 87*, 113–132.

Fisher, P. (1993). Visualizing uncertainty in soil maps by animation. *Cartographica, 30*(2 & 3), 20–27.

Fisher, P. (1994). Randomization and sound for the visualization of uncertain spatial information. In D. Unwin and H. Hearnshaw (Eds.), *Visualization in Geographic Information Systems* (pp. 181–185). London: Wiley.

Ford, K. L. (1985). Map reading, divided attention, and limited viewing area (abstract). *Conference Program and Abstracts, 5th Annual Meeting of the North American Cartographic Information Society*, November 10–13, Chicago, p. 10.

Forrest, D., and Castner, H. W. (1985). The design and perception of point symbols for tourist maps. *Cartographic Journal, 22*, 11–19.

Friedhoff, R. M., and Benzon, W. (1989). *Visualization: The Second Computer Revolution*. New York: Harry Abrams.

Frutiger, A. (1989). *Signs and Symbols: Their Design and Meaning*. New York: Van Nostrand Reinhold.

Ganter, J. H., and MacEachren, A. M. (1989). *Cognition and the Design of Scientific Visualization Systems*. Unpublished manuscript, Department of Geography, Pennsylvania State University.

Garling, T., Lindberg, E., and Mantyla, T. (1983). Orientation in buildings: Effects of familiarity, visual access, and orientation aids. *Journal of Applied Psychology, 68*, 177–186.

Garner, W. R. (1976). Interaction of stimulus dimensions in concept and choice processes. *Cognitive Psychology, 8*, 98–123.

Gerber, R. V. (1981). Competence and performance in cartographic language. *Cartographic Journal*, 18(2), 104–111.

Gerber, R., Burden, P., and Stanton, G. (1990). Development of public information symbols for tourism and recreation mapping. *Cartographic Journal*, 27(2), 92–103.

Gershon, N. D. (1992). Visualization of fuzzy data using generalized animation. *Proceedings, Visualization '92*, October 19–23, Boston, IEEE Computer Society Technical Committee on Computer Graphics, pp. 268–273.

Gersmehl, P. J. (1977). Soil taxonomy and mapping. *Annals of the Association of American Geographers*, 67(3), 419–428.

Gersmehl, P. J. (1990, Spring). Choosing tools: Nine metaphors of four-dimensional cartography. *Carographic Perspectives*, No. 5, 3–17.

Gibson, A. E. (1987). A model describing options for parallel color/data structuring. *Technical Papers, ACSM & ASPRS*, 4, 97–106.

Gibson, J. J. (1950). *Perception of the Visual World*. New York: Houghton Mifflin.

Gibson, J. J. (1979). *An Ecological Approach to Visual Perception*. Boston: Houghton Mifflin.

Giere, R. N. (1988). *Explaining Science: A Cognitive Approach*. Chicago: University of Chicago Press.

Gilmartin, P. P. (1978). Evaluation of thematic maps using the semantic differential test. *American Cartographer*, 5, 133–139.

Gilmartin, P. P. (1981). Influence of map content on circle perception. *Annals of the Association of American Geographers*, 71, 253–258.

Gilmartin, P. P. (1988). The design of choropleth shadings for maps on 2- and 4-bit color graphics monitors. *Cartographica*, 25(4), 1–10.

Gilmartin, P. P., and Shelton, E. (1989). Choropleth maps on high resolution CRTs: The effects of number of classes and hue on communication. *Cartographica*, 26(2), 40–52.

Goldberg, J. H., MacEachren, A. M., and Kotval, X. P. (1992). Mental image transformations in terrain map comparison. *Cartographica*, 29(2), 46–59.

Goldhamer, H. (1934). The influence of area, position, and brightness in the visual perception of a reversible configuration. *American Journal of Psychology*, 46(2), 189–206.

Golledge, R. G., Gale, N., Pellegrino, J. W., and Doherty, S. (1992). Spatial knowledge acquisition by children: Route learning and relational distances. *Annals of the Association of American Geographers*, 82(2), 223–244.

Golledge, R. G., and Stimson, R. J. (1987). Spatial cognition. In *Analytical Behavioral Geography* (pp. 52–83). New York: Croom Helm.

Gombrich, E. (1990). Pictorial instructions. In C. B. H. Barlow and M. Weston-Smith (Eds.), *Images and Understanding* (pp. 26–45). Cambridge: Cambridge University Press.

Goodman, N. (1976). *Languages of Art: An Approach to a Theory of Symbols* (2nd ed.). Indianapolis: Hackett.

Gould, P., Kabel, J., Gorr, W., and Golub, A. (1991). AIDS: Predicting the next map. *Interfaces*, 21(3), 80–92.

Gratzer, M. A., and McDowell, R. D. (1971). *Adaptation of an Eye Movement*

Recorder to Aesthetic Environment Mensuration. Storrs, CT: Storrs Agricultural Station, University of Connecticut.

Gregory, R. (1990). How do we interpret images? In C. M. H. Barlow and M. Weston-Smith (Eds.), *Images and Understanding* (pp. 310–330). Cambridge: Cambridge University Press.

Griffin, T. L. C. (1985). Group and individual variations in judgment and their relevance to the scaling of graduated circles. *Cartographica, 22*(1), 21–37.

Grinstein, G., Sieg, J. C. J., Smith, S., and Williams, M. G. (1992). Visualization for knowledge discovery. *International Journal of Intelligent Systems, 7*, 637–648.

Groop, R. E., and Cole, D. (1978). Overlapping graduated circles: Magnitude estimation and method of portrayal. *Canadian Cartographer, 15*, 114–122.

Guiraud, P. (1975). *Semiology*. London: Routledge & Kegan Paul.

Hadamard, J. (1945). *An Essay on the Psychology of Invention in the Mathematical Field*. New York: Dover.

Haining, R. P. (1978). A spatial model for high plains agriculture. *Annals of the Association of American Geographers, 68*, 493–504.

Hall, S. S. (1992). *Mapping the Next Millenium: The Discovery of New Geographies*. New York: Random House.

Handel, S., Imai, S., and Spottswood, P. (1980). Dimensional similarity and configural classification of integral and separable stimuli. *Perception and Psychophysics, 28*, 205–212.

Hannah, M. (1993). *Foucault Deinstitutionalized: Spatial Prerequisites for Modern Social Control*. Unpublished Ph.D. dissertation, Department of Geography, Pennsylvania State University.

Harley, J. B. (1988). Maps, knowledge, and power. In D. Cosgrove and S. Daniels (Eds.), *The Iconography of Landscape: Essays on the Symbolic Representation, Design and Use of Past Environments* (pp. 277–311). Cambridge: Cambridge University Press.

Harley, J. B. (1989). Deconstructing the map. *Cartographica, 26*(2), 1–20.

Harley, J. B. (1990). Cartography, ethics, and social theory. *Cartographica, 27*(2), 1–23.

Harrower, M. (1936). Some factors determining figure–ground articulation. *British Journal of Psychology, 4*, 407–424.

Hartshorne, C., and Weiss, P. (Eds.). (1931). *The Collected Papers of Charles Sanders Peirce, Volumes I and II*. Cambridge, MA: Belknap Press of Harvard University Press.

Harvey, D. (1969). *Explanation in Geography*. London: Edward Arnold.

Harwood, K. (1989). Cognitive perspectives on map displays for helicopter flight. *Proceedings, 33rd Annual Meeting of the Human Factors Society*, October 16–18, Denver, CO, pp. 13–17.

Head, C. G. (1984). The map as natural language: A paradigm for understanding. *Cartographica, 21*(1), 1–31.

Hervey, S. (1982). *Semiotic Perspectives*. London: Allen & Unwin.

Hochberg, J. (1980). Pictorial functions and perceptual structures. In M. A. Hagen (Eds.), *The Perception of Pictures* (pp. 47–93). New York: Academic Press.

Hoffman, D. D., and Richards, W. A. (1984). Parts of recognition. *Cognition, 18,* 65–96.
Hoffman, H. S. (1989). *Vision and the Art of Drawing.* Englewood Cliffs, NJ: Prentice-Hall.
Hoffman, R. R., Detweiler, M., Conway, J. A., and Lipton, K. S. (1993). Some considerations in using color in meteorological displays. *Weather and Forecasting, 8,* 505–518.
Holyoak, K. J., and Nisbett, R. E. (1988). Induction. In R. J. Sternberg and E. E. Smith (Eds.), *The Psychology of Human Thought* (pp. 50–91). Cambridge: Cambridge University Press.
Howard, V. A. (1980). Theory of representation: Three questions. In P. A. Kolers, M. E. Wrolstad, and H. Bouma (Eds.), *Processing of Visible Language, Volume 2* (pp. 501–515). New York: Plenum Press.
Hsu, M.-L. (1979). The cartographer's conceptual process and thematic symbolization. *American Cartographer, 6,* 117–127.
Hubel, D. H. (1988). *Eye, Brain and Vision.* New York: Scientific American Library.
Humphreys, G. W., and Bruce, V. (1989). *Visual Cognition: Computational, Experimental and Neuropsychological Perspectives.* London: Erlbaum.
Humphreys, G. W., and Quinlan, P. T. (1987). Normal and pathological processes in visual object constancy. In G. W. Humphreys and M. J. Riddoch (Eds.), *Visual Object Processing: A Cognitive Neurological Approach* (pp. 43–105). London: Erlbaum.
Hurvich, L. M., and Jameson, D. (1957). Opponent-process theory of color vision. *Psychological Review, 64*(6), 384–404.
Imhof, E. (1982). *Cartographic Relief Presentation.* Berlin: Walter de Gruyter. (German edition, 1965)
Innis, R. E. (Ed.). (1985). *Semiotics: An Introductory Anthology.* Bloomington: Indiana University Press.
Jakobson, R. (1960). Linguistics and poetics. In T. A. Sebeok (Ed.), *Style in Language* (pp. 350–377). Cambridge, MA: MIT Press.
James, W. (1950). *The Principles of Psychology, Volume 1.* New York: Dover. (Original work published 1890)
Japan Cartographers Association. (1980). Examples of methods used for oceanic cartography in Japan. *The Dynamics of Oceanic Cartography, Monograph 25, Cartographica, 17*(2), 157–181.
Jenks, G. F. (1963). Generalization in statistical mapping. *Annals of the Association of American Geographers, 53*(1), 15–26.
Jenks, G. F. (1967). The data model concept in statistical mapping. *International Yearbook of Cartography, 7,* 186–190.
Jenks, G. F. (1973). Visual integration in thematic mapping: Fact or fiction? *International Yearbook of Cartography, 13,* 27–34.
Jenks, G. F. (1975). The evaluation and prediction of visual clustering in maps symbolized with proportional circles. In J. Davis and M. McCullaugh (Eds.), *Display and Analysis of Spatial Data* (pp. 311–327). New York: Wiley.
Jenks, G. F. (1977). *Optimal Data Classification for Choropleth Maps.* Lawrence: Department of Geography, University of Kansas.

Jenks, G. F., and Brown, D. A. (1966). Three-dimensional map construction. *Science, 154*, 857–864.

Jenks, G. F., and Caspall, F. C. (1971). Error on choroplethic maps: Definition, measurement, reduction. *Annals of the Association of American Geographers, 61*, 217–244.

Johnson, G. (1983). *Qualitative Symbology: An Evaluation of the Pictorial Signs of the National Park Service as Cartographic Symbols.* Unpublished master's thesis, Virginia Polytechnic Institute and State University.

Johnson, M. (1987). *The Body in the Mind: The Bodily Basis of Meaning, Imagination, and Reason.* Chicago: University of Chicago Press.

Judson, H. F. (1987). *The Search for Solutions.* Baltimore: Johns Hopkins University Press.

Julesz, B. (1965). Texture and visual perception. *Scientific American, 212*, 38–48.

Julesz, B. (1975). Experiments in the visual perception of texture. *Scientific American, 232*, 34–43.

Julesz, B. (1981). Textons, the elements of texture perception, and their interactions. *Nature, 290*, 91–97.

Kaiser, M. K., and Proffitt, D. R. (1992). Using the stereokinetic effect to convey depth: Computationally efficient depth-from-motion displays. *Human Factors, 34*(5), 571–581.

Kay, P., and McDaniel, C. (1978). The linguistic significance of the meanings of basic color terms. *Language, 54*(3), 610–646.

Keates, J. S. (1982). *Understanding Maps.* New York: Halsted Press.

Keates, J. S. (1984). The cartographic art. *New Insights in Cartographic Communication, Monograph 31, Cartographica, 23*(1), 37–43.

Kienker, P. K., and Sejnowski, T. J. (1986). Separating figure from ground with a parallel network. *Perception, 15*, 197–216.

Kimchi, R. (1988). Selective attention to global and local levels in the comparison of hierarchical patterns. *Perception and Psychophysics, 43*, 189–198.

Kimerling, A. J. (1985). The comparison of equal-value gray scales. *American Cartographer, 12*, 132–142.

Kinchla, R. A., and Wolfe, J. M. (1979). The order of visual processing: "Top-down," "bottom-up" or "middle-out." *Perception and Psychophysics, 25*, 225–231.

Kishto, B. N. (1965). The color stereoscopic effect. *Vision Research, 5*, 313–329.

Klymenko, V., and Weisstein, N. (1986). Spatial frequency differences can determine figure–ground organization. *Journal of Experimental Psychology: Human Perception and Performance, 12*(3), 324–330.

Klymenko, V., and Weisstein, N. (1989). Figure and ground in space and time: 1. Temporal response surfaces of perceptual organization. *Perception, 18*(5), 627–637.

Klymenko, V., Weisstein, N., Topolski, R., and Hsieh, C. (1989). Spatial and temporal frequency in figure–ground organization. *Perception and Psychophysics, 45*, 395–403.

Knowlton, J. Q. (1966). On the definition of "picture." *AV Communication Review, 14*, 157–183.

Koffka, K. (1935). *Principles of Gestalt Psychology*. New York: Harcourt, Brace & Co.
Kohler, W. (1947). *Gestalt Psychology*. New York: Liveright.
Koláčný, A. (1969). Cartographic information—a fundamental concept and term in modern cartography. *Cartographic Journal, 6*, 47–49.
Korac, N. (1988). Functional, cognitive and semiotic factors in the development of audiovisual comprehension. *ECTJ, 36*(2), 67–91.
Kosslyn, S. M. (1980). *Images and Mind*. Cambridge, MA: Harvard University Press.
Kosslyn, S. M. (1981). The medium and the message in mental imagery. *Psychological Review, 88*, 46–66.
Kosslyn, S. M. (1989). Understanding charts and graphs. *Applied Cognitive Psychology, 3*, 185–226.
Kosslyn, S. M., and Koenig, O. (1992). *Wet Mind: The New Cognitive Neuroscience*. New York: Free Press.
Kraak, M.-J. (1988). *Computer-Assisted Cartographical Three-Dimensional Imaging Techniques*. Delft: Delft University Press.
Kraak, M.-J. (1989). Computer-assisted cartographical 3D imaging techniques. In J. Raper (Ed.), *Three-Dimensional Applications in Geographic Information Systems* (pp. 99–113). London: Taylor & Francis.
Kraak, M.-J. (1993). Cartographic terrain modeling in a three-dimensional GIS environment. *Cartography and Geographic Information Systems, 20*(1), 13–18.
Krampen, M. (1965). Signs and symbols in graphic communication. *Design Quarterly, 62*, 1–31.
Kröse, B. (1987). Local structure analyzers as determinants of preattentive pattern discrimination. *Biological Cybernetics, 55*, 289–298.
Krygier, J. (1991). *An elemental approach to animation and sound in information graphics*. Electronically published on the INGRAPHX listserve.
Krygier, J. (1994). Sound and geographic visualization. In A. M. MacEachren and D. R. F. Taylor (Eds.), *Visualization in Modern Cartography* (pp. 149–166). Oxford, UK: Elsevier.
Krygier, J. (forthcoming). *Visualization, Geography, and Derelict Landscapes*. Unpublished Ph.D. dissertation, Pennsylvania State University.
Kubovy, M. (1981). Concurrent pitch segregation and the theory of indispensable attributes. In M. Kubovy and J. Pomerantz (Eds.), *Perceptual Organization* (pp. 55–98). Hillsdale, NJ: Erlbaum.
Kuhn, T. S. (1970). *The Structure of Scientific Revolutions* (2nd ed.). Chicago: University of Chicago Press.
Labov, W. (1973). The boundaries of words and their meanings. In J. Fishman (Ed.), *New Ways of Analyzing Variation in English* (pp. 340–373). Washington, DC: Georgetown University Press.
Lakoff, G. (1987). *Woman, Fire, and Dangerous Things: What Categories Reveal about the Mind*. Chicago: University of Chicago Press.
Lakoff, G., and Johnson, M. (1980). *Metaphors We Live By*. Chicago: University of Chicago Press.
Lamb, M., and Robertson, L. (1988). The processing of hierarchical stimuli: Ef-

fects of retinal locus, locational uncertainty, and stimulus identity. *Perception and Psychophysics*, 44(2), 172–181.

Land, E. H., and McCann, J. J. (1971). Lightness and retinex theory. *Journal of the Optical Society of America*, 61, 1–11.

Larkin, J., McDermott, J., Simon, D. P., and Simon, H. A. (1980). Expert and novice performance in solving physics problems. *Science*, 208, 1335–1342.

Larkin, J. H., and Simon, H. A. (1987). Why a diagram is (sometimes) worth ten thousand words. *Cognitive Science*, 11, 65–99.

Lavin, S. (1986). Mapping continuous geographical distributions using dot-density shading. *American Cartographer*, 13(2), 140–150.

Leach, E. (1964). Anthropological aspects of language: Animal categories and verbal abuse. In E. H. Lenneberg (Ed.), *New Directions in the Study of Language* (pp. 23–64). Cambridge, MA: MIT Press.

Leonard, J. J., and Buttenfield, B. P. (1989). An equal value gray scale for laser printer mapping. *American Cartographer*, 16(2), 97–107.

Levine, M., Jankovic, L., and Palij, M. (1982). Principles of spatial problem solving. *Journal of Experimental Psychology: General*, 111, 157–175.

Lewandowsky, S., and Spence, I. (1989). Discriminating strata in scatterplots. *Journal of the American Statistical Association*, 84(407), 682–688.

Lewis, P. (1992). Introducing a cartographic masterpiece: A review of the U.S. Geological Survey's digital terrain map of the United States, by Gail Thelin and Richard Pike. *Annals of the Association of American Geographers*, 82(2), 289–304.

Liben, L. S., and Downs, R. M. (1989). Understanding maps as symbols: The development of map concepts in children. *Advances in Child Development and Behavior*, 22, 145–201.

Lindauer, M., and Lindauer, J. (1970). Brightness differences and the perception of figure–ground. *Journal of Experimental Psychology*, 84(2), 291–295.

Liu, Y., and Wickens, C. D. (1992). Use of computer graphics and cluster analysis in aiding relational judgment. *Human Factors*, 34(2), 165–178.

Liverman, D. M., and O'Brian, K. L. (1991). Global warming and climate change in Mexico. *Global Environmental Change*, 1(4), 351–364.

Lloyd, R. (1982). A look at images. *Annals of the Association of American Geographers*, 72, 532–548.

Lloyd, R. (1988). Searching for map symbols: The cognitive processes. *American Cartographer*, 15(4), 363–377.

Lloyd, R. (1989). Cognitive maps: Encoding and decoding information. *Annals of the Association of American Geographers*, 79(1), 101–124.

Lloyd, R., and Steinke, T. (1976). The decision-making process for judging the similarity of choropleth maps. *American Cartographer*, 3, 177–184.

Lloyd, R., and Steinke, T. (1985). Comparison of qualitative point symbols: The cognitive process. *American Cartographer*, 12, 156–168.

Lobeck, A. K. (1958). *Block Diagrams* (2nd ed.). Amherst, MA: Emerson–Trussel.

Lockhead, G. R. (1970). Identification and the form of multidimensional discrimination space. *Journal of Experimental Psychology*, 85, 1–10.

Luckiesh, M. (1918). On "retiring" and "advancing" colors. *American Journal of Psychology*, 29, 182–186.

Luria, S. M., Neri, D. F., and Jacobsen, A. R. (1986). The effects of set size on color matching using CRT displays. *Human Factors, 28*(1), 49–61.

Lynch, K. (1960). *The Image of the City*. Cambridge, MA: MIT Press.

Lyutyy, A. A. (1985). On the essence of the language of the map. *Mapping Sciences and Remote Sensing, 23*, 127–139.

MacDougall, E. B. (1992). Exploratory analysis, dynamic statistical visualization, and geographic information systems. *Cartography and Geographic Information Systems, 19*(4), 237–246.

MacEachren, A. M. (1979). *Communication Effectiveness of Choropleth and Isopleth Maps: The Influence of Visual Complexity*. Unpublished Ph.D. dissertation, University of Kansas.

MacEachren, A. M. (1982). The role of complexity and symbolization method in thematic map effectiveness. *Annals of the Association of American Geographers, 72*, 495–513.

MacEachren, A. M. (1989, March). Review of [Yamahira, T., Kasahara, Yutaka, and Tsurutani, Tateyuki (1985). How map designers can represent their ideas in thematic maps. *The Visual Computer, 1*, 174–184.] *Cartographic Perspectives*, No. 11, 5–16.

MacEachren, A. M. (1991a). The role of maps in spatial knowledge acquisition. *Cartographic Journal, 28*, 152–162.

MacEachren, A. M. (1991b). *Visualization Quality and the Representation of Uncertainty*. Orono, ME: National Center for Geographic Information and Analysis.

MacEachren, A. M. (1992a). Application of environmental learning theory to spatial knowledge acquisition from maps. *Annals of the Association of American Geographers, 82*(2), 245–274.

MacEachren, A. M. (1992b, Fall). Visualizing uncertain information. *Cartographic Perspectives*, No. 13, 10–19.

MacEachren, A. M. (1994a). *Some Truth with Maps: A Primer on Design and Symbolization*. Washington, DC: Association of American Geographers.

MacEachren, A. M. (1994b). Viewing time as a cartographic variable. In H. Hearnshaw and D. Unwin (Eds.), *Visualization in GIS* (pp. 115–130). London: Wiley.

MacEachren, A. M. (1994c). Visualization in modern cartography: Setting the agenda. In A. M. MacEachren and D. R. F. Taylor (Eds.), *Visualization in Modern Cartography* (pp. 1–12). Oxford, UK: Elsevier.

MacEachren, A. M. (in collaboration with Buttenfield, B., Campbell, J., DiBiase, D., and Monmonier, M.) (1992). Visualization. In R. Abler, M. Marcus, and J. Olson (Eds.), *Geography's Inner Worlds: Pervasive Themes in Contemporary American Geography* (pp. 99–137). New Brunswick, NJ: Rutgers University Press.

MacEachren, A. M., and DiBiase, D. W. (1991). Animated maps of aggregate data: Conceptual and practical problems. *Cartography and Geographic Information Systems, 18*(4), 221–229.

MacEachren, A. M., and Ganter, J. H. (1990). A pattern identification approach to cartographic visualization. *Cartographica, 27*(2), 64–81.

MacEachren, A. M., Howard, D., von Wyss, M., Askov, D., and Taormino, T. (1993). Visualizing the health of Chesapeake Bay: An uncertain endeavor.

Proceedings, GIS/LIS '93, Minneapolis, MN, November 2–4, ACSM & ASPRS, pp. 449–458.

MacEachren, A. M., and Mistrick, T. A. (1992). The role of brightness differences in figure–ground: Is darker figure? *Cartographic Journal,* 29(2), 91–100.

Mackworth, N. H., and Morandi, A. J. (1967). The gaze selects the information details within pictures. *Perception and Psychophysics,* 2, 547–552.

Maclean, A. L., D'Avello, T. P., and Shetron, S. G. (1993). The use of variability diagrams to improve the interpretation of digital soil maps in a GIS. *Photogrammetric Engineering and Remote Sensing,* 59(2), 223–228.

Makkonen, K., and Sainio, R. (1991). Computer assisted cartographic communication. *Proceedings, 15th Conference of the International Cartographic Association,* September 23–October 1, Bournemouth, UK, pp. 211–222.

Malik, J., and Perona, P. (1990). Preattentive texture discrimination with early vision mechanisms. *Journal of the Optical Society of America,* A7, 923–932.

Margolis, H. (1987). *Patterns, Thinking, and Cognition: A Theory of Judgment.* Chicago: University of Chicago Press.

Mark, D., and Frank, A. (Eds.). (1991). *Cognitive and Linguistic Aspects of Geographic Space, NATO ASI Series D: Behavioral and Social Sciences, Volume 63.* Dordrecht: Kluwer Academic.

Mark, D. M., and Gold, M. D. (1991). Interacting with geographic information: A commentary. *Photogrammetric Engineering and Remote Sensing,* 57(11), 1427–1430.

Marr, D. (1982). *Vision: A Computational Investigation into the Human Representation and Processing of Visual Information.* San Francisco: W. H. Freeman.

Marr, D. (1985). Vision: The philosophy and the approach. In A. M. Aitkenhead and J. M. Slack (Eds.), *Issues in Cognitive Modeling* (pp. 103–126). London: Erlbaum.

Marr, D., and Nishihara, H. K. (1978). Representation and recognition of the spatial organization of three-dimensional shapes. *Proceedings of the Royal Society,* 200, 269–294.

Marshall, R., Kempf, J., and Dyer, S. (1990). Visualization methods and simulation steering for a 3D turbulence model of Lake Erie. *Computer Graphics,* 24(2), 89–97.

Martinet, J. (1973). *Clefs pour la Sémiologie.* Paris: Editions Seghers.

Martinet, J. (1990). The semiotics of Luis Prieto. In T. A. Sebeok and Umiker-Sebeok (Eds.), *The Semiotic Web 1989* (pp. 89–108). Berlin: Mouton de Gruyter.

McCleary, G. F. (1970). Beyond simple psychophysics: Approaches to the understanding of map perception. *Technical Papers,* ACSM, 189–209.

McCleary, G. F. (1975). In pursuit of the map user. *Proceedings, Auto-Carto II,* September 21–25, Washington, DC, U.S. Bureau of the Census & ACSM, pp. 238–250.

McCleary, G. F. (1981). How to design an effective graphics presentation. *Harvard Library of Computer Graphics 1981 Mapping Collection,* 17, 15–64.

McGranaghan, M. (1989). Ordering choropleth map symbols: The effect of background. *American Cartographer,* 16(4), 279–285.

McGranaghan, M., Mark, D., and Gould, M. D. (1987). Automated provision of navigation assistance to drivers. *American Cartographer*, 14(2), 121–138.

McGuigan, F. J. (1957). An investigation of several methods of teaching contour interpretation. *Journal of Applied Psychology*, 41, 53–57.

McGuinness, C., van Wersch, A., and Stringer, P. (1993). User differences in a GIS environment: A protocol study. *Proceedings, 16th Conference of the International Cartographic Association*, May 3–9, Cologne, Germany, pp. 478–485.

McLay, W. J. (1987). *Two-Variable Mapping: A Practical Case for the Soil Map*. Unpublished master's thesis, Pennsylvania State University.

Medyckyj-Scott, D., and Board, C. (1991). Cognitive cartography: A new heart for a lost soul. In J.-C. Muller (Ed.), *Advances in Cartography* (pp. 201–230). London: Elsevier.

Mersey, J. E. (1990). *Colour and Thematic Map Design: The Role of Colour Scheme and Map Complexity in Choropleth Map Communication, Monograph 41, Cartographica*, 27(3).

Metz, C. (1974). *Film Language: A Semiotics of the Cinema*. New York: Oxford University Press. (French edition, 1968)

Metzger, P. (1992). *Perspective without Pain*. Cincinnati: North Light Books.

Meyers, G. (1988). Every picture tells a story: Illustrations in E. O. Wilson's *Sociobiology*. In M. Lynch and S. Woolgar (Eds.), *Representation in Scientific Practice* (pp. 231–265). Cambridge, MA: MIT Press.

Mistrick, T. (1990). *The Effects of Brightness Contrast on Figure–Ground Discrimination for Black and White Maps*. Unpublished master's thesis, Pennsylvania State University.

Moellering, H. (1976). The potential uses of a computer animated film in the analysis of geographical patterns of traffic crashes. *Accident Analysis and Prevention*, 8, 215–227.

Moellering, H. (1980a). The real-time animation of three dimensional maps. *American Cartographer*, 7, 67–75.

Moellering, H. (1980b). Strategies of real-time cartography. *Cartographic Journal*, 17, 12–15.

Moellering, H. (1993). MKS-Aspect™—a new way of rendering cartographic Z surfaces. *Proceedings, 16th Conference of the International Cartographic Association*, May 3–9, Cologne, Germany, pp. 675–681.

Moellering, H., and Kimerling, J. (1990). A new digital slope–aspect display process. *Cartography and Geographic Information Systems*, 17(2), 151–159.

Mollon, J. (1990). The tricks of colour. In C. B. H. Barlow and M. Weston-Smith (Eds.), *Images and Understanding* (pp. 61–78). Cambridge: Cambridge University Press.

Monmonier, M. (1972). Contiguity-biased class-interval selection: A method for simplifying patterns on statistical maps. *Geographical Review*, 62, 203–228.

Monmonier, M. (1989a). Geographic brushing: Enhancing exploratory analysis of the scatterplot matrix. *Geographical Analysis*, 21(1), 81–84.

Monmonier, M. (1989b). Graphic scripts for the sequenced visualization of geographic data. *Proceedings, GIS/LIS '89*, ASPRS & ACSM, pp. 381–389.

Monmonier, M. (1990). Strategies for the visualization of geographic time-series data. *Cartographica, 27*(1), 30–45.

Monmonier, M. (1991a). *How to Lie with Maps*. Chicago: University of Chicago Press.

Monmonier, M. (1991b). On the design and application of biplots in geographic visualization. *Journal of the Pennsylvania Academy of Science, 65*(1), 40–47.

Monmonier, M. (1992). Authoring graphics scripts: Experiences and principles. *Cartography and Geographic Information Systems, 19*(4), 247–260.

Monmonier, M., and Gluck, M. (1994). Focus groups for design improvement in dynamic cartography. *Cartography and Geographic Information Systems, 21*(1), 37–47.

Morita, T. (1991). *The Measurement of Eye Movements for Map Design Evaluation: Legibility of Quantitative Symbols*. Paper presented at the 15th Meeting of the International Cartographic Association, September 23–October 1, Bournemouth, UK.

Morris, C. W. (1938). *Foundations of the Unity of Science: Towards an International Encyclopedia of Unified Science, Volumes 1 and 2*. Chicago: University of Chicago Press.

Morris, C. W. (1971). Esthetics and the theory of signs. In C. W. Morris (Ed.), *Writings on the General Theory of Signs* (pp. 415–433). The Hague: Mouton. (Original paper published 1939)

Morris, C. W. (1971). Signs, language, and behavior. In C. Morris (Ed.), *Writings on the General Theory of Signs* (pp. 73–398). The Hague: Mouton. (Original paper published 1946)

Morris, C. W. (1964). *Signification and Significance*. Cambridge, MA: MIT Press.

Morrison, J. L. (1974). A theoretical framework for cartographic generalization with the emphasis on the process of symbolization. *International Yearbook of Cartography, 14*, 115–127.

Morrison, J. L. (1984). Applied cartographic communication: Map symbolization for atlases. *New Insights in Cartographic Communication, Monograph 31, Cartographica, 21*, 44–84.

Movshon, A. (1990). Visual processing of moving images. In C. B. H. Barlow and M. Weston-Smith (Eds.), *Images and Understanding* (pp. 122–137). Cambridge: Cambridge University Press.

Mowafy, L., Blake, R., and Lappin, J. S. (1990). Detection and discrimination of coherent motion. *Perception and Psychophysics, 48*(6), 583–592.

Muehrcke, P. (1973). Visual pattern comparison in map reading. *Proceedings, Association of American Geographers*, April 14–18, Atlanta, GA, pp. 190–194.

Muehrcke, P. (1974). Beyond abstract map symbols. *Journal of Geography, 73*(8), 35–52.

Muehrcke, P. (1990). Cartography and geographic information systems. *Cartography and Geographic Information Systems, 17*(1), 7–15.

Muehrcke, P., and Muehrcke, J. O. (1974). Maps in literature. *Geographical Review, 64*, 317–338.

Muehrcke, P., and Muehrcke, J. O. (1978). *Map Use: Reading, Analysis, and Interpretation*. Madison, WI: JP Publications.

Mulder, J., and Hervey, S. (1990). *The Strategy of Linguistics*. Edinburgh: Scotish Academic Press.

Muller, J.-C. (1979). Perception of continuously shaded maps. *Annals of the Association of American Geographers, 69,* 240–249.

Murphy, G. L., and Medin, D. L. (1985). The role of theories in conceptual coherence. *Psychological Review, 92*(3), 289–316.

Nakayama, K., and Silverman, G. H. (1986). Serial and parallel processes of visual feature conjunctions. *Nature, 320,* 264–265.

Navon, D. (1977). Forest before trees: The precedence of global features in visual perception. *Cognitive Psychology, 9,* 353–383.

Neisser, U. (1976). *Cognition and Reality: Principles and Implications of Cognitive Psychology*. San Francisco: W. H. Freeman.

Neisser, U. (1987). A sense of where you are: Functions of the spatial module. In *Proceedings of the NATO Advanced Study Institute on Cognitive Processes and Spatial Orientation in Animal and Man, 1985* (pp. 293–311). La-Baume-les-Aix, France: Martinus Nijhoff.

Nöth, W. (1990). *Handbook of Semiotics*. Bloomington: Indiana University Press.

Nothdurft, H.-C. (1992). Feature analysis and the role of similarity in preattentive vision. *Perception and Psychophysics, 52*(4), 355–375.

Nyerges, T. L. (1991a). Geographic information abstractions: Conceptual clarity for geographic modeling. *Environment and Planning A, 23,* 1483–1499.

Nyerges, T. L. (1991b). Representing geographical meaning. In B. P. Buttenfield and R. B. McMaster (Eds.), *Map Generalization: Making Rules for Knowledge Representation* (pp. 59–85). Essex, UK: Longman.

Ogden, C. K., and Richards, I. A. (1923). *The Meaning of Meaning*. New York: Harcourt, Brace & Co.

Olson, J. M. (1972). Class interval system on maps of observed correlation distribution. *Canadian Cartographer, 9,* 122–131.

Olson, J. M. (1976). A coordinated approach to map communication improvement. *American Cartographer, 3,* 151–159.

Olson, J. M. (1979). Cognitive cartographic experimentation. *Canadian Cartographer, 16,* 34–44.

Olson, J. M. (1981). Spectrally encoded two-variable maps. *Annals of the Association of American Geographers, 71*(2), 259–276.

Olson, J. M. (1983). Future research directions in cartographic communication and design. In D. R. F. Taylor (Ed.), *Graphic Communication and Design in Contemporary Cartography, Volume 2, Progress in Contemporary Cartography* (pp. 257–284). New York: Wiley.

Olson, J. M. (1989, February). *Maps for People with Defective Color Vision: Experimental Results*. Paper presented at the Pennsylvania State University Department of Geography Coffee Hour.

Olson, J. M. (1994). Problems and uses of color in cartography: Examples from Michigan State University. In J. Bares (Chair/Ed.), *Color Hardcopy and Graphic Arts III, Proceedings, SPIE—The International Society for Optical Engineering, Volume 2171,* pp. 46–53.

Olson, R. K., and Attneave, F. (1970). What variables produce similarity grouping? *American Journal of Psychology, 83,* 1–21.

Oyama, T. (1960). Figure–ground dominance as a function of sector angle, brightness, hue, and orientation. *Journal of Experimental Psychology*, 60(5), 299–305.

Paivio, A. (1969). Mental imagery in associative learning and memory. *Psychological Review*, 76, 241–263.

Palmer, S. E. (1975). Visual perception and world knowledge: Notes on a model of sensory-cognitive interaction. In D. A. Norman and D. E. Rumelhart (Eds.), *Explorations in Cognition* (pp. 279–307). San Francisco: W. H. Freeman.

Palmer, S. E. (1977). Hierarchical structure in perceptual representation. *Cognitive Psychology*, 9, 441–474.

Papathomas, T. V., and Julesz, B. (1988). The application of depth separation to the display of large data sets. In W. S. Cleveland and M. E. McGill (Eds.), *Dynamic Graphics for Statistics* (pp. 353–377). Belmont, CA: Wadsworth.

Paquet, L., and Merikle, P. (1988). Global precedence in attended and nonattended objects. *Journal of Experimental Psychology: Human Perception and Performance*, 14(1), 89–100.

Patton, J. C., and Crawford, P. V. (1978). The perception of hyposometric colours. *Cartographic Journal*, 15, 115–127.

Peirce, C. S. (1985). Logic as semiotic: The theory of signs. In R. E. Innis (Ed.), *Semiotics: An Introductory Anthology* (pp. 4–23). Bloomington: Indiana University Press.

Penn, A. (1993). Intelligent analysis of urban space patterns: Graphical interfaces to precedent databases for urban design. *Proceedings, Auto-Carto 11*, October 30–November 1, Minneapolis, MN, ASPRS & ACSM, pp. 53–62.

Pentland, A. P. (1985). The focal gradient: Optics ecologically salient (abstract). *Ophthalmology and Visual Science, Supplement*, 26(3), 243.

Perkins, D. N. (1981). *The Mind's Best Work*. Cambridge, MA: Harvard University Press.

Perkins, D. N. (1988). Creativity and the quest for mechanism. In R. J. Sternberg and E. E. Smith (Eds.), *The Psychology of Human Thought* (pp. 309–336). Cambridge: Cambridge University Press.

Petchenik, B. B. (1975). Cognition in cartography. *Proceedings, Auto-Carto II*, September 21–25, Washington, DC, U.S. Bureau of the Census & ACSM, pp. 183–193.

Peterson, M. A., and Gibson, B. S. (1991). The initial identification of figure–ground relationships: Contributions from shape recognition processes. *Journal of Experimental Psychology: Human Learning and Memory*, 29(3), 199–202.

Peterson, M. A., Harvey, E. M., and Weidenbacher, H. J. (1991). Shape recognition contributions to figure–ground reversal: Which route counts? *Journal of Experimental Psychology: Human Perception and Performance*, 17(4), 1075–1089.

Peterson, M. J., and Graham, S. E. (1974). Visual detection and visual imagery. *Journal of Experimental Psychology*, 103(3), 509–514.

Peterson, M. P. (1979). An evaluation of unclassed crossed-line choropleth mapping. *American Cartographer*, 6(1), 21–37.

Peterson, M. P. (1985). Evaluating a map's image. *American Cartographer*, 12, 41–55.

Peterson, M. P. (1987). The mental image in cartographic communication. *Cartographic Journal*, 24(1), 35–41.

Peuquet, D. J. (1988). Representations of geographic space: Toward a conceptual synthesis. *Annals of the Association of American Geographers*, 78(3), 373–394.

Peuquet, D. J. (1994). It's about time: A conceptual framework for the representation of temporal dynamics in geographic information systems. *Annals of the Association of American Geographers*, 84(3), 441–461.

Phillips, R. J. (1984). Experimental method in cartographic communication: Research on relief maps. *New Insights in Cartographic Communication, Monograph 31, Cartographica*, 21(4), 120–128.

Phillips, R. J., and Noyes, L. (1977). Searching for names in two city street maps. *Applied Ergonomics*, 8(2), 73–77.

Phillips, R. J., Noyes, L., and Audley, R. J. (1978). Searching for names on maps. *Cartographic Journal*, 15, 72–77.

Phillips, W. A. (1974). On the distinction between sensory storage and short-term visual memory. *Perception and Psychophysics*, 16, 283–290.

Phillips, W. A. (1983). Short-term visual memory. *Philosophical Transactions of the Royal Society of London*, B302, 295–309.

Pike, R. J., and Thelin, G. P. (1989). Shaded relief map of U.S. topography from digital elevations. *Eos*, 70(38), cover, 843, 853.

Pinker, S. (1984). Visual cognition: An introduction. *Cognition*, 18, 1–63.

Pinker, S. (1990). A theory of graph comprehension. In R. Friedle (Ed.), *Artificial Intelligence and the Future of Testing* (pp. 73–126). Norwood, NJ: Ablex.

Pittman, K. (1992). A laboratory for the visualization of virtual environments. *Landscape and Urban Planning*, 21, 327–331.

Plumb, G. A. (1988). Displaying GIS data sets using cartographic classification techniques. *Proceedings, GIS/LIS '88: Assessing the World, Volume 1*, November 30–December 2, San Antonio, TX, pp. 340–349.

Pomerantz, J. R. (1985). Perceptual organization in information processing. In A. M. Aitkenhead and J. M. Slack (Eds.), *Issues in Cognitive Modelling*. London: Erlbaum.

Pomerantz, J. R., and Garner, W. R. (1973). Stimulus configuration in selective attention tasks. *Perception and Psychophysics*, 14, 565–569.

Pomerantz, J. R., and Schwaitzberg, S. D. (1975). Grouping by proximity: Selective attention measures. *Perception and Psychophysics*, 18, 335–361.

Pomerantz, S. (1983). Global and local precedence: Selective attention in form and motion perception. *Journal of Experimental Psychology: General*, 112(4), 516–540.

Posner, M. I. (1980). Orienting attention. *Quarterly Journal of Experimental Psychology*, 32, 3–25.

Pylyshyn, Z. W. (1981). The imagery debate: Analogue media versus tacit knowledge. *Psychological Review*, 88, 16–45.

Quinlan, P. T., and Humphreys, G. W. (1987). Visual search for targets defined by combinations of color, shape, and size: An examination of the task constraints on feature and conjunction searches. *Perception and Psychophysics, 41*(5), 455–472.

Rabenhorst, D. A., Farrell, E. J., Jameson, D. H., Linton, T. D., and Mandelman, J. A. (1990). Complementary visualization and sonification of multi-dimensional data. *Proceedings, Extracting Meaning from Complex Data: Processing, Display, Interaction*, February 14–16, Santa Clara, CA, SPIE—The International Society for Optical Engineering, pp. 147–153.

Rader, C. P. (1989). *A Functional Model of Color in Cartographic Design*. Unpublished master's thesis, University of Washington.

Raichle, M. (1991). Plate 3.1: Computerized PET images showing the changes in local blood flow in the brain, associated with local changes in neuronal activity, that occur during different states of information processing. In C. M. Pechura and J. B. Martin (Eds.), *Mapping the Brain and Its Functions: Integrating Enabling Technologies into Neuroscience Research*. Washington, DC: National Academy Press.

Ramachandran, V. S., and Anstis, S. M. (1986). The perception of apparent motion. *Scientific American, 254*(6), 102–109.

Raper, J. (Ed.). (1989). *Three-Dimensional Applications in Geographic Information Systems*. London: Taylor & Francis.

Ratajski, L. (1971). The methodical basis of the standardization of signs on economic maps. *International Yearbook of Cartography, 11*, 137–159.

Ratner, C. (1989). A sociohistorical critique of naturalistic theories of color perception. *Journal of Mind and Behavior, 10*, 361–372.

Rheingans, P. (1992). Color, change, and control of quantitative data display. *Proceedings, Visualization '92*, October 19–23, Boston, IEEE Computer Society Technical Committee on Computer Graphics, pp. 252–259.

Rheingans, P., and Tebbs, B. (1990). A tool for dynamic explorations of color mappings. *Computer Graphics, 24*(2), 145–146.

Rhind, D. (1993). Mapping for the new millenium. *Proceedings, 16th Conference of the International Cartographic Association*, May 3–9, Cologne, Germany, pp. 3–14.

Rice, K. (1990). Disoriented prism maps: A recognition experiment (abstract). *Cartographic Perspectives, 4*, 32.

Robinson, A. H. (1952). *The Look of Maps*. Madison: University of Wisconsin Press.

Robinson, A. H. (1953). *Elements of Cartography*. New York: Wiley.

Robinson, A. H. (1960). Symbolization and processing map data. In *Elements of Cartography* (2nd ed, pp. 136–145). New York: Wiley.

Robinson, A. H. (1967). Psychological aspects of color in cartography. *International Yearbook of Cartography, 7*, 50–61.

Robinson, A. H. (1973). An international standard symbolism for thematic maps: Approaches and problems. *International Yearbook of Cartography, 13*, 19–26.

Robinson, A. H., and Petchenik, B. B. (1976). *The Nature of Maps*. Chicago: University of Chicago Press.

Robinson, A. H., Sale, R. D., Morrison, J. L., and Muehrcke, P. C. (1984). *Elements of Cartography* (5th ed.). New York: Wiley.

Rogers, J. E., and Groop, R. E. (1981). Regional portrayal with multi-pattern color dot maps. *Cartographica*, 18(4), 51–64.
Rollins, M. (1989). *Mental Imagery: On the Limits of Cognitive Science*. New Haven: Yale University Press.
Rosch, E. (1973). Natural categories. *Cognitive Psychology*, 4, 328–350.
Rosch, E. (1975a). Cognitive reference points. *Cognitive Psychology*, 7, 532–547.
Rosch, E. (1975b). Cognitive representations of semantic concepts. *Journal of Experimental Psychology: General*, 104(3), 192–233.
Rosch, E. (1977). Human categorization. In N. Warren (Ed.), *Studies in Cross-Cultural Psychology* (pp. 27–48). London: Academic Press.
Rosch, E. (1978). Principles of categorization. In E. Rosch and B. B. Lloyd (Eds.), *Cognition and Categorization* (pp. 27–48). Hillsdale, NJ: Erlbaum.
Rosch, E., and Lloyd, B. B. (Eds.). (1975). *Cognition and Categorization*. Hillsdale, NJ: Erlbaum.
Rosch, E., Mervis, C., Gray, W., Johnson, D., and Boyes-Braem, P. (1976). Basic objects in natural categories. *Cognitive Psychology*, 8, 382–439.
Roth, I., and Frisby, J. P. (1986). *Perception and Representation: A Cognitive Approach*. Philadelphia: Open University Press.
Rowles, R. A. (1978). Perception of perspective block diagrams. *American Cartographer*, 5, 331–344.
Rumelhart, D. E., and Norman, D. A. (1985). Representation of knowledge. In A. M. Aitkenhead and J. M. Slack (Eds.), *Issues in Cognitive Modeling* (pp. 15–62). London: Erlbaum.
Rumelhart, D. E., and Ortony, A. (1977). The representation of knowledge in memory. In R. C. Anderson, R. J. Spiro, and W. E. Montague (Eds.), *Schooling and the Acquisition of Knowledge* (pp. 99–135). Hillsdale, NJ: Erlbaum.
Rundstrom, R. (1993). The role of ethics, mapping, and the meaning of place in relations between Indians and whites in the United States. *Introducing Cultural and Social Cartography, Monograph 44, Cartographica*, 30(1), 21–28.
Saint-Martin, F. (1989). From visible to visual language: Artificial intelligence and visual semiology. *Semiotica*, 77(1/3), 303–316.
Salichtchev, K. A. (1983). Cartographic communication: A theoretical survey. In D. R. F. Taylor (Ed.), *Graphic Communication and Design in Contemporary Cartography, Volume 2, Progress in Contemporary Cartography* (pp. 11–36). New York: Wiley.
Saushkin, Y. G., and Il'yina, L. N. (1985). Conversations in the language of maps. *Soviet Geography*, 26, 1–10.
Saussure, F. de. (1986). *Cours de Linguistique Générale*, 25th ed. (C. Bally and A. Sechehaye, Eds.). Paris: Payot. (Original work published 1916)
Schank, R. C., and Abelson, R. P. (1977). *Scripts, Plans, Goals and Understanding: An Inquiry into Human Knowledge Structures*. Hillsdale, NJ: Erlbaum.
Schlichtmann, H. (1979). Codes in map communication. *Canadian Cartographer*, 16, 81–97.
Schlichtmann, H. (1984). Discussion of C. Grant Head, "The map as natural language: A paradigm for understanding." *Cartographica*, 21, 33–36.

Schlichtmann, H. (1985). Characteristic traits of the semiotic system "map symposium." *Cartographic Journal*, 22, 23–30.

Schlichtmann, H. (1991). Plan information and its retrieval in map interpretation: The view from semiotics. In D. M. Mark and A. U. Frank (Eds.), *Cognitive and Linguistic Aspects of Geographic Space* (pp. 263–284). Amsterdam: Kluwer Academic.

Schneider, W., and Shiffrin, R. M. (1977). Controlled and automatic human information processing: 1. Detection, search, and attention. *Psychological Review*, 84, 1–66.

Schustack, M. W. (1988). Thinking about causality. In R. J. Sternberg and E. E. Smith (Eds.), *The Psychology of Human Thought* (pp. 92–115). Cambridge: Cambridge University Press.

Schweizer, D. M., and Goodchild, M. F. (1992). Data quality and choropleth maps: An experiment with the use of color. *Proceedings, GIS/LIS '92*, November 10–12, San Jose, CA, ACSM & ASPRS, pp. 686–699.

Sebeok, T. A. (1976). Problems in the classification of signs. In T. Sebeok (Ed.), *Contributions to the Doctrine of Signs, Volume 5, Studies in Semiotics* (pp. 71–143). Bloomington: Indiana University Press.

Shannon, C., and Weaver, W. (1949). *The Mathematical Theory of Communication*. Urbana: University of Illinois Press.

Shapley, R., Caelli, T., Grossberg, S., Morgan, M., and Rentschler, I. (1990). Computational theories of visual perception. In L. Spillmann and J. S. Werner (Eds.), *Visual Perception: The Neurophysiological Foundations* (pp. 417–447). New York: Academic Press.

Shepard, R. N. (1978). The mental image. *American Psychologist*, 33, 125–137.

Shepard, R. N., and Cooper, L. (1982). *Mental Images and Their Transformations*. Cambridge, MA: MIT Press.

Shortridge, B. G. (1979). Map reader discrimination of lettering size. *American Cartographer*, 6, 13–20.

Shortridge, B. G. (1982). Stimulus processing models from psychology: Can we use them in cartography? *American Cartographer*, 9, 155–167.

Shortridge, B. G., and Welch, R. B. (1980). Are we asking the right questions? Comments on instructions in cartographic psychophysical studies. *American Cartographer*, 7, 19–23.

Shortridge, B. G., and Welch, R. B. (1982). The effect of stimulus redundancy on the discrimination of town size on maps. *American Cartographer*, 9, 69–80.

Simcox, W. A. (1983). *A Perceptual Analysis of Graphic Information Processing*. Unpublished Ph.D. dissertation, Tufts University, Medford, MA.

Singh, G., and Chignell, M. H. (1992). Components of the visual computer: A review of relevant technologies. *Visual Computer*, 9, 115–142.

Slocum, T. A. (1983). Predicting visual clusters on graduated circle maps. *American Cartographer*, 10, 59–72.

Slocum, T. A., and McMaster, R. B. (1986). Gray tone versus line plotter area symbols: A matching experiment. *American Cartographer*, 13(2), 151–164.

Slocum, T. A., Roberson, S. H., and Egbert, S. L. (1990). Traditional versus sequenced choropleth maps: An experimental investigation. *Cartographica*, 27(1), 67–88.

Smith, E. E., and Medin, D. L. (1981). *Categories and Conceps*. Cambridge, MA: Harvard University Press.
Smith, S., Grinstein, G., and Pickett, R. (1991). Global geometric, sound, and color controls for iconographic displays of scientific data. *Proceedings, Extracting Meaning from Complex Data: Processing, Display, Interaction II*, February 26–28, Santa Clara, CA, SPIE—The International Society for Optical Engineering, pp. 192–206.
Speeth, S. D. (1961). Seismometer sounds. *Journal of the Acoustical Society of America, 33*, 909–916.
Spiess, E. (1978). *Some Graphic Means to Establish Visual Levels in Map Design*. Paper presented at the 9th International Conference on Cartography, July 26–August 2, College Park.
Spiess, E. (1988). Map compilation. In R. W. Anson (Ed.), *Basic Cartography for Students and Technicians, Volume 2* (pp. 35–69). London: International Cartographic Association.
Steinke, T. R. (1987). Eye movement studies in cartography and related fields. *Studies in Cartography, Monograph 37, Cartographica, 24*(2), 40–73.
Steinke, T. R., and Lloyd, R. E. (1983a). Images of maps: A rotation experiment. *Professional Geographer, 35*(4), 455–461.
Steinke, T. R., and Lloyd, R. E. (1983b). Judging the similarity of choropleth map images. *Cartographica, 20*, 35–42.
Suchan, T. (1991). *Useful Categories: A Cognitive Approach to Land Use Categorization Systems*. Unpublished master's thesis, University of Washington.
Szegö, J. (1987). *Human Cartography: Mapping the World of Man*. Stockholm: Swedish Council for Building Research.
Tanaka, K. (1932). The orthographical relief method of representing hill features on a topographical map. *Geographical Journal, 79*, 213–219.
Tatham, A. F. (1991). The design of tactile maps: Theoretical and practical considerations. *Proceedings, 15th Conference of the International Cartographic Association*, September 23–October 1, Bournemouth, UK, pp. 157–166.
Tatham, A. F., and Dodds, A. G. (Eds.). (1988). *Proceedings of the Second International Symposium on Maps and Graphics for Visually Handicapped People*. London: University of Nottingham.
Taylor, D. R. F. (1991). Geographic information systems: The microcomputer and modern cartography. In D. R. F. Taylor (Ed.), *Geographic Information Systems: The Microcomputer and Modern Cartography* (pp. 1–20). Oxford, UK: Elsevier.
Thomas, E. L., and Lansdown, E. L. (1963). Visual search patterns of radiologists in training. *Radiology, 81*, 288–292.
Tobler, W. R. (1981). Depicting federal fiscal transfers. *Professional Geographer, 33*(4), 419–422.
Travis, D. S. (1990). Applying visual psychophysics to user interface design. *Behaviour and Information Technology, 9*(5), 425–438.
Treinish, L. (1993). Visualization techniques for correlative data analysis in the earth and space sciences. In R. A. Earnshaw and D. Watson (Eds.), *Animation and Scientific Visualization: Tools and Applications* (pp. 193–204). New York: Academic Press.

Treisman, A. (1988). Features and objects: The fourteenth Bartlett Memorial Lecture. *Quarterly Journal of Experimental Psychology, 40A,* 201–237.
Treisman, A., Cavenagh, P., Fischer, B., Ramachandran, V. S., and von der Heydt, R. (1990). Form, perception, and attention. In L. Spillmann and J. S. Werner (Eds.), *Visual Perception: The Neurophysiological Foundations* (pp. 273–316). New York: Academic Press.
Treisman, M. (1985). The magical number seven and some other features of category scaling: Properties of a model for absolute judgement. *Journal of Mathematical Psychology, 29,* 175–230.
Tsal, Y., and Kolbert, L. (1985). Disambiguating ambiguous figures by selective attention. *Quarterly Journal of Experimental Psychology, 37A,* 25–37.
Tsal, Y., and Lavie, N. (1988). Attending to color and shape: The special role of location in selective visual processing. *Perception and Psychophysics, 44,* 15–21.
Tufte, E. (1983). *The Visual Display of Quantitative Information.* Cheshire, CT: Graphics Press.
Tufte, E. (1990). *Envisioning Information.* Cheshire, CT: Graphics Press.
Tukey, J. W. (1977). *Exploratory Data Analysis.* Reading, MA: Addison-Wesley.
Tversky, B., and Hemenway, K. (1984). Objects, parts, and categories. *Journal of Experimental Psychology: General, 113*(2), 169–191.
Tyner, J. A. (1974). *Persuasive Cartography.* Unpublished Ph.D. dissertation, University of California at Los Angeles.
Ucar, D. (1993). A semiological approach to typology of the map signs. *Proceedings, 16th Conference of the International Cartographic Association,* May 3–9, Cologne, Germany, pp. 768–781.
Ullman, S. (1984). Visual routines. *Cognition, 18,* 97–159.
U.S. Environmental Protection Agency. (1990). *Region 6 Comparative Risk Project, Appendix A: Ecological Report.* EPA Region 6 Office of Planning and Analysis, Dallas, TX.
Usery, L. (1993). Category theory and the structure of features in geographic information systems. *Cartography and Geographic Information Systems, 20*(1), 5–12.
Uttal, W. R. (1988). *On Seeing Forms.* Hillsdale, NJ: Erlbaum.
van der Wel, F. J. M. (1993). Visualization of quality information as an indispensable part of optimal information extraction from a GIS. *Proceedings, 16th Conference of the International Cartographic Association,* May 3–9, Cologne, Germany, pp. 881–897.
Vasconcellos, R. (1992). Knowing the Amazon through tactual graphics. *Proceedings, 15th Conference of the International Cartographic Association,* September 23–October 1, Bournemouth, UK, pp. 206–210.
Vasilev, I., Freundschuh, S., Mark, D. M., Theisen, G. D., and McAvoy, J. (1990). What is a map? *Cartographic Journal, 27*(2), 119–123.
Vernon, M. D. (1962). *The Psychology of Perception.* Baltimore: Penguin Books.
Wade, N. J., and Swanston, M. (1991). *Visual Perception.* New York: Routledge.
Wanger, L. R., Ferwerda, J. A., and Greenburg, D. P. (1992, May). Perceiving spatial relationships in computer-generated images. *IEEE Computer Graphics and Applications,* pp. 44–55.

Watt, R. J. (1988). *Visual Processing: Computational, Psychological and Cognitive Research*. London: Erlbaum.

Wever, E. (1927). Figure and ground in the visual perception of form. *American Journal of Psychology, 38*, 194–226.

Wiedel, J. (Ed.). (1983). *Proceedings of the First International Symposium on Maps and Graphics for the Visually Handicapped*, International Cartographic Association.

Williams, L. G. (1967). The effects of target specification on objects fixated during visual search. *Acta Psychologica, 27*, 355–360.

Wilson, H. R., Levi, D., Maffei, L., Rovamo, J., and DeValois, R. (1990). The perception of form. In L. Spillmann and J. S. Werner (Eds.), *Visual Perception: The Neurophysiological Foundations* (pp. 231–272). New York: Academic Press.

Wolodtschenko, A., and Pravda, J. (1993). Cartosemiotics—ideas and inducements. *Proceedings, 16th Conference of the International Cartographic Association*, May 3–9, Cologne, Germany, pp. 1235–1237.

Wong, E., and Weisstein, N. (1983). Sharp targets are detected better against a figure, and blurred targets are detected better against a ground. *Journal of Experimental Psychology: Human Learning and Memory, 9*, 194–202.

Wong, E., and Weisstein, N. (1984). Flicker induces depth: Spatial and temporal factors in the perceptual segregation of flickering and nonflickering regions in depth. *Perception and Psychophysics, 35*, 229–236.

Wood, C. H. (1976). Brightness gradients operant in the cartographic context of figure–ground relationship. *Proceedings, American Congress on Surveying and Mapping*, pp. 5–34.

Wood, D. (1977). Now and then: Comparisons of ordinary American's symbol conventions with those of past cartographers. *Prologue, 9*, 151–161.

Wood, D. (1978). *Cultured Symbols: Thoughts on the Cultural Context of Cartographic Symbols*. Unpublished manuscript, Department of Landscape Architecture, University of North Carolina at Chapel Hill.

Wood, D. (1992). *The Power of Maps*. New York: Guilford Press.

Wood, D., and Fels, J. (1986). Designs on signs: Myth and meaning in maps. *Cartographica, 23*(3), 54–103.

Wood, M. (1968). Visual perception and map design. *Cartographic Journal, 5*, 54–64.

Wood, M. (1972). Human factors in cartographic communication. *Cartographic Journal, 9*, 123–132.

Woods Hole Oceanographic Institute. (1982). *Georges Bank* (map). Cambridge, MA: MIT Press.

Woodsworth, R. S. (1938). *Experimental Psychology*. New York: Holt.

Woodward, D. (1985). Reality, symbolism, time, and space in medieval world maps. *Annals of the Association of American Geographers, 75*(4), 510–521.

Wright, J. K. (1942). Map makers are human: Comments on the subjective in maps. *Geographical Review, 32*(4), 527–544.

Wright, J. K. (1944). The terminology of certain map symbols. *Geographical Review, 34*, 653–654.

Yamahira, T., Kasahara, Y., and Tsurutani, T. (1985). How map designers can represent their ideas in thematic maps. *Visual Computing, 1*, 174–184.

Yapa, L. S. (1992). Why do they map GNP per capita? In S. K. Majumdar, G. S. Forbes, E. W. Miller, and R. F. Schmalz (Eds.), *Natural and Technological Disasters: Causes, Effects and Preventative Measures* (pp. 494–510). Easton: Pennsylvania Academy of Science.

Zadeh, L. (1965). Fuzzy sets. *Information and Control, 8*, 338–353.

Author Index

Abelson, R. P., 191–193
Ahmad-Taylor, T., 225
Allis, S., 2
Allman, J., 171, 185
Anderson, J. R., 211
Andrews, S. K., 276
Anstis, S. M., 383
Antes, J. R., 40, 99, 100, 203, 380
Antonietti, A., 422
Armstrong, S. L., 169
Arnheim, R., 48, 108, 148, 176
Attneave, F., 83
Axelsen, B., 242, 334, 338, 345, 353

Baddeley, A. D., 35, 50, 211
Balchin, W. G. V., 339
Barlow, H., 116
Barsalou, L. W., 157, 210, 212
Barthes, R., 230, 231, 331, 333, 345, 353
Bastide, F., 453
Bates, K., 135
Beck, J., 130
Becker, R. A., 418
Bemis, D., 135
Benzon, W., 356, 369, 392, 451, 452
Berlin, B., 159, 162
Bertin, J., 34, 36, 38, 82–85, 91, 92, 120, 134, 147, 182, 206, 209, 210, 229, 240, 245, 259, 269–279, 281, 288, 291, 307, 308, 318, 324, 325, 329, 368, 374, 387, 388, 402–404, 414
Berton, J. A. J., 381, 391
Beshers, C., 430
Beveridge, M., 48
Beyls, P., 356

Biederman, I., 50
Bishop, I. D., 371, 376
Bjorklund, E. M., 152
Board, C., 4, 12, 18, 19, 25, 156, 234, 434
Bowman, W. J., 121, 122
Brachman, R. J., 261
Brewer, C. A., 62–64, 144, 292, 295–297, 309, 389, 391, 392, 411, 414–416
Brown, A., 440
Brown, D. A., 399
Brown, M. H., 389, 428
Brown, R., 162
Bruce, V., 25, 70, 84, 94–96, 109, 112
Buja, A., 384, 434
Bunn, J. H., 300
Buttenfield, B. P., 36, 37, 458
Buxton, B., 379

Caivano, J. L., 272–275, 386
Campbell, J., 87
Carr, D. B., 90, 91, 409, 440
Carroll, L., 151
Carswell, M. C., 89, 90, 452
Caspall, F. C., 163, 265
Castner, H. W., 23, 39, 42, 102, 103, 125, 130, 132
Cave, K. R., 106, 107
Cavenagh, P., 105
Chang, K.-T., 200, 201, 395
Chase, W. G., 42, 365, 395
Cheal, M., 107
Cheng, P. W., 89, 394
Chignell, M. H., 428
Cleveland, W. S., 131, 206–208, 418, 439
Cole, D., 75

Cooper, L., 187
Coulson, M. R. C., 156, 276
Cox, C. W., 136
Crampton, J., 173, 331, 348, 395–397, 400
Crawford, P. V., 146, 319, 320
Cuff, D. J., 134, 135, 195, 318–320

Dacey, M., 214
Darius, J., 378, 449
Davis, J., 87, 167
De Mey, M., 448
DeLucia, A. A., 204, 205, 320
Dennett, D., 425, 426
Dent, B. D., 71, 77, 78, 114, 121, 122, 334
Detweiler, M., 324, 326, 327, 352
DeValois, R., 160
DeYoe, E., 83
DiBiase, D., 7, 73, 127, 134, 238, 241, 253, 278, 281, 283, 284, 302, 303, 356, 357, 362, 377, 405, 431, 437–439
Dobson, M. W., 11, 41, 68, 88, 89, 102, 126, 211
Dodds, A. G., 276
Dorling, D., 301, 356, 369, 409, 426, 429, 430, 438
Downing, C. J., 95
Downs, R. M., 154, 157, 161, 196–199, 308
Duncan, J., 95, 107

Eastman, J. R., 23, 25, 40–44, 46, 62, 78, 79, 102, 103, 148, 174, 194, 195, 203, 204, 209, 292–296, 455
Eco, U., 211, 222, 224, 231, 262, 311, 336
Edney, M. H., 331, 338, 341
Edwards, G., 394, 395, 400
Ellis, W., 69–72, 74–76, 93, 108
Engel, F., 126
Eriksen, C. W., 95, 132
Eyton, J. R., 145, 146

Familant, M. E., 324, 327, 352
Feiner, S., 430
Fels, J., 7, 10, 220, 242, 243, 298–301, 312, 313, 319, 325, 332–335, 337, 352, 353
Ferreia, J., 384
Finke, R., 46
Fisher, P., 186, 442, 444
Ford, K. L., 173
Forrest, D., 130, 132
Frank, A., 152
Friedhoff, R. M., 356, 369, 392, 451, 452
Frisby, J. P., 70, 154, 158, 168, 169
Frutiger, A., 214, 329, 330

Ganter, J., 44–46, 48, 74, 120, 161, 170, 176, 255, 263, 308, 356, 361–363, 385, 393, 436, 445, 453

Garling, T., 173
Garner, W. R., 82, 88, 90
Gerber, R., 305, 306, 309, 317
Gershon, N. D., 383, 417
Gersmehl, P. J., 38, 301
Gibson, A. E., 292–296
Gibson, B. S., 119
Gibson, J. J., 23, 26, 121
Giere, R. N., 393, 447, 450, 451, 457
Gilmartin, P. P., 134, 136, 195, 350
Gluck, M., 378, 428, 434
Gold, M. D., 451
Goldberg, J. H., 173, 371
Goldhamer, H., 112
Golledge, R. G., 171–173
Gombrich, E., 426
Goodchild, M. F., 440, 441
Goodman, N., 13, 199
Gould, M. D., 430
Graham, S. E., 171
Gratzer, M. A., 102
Green, P. R., 25, 70, 84, 109, 112
Gregory, R., 117, 120, 363, 444
Griffin, T. L. C., 136
Grinstein, G., 387, 388, 409, 410, 421, 433
Groop, R. E., 75, 93, 280, 413, 418
Guiraud, P., 227, 229, 231, 232, 243, 247, 330–332, 334, 337, 345

Hadamard, J., 48
Haining, R. P., 380
Hall, S. S., 155, 157, 171, 198
Handel, S., 151, 210
Hannah, M., 346
Harley, J. B., 10, 13, 14, 190, 227, 230, 243, 245, 311, 331, 335, 336, 338, 339, 342–345, 347, 350, 351
Harrower, M., 113
Hartshorne, C., 215, 243
Harwood, K., 372
Head, C. G., 214, 215
Hemenway, K., 162, 164, 165
Hershberger, J., 389, 428
Hervey, S., 214, 217, 220, 224, 226, 228, 230, 232, 233, 238–241, 243
Hiller, D. W., 204, 205, 320
Hitch, G., 35
Hochberg, J., 107
Hoffman, D. D., 50
Hoffman, H. S., 67, 123
Hoffman, R. R., 400
Holyoak, K. J., 394, 450
Howard, V. A., 12–15, 214
Hsu, M.-L., 252
Hubel, D. H., 57, 58, 64–68
Humphreys, G. W., 93, 95, 96, 107, 129, 132
Hurvich, L. M., 62

Idol, T., 61
Il'yina, L. N., 214
Imhof, E., 9, 21, 142
Innis, R. E., 213, 219, 220, 222, 223, 330

Jacobs, G. H., 160
James, W., 95
Jameson, D., 62
Japan Cartographers Association, 146, 147
Jenks, G. F., 78, 102–104, 151, 156, 163–166, 210, 252, 265, 266, 302, 384, 399
Johnson, G., 129, 317, 318
Johnson, M., 185, 186, 223
Jones, M., 242, 334, 338, 345, 353
Judson, H. F., 39, 366, 449, 450
Julesz, B., 84, 85, 130, 131, 389, 410, 418

Kaiser, M. K., 372, 417, 432
Kay, P., 159, 160, 168
Keates, J. S., 9, 125, 223, 228, 234, 250, 256–258, 309, 336, 337
Kienker, P. K., 117, 118
Kimchi, R., 99, 100, 380
Kimerling, A. J., 36, 37, 136, 143, 144, 149
Kinchla, R. A., 99
Kishto, B. N., 145
Klymenko, V., 117, 377, 378
Knowlton, J. Q., 222, 232, 258–260, 262–264, 267–269, 457
Koenig, O., 187, 211
Koffka, K., 76, 109
Kohler, W., 69–71
Koláčný, A., 4
Kolbert, L., 118
Korac, N., 237, 238
Kosslyn, S. M., 43, 46, 77, 171, 187, 199, 211
Kraak, M.-J., 137, 138, 149, 307, 371, 373–376
Krampen, M., 259, 261, 262, 267, 291, 326, 335
Kröse, B., 131
Krygier, J., 72, 152, 169, 210, 288–290, 309, 379, 421, 431
Kubovy, M., 35, 36, 48, 92, 368
Kuhn, T. S., 448, 457

Labov, W., 157
Lakoff, G., 149, 151, 152, 155, 160, 162, 163, 168, 175, 176, 185–190, 193, 194, 201, 202, 211, 223, 339, 436, 449, 455
Lamb, M., 99
Land, E. H., 133
Lansdown, E. L., 102
Larkin, J., 48, 209, 366, 450
Lavie, N., 95
Lavin, S., 273
Leach, E., 167

Lederberg, J., 367
Leonard, J. J., 36, 37
Leous, J., 390
Levine, M., 173
Lewandowsky, S., 132
Lewis, P., 142
Liben, L. S., 157, 196, 197, 308
Lindauer, J., 113
Lindauer, M., 113
Liu, Y., 434
Liverman, D. M., 437
Lloyd, B. B., 158
Lloyd, R., 46, 104, 105, 171–173, 187, 402, 446
Lobeck, A. K., 399
Lockhead, G. R., 87
Luckiesh, M., 145
Luria, S. M., 132
Lynch, K., 11
Lyon, D. R., 107
Lyutyy, A. A., 214, 251, 259

MacDougall, E. B., 357, 384
MacEachren, A. M., 44–46, 48, 61, 71, 74, 115–117, 120, 127, 134, 161, 164–166, 170, 172, 173, 176, 186, 187, 202, 253, 263, 275, 278, 279, 284, 285, 302–304, 308, 320, 340, 356–358, 361–363, 380, 384, 385, 388, 393, 417, 436, 437, 440, 441, 445, 447, 453
Mackworth, N. H., 102
Maclean, A. L., 342
Makkonen, K., 356
Malik, J., 130, 131
Mann, S. W., 40, 99, 100, 203, 380
Margolis, H., 45, 362, 363, 381–383, 393, 394, 398, 450, 458
Mark, D., 152, 451
Marks, R. A., 146
Marr, D., 12, 24, 26–33, 35, 39, 43, 45, 51, 54, 64, 70, 71, 80, 95, 101, 118, 120, 130–132
Marshall, R., 427, 432
Martinet, J., 224, 243
McCann, J. J., 133
McCleary, G. F., 38, 77, 79, 80, 116, 121, 136, 203
McDaniel, D., 160, 168
McDowell, R. D., 102
McGill, R., 439
McGranaghan, M., 135, 173, 195, 318, 319
McGuigan, F. J., 201
McGuinness, C., 397
McLay, W. J., 415
McMaster, R. B., 273
Medin, D. L., 154, 212, 260
Medyckyj-Scott, D., 11, 18, 19, 25

Merikle, P., 97, 98
Mersey, J. E., 292
Metz, C., 236–238, 243
Metzger, P., 137
Meyers, G., 452–455
Mistrick, T. A., 71, 101, 108, 114–116
Moellering, H., 143, 144, 149, 348, 371, 426
Mollon, J., 390
Monmonier, M., 73, 127, 167, 193, 282, 284, 304, 317, 334, 347, 348, 356, 357, 377, 378, 385, 386, 399, 400, 407, 408, 417–420, 428, 430, 433, 434
Montesino, D., 225
Morandi, A. J., 102
Morita, T., 42, 44, 103
Morris, C. W., 219, 225–228, 234, 235, 238, 246, 247, 256
Morrison, J. L., 234, 272–276, 291–294
Movshon, A., 368
Mowafy, L., 126, 133
Muehrcke, J. O., 151, 340, 341, 345, 347
Muehrcke, P., 151, 339–341, 343, 345, 347, 402, 445, 446
Mulder, J., 239
Muller, J.-C., 79, 164, 165, 211
Murphy, G. L., 154, 212
Murphy, T. D., 95

Nakayama, K., 93
Navon, D., 42, 95, 97–99
Neisser, U., 40, 174, 175, 177, 419
Nisbett, R. E., 450
Nishihara, H. K., 29, 39, 101
Norman, D. A., 171–173, 175–177
Nöth, W., 217–219, 222, 230, 231, 235, 243, 336
Nothdurft, H.-C., 83
Noyes, L., 95, 104, 107
Nyerges, T. L., 251, 252, 260, 261, 322, 323, 352

O'Brian, K. L., 437
Ogden, C. K., 221, 261
Olson, J. M., 11, 22, 23, 28, 39–41, 63, 68, 77, 126, 135, 149, 402, 440
Olson, R. K., 83
Openshaw, S., 426, 429, 430
Orland, B., 456
Ortony, A., 177
Oyama, T., 112, 113

Pachella, R. G., 89
Paivio, A., 171
Palmer, S. E., 42, 100
Papathomas, T. V., 389, 410, 418
Paquet, L., 97, 98
Parkins, E., 48

Patton, J. C., 319, 320
Peirce, C. S., 213, 215, 217, 219–224, 232, 242, 243, 247, 256–258, 263, 330, 352
Penn, A., 450
Pentland, A. P., 117
Perkins, D. N., 363, 367
Perona, P., 130, 131
Petchenik, B. B., 7, 236, 258, 259, 263
Peterson, M. A., 118–120
Peterson, M. J., 171
Peterson, M. P., 23, 28, 46–48, 71, 173, 211
Peterson, W., 390
Peucker, K., 21, 145
Peuquet, D. J., 152, 212, 301, 352
Phillips, R. J., 68, 95, 104, 107, 208
Phillips, W. A., 35, 211
Pike, R. J., 142
Pinker, S., 33–42, 45, 46, 51, 92, 95, 176, 178–185, 198, 205–212, 363, 366, 393, 455
Pittman, K., 451
Plumb, G. A., 384
Pomerantz, J. R., 70, 77, 81, 82, 87, 88, 92, 96, 148
Pomerantz, S., 99
Posner, M. I., 94, 235
Pravda, J., 234
Proffitt, D. R., 372, 417, 432
Pylyshyn, Z. W., 171

Quinlan, P. T., 96, 129, 132

Rabenhorst, D. A., 420
Rader, C. P., 292, 296–299, 302, 320, 334
Ramachandran, V. S., 383
Raper, J., 149
Ratajski, L., 2, 32, 34, 290–292, 329
Ratner, C., 160
Rheingans, P., 391, 416
Rice, K., 173
Richards, I. A., 221, 261
Richards, W. A., 50
Robertson, L., 99
Robinson, A. H., 2–4, 6, 9, 22, 120, 124–126, 136, 236, 252, 257–261, 263
Rogers, J. E., 93, 280, 413, 418
Rollins, M., 211
Rosch, E., 152–154, 158, 162, 167, 168, 211
Roth, I., 70, 154, 158, 168, 169
Rowles, R. A., 141, 142
Rumelhart, D. E., 171–173, 175–177
Rundstrom, R., 331

Sainio, R., 356
Saint-Martin, P., 295
Salichtchev, K. A., 213
Saushkin, Y. G., 214
Saussure, F. de, 217, 219, 243

Schank, R. C., 191–193
Schlichtmann, H., 214, 223, 251, 252, 267, 290
Schneider, W., 39
Schustack, M. W., 448, 449
Schwaitzberg, S. D., 87, 92, 96, 148
Schweizer, D. M., 440, 441
Sebeok, T. A., 224, 225, 232, 246, 247
Sejnowski, T. J., 117, 118
Shannon, C., 8
Shapley, R., 116, 133, 134
Shelton, E., 195
Shepard, R. N., 48, 187
Shiffrin, R. M., 39
Shortridge, B. G., 22, 41, 85, 86, 88, 89, 92, 127, 128
Silverman, G. H., 93
Simcox, W. A., 206
Simon, H. A., 42, 48, 209, 365, 395
Singh, G., 428
Slocum, T. A., 72, 77, 78, 81, 93, 123, 148, 273, 284, 416
Smith, E. E., 154, 260
Smith, S., 410–413
Speeth, S. D., 379
Spence, I., 132
Spiess, E., 116, 121, 122, 125
Steinke, T., 101–104, 173, 187, 402, 446
Stimson, R. J., 171, 172
Suchan, T., 152
Szegö, J., 192, 193, 282, 301

Tanaka, K., 146, 147
Tatham, A. F., 276
Taylor, D. R. F., 156, 356
Tebbs, B., 391, 416
Thelin, G. P., 142
Thomas, E. L., 102
Tobler, W. R., 283
Travis, D. S., 126, 134
Treinish, L., 276, 405
Treisman, A., 105, 106, 127, 129
Treisman, M., 107, 130
Tsal, Y., 95, 118
Tufte, E., 36, 38, 402, 403

Tukey, J. W., 7
Tversky, B., 162, 164, 165
Tyner, J. A., 348

Ucar, D., 259, 260
Ullman, S., 117, 118
Usery, L., 152
Uttal, W. R., 70, 127, 176

Van der Wel, F. J. M., 441, 443
Van Elzakker, C. P. J. M., 440
Vasconcellos, R., 276–278, 280
Vasiliev, I., 154, 157, 161
Vernon, M. D., 121

Wanger, L. R., 137, 149, 373, 374, 376
Watt, R. J., 96
Weaver, W., 8
Weiss, P., 215, 243
Weisstein, N., 117, 377, 378
Welch, R. B., 127, 128
Wever, E., 111, 112
Wickens, C. D., 90, 434
Wiedel, J., 276
Wiggins, L., 384
Williams, L. G., 132
Wilson, H. R., 96
Wolfe, J. M., 99, 106, 107
Wolodtschenko, A., 234
Wong, E., 116, 117, 378
Wood, C. H., 114
Wood, D., 7, 10, 13, 14, 197, 198, 220, 227, 242, 243, 298–301, 311–313, 319, 325, 329–341, 345, 346, 348, 352, 353, 456
Wood, M., 77, 116, 120–122
Woods Hole Oceanographic Institute, 126
Woodsworth, R. S., 148
Woodward, D., 308, 343
Wright, J. K., 252, 335, 338, 339, 341

Yamahira, T., 384
Yapa, L. S., 7, 169

Zadeh, L., 158

Subject Index

Abstractness continuum, 161, 259, 261, 262, 453, 454, 456
Accommodation of eyes, 137
Action patterns, 365
Advance-and-retreat theory, 21, 134, 135, 138, 145, 194
Advertising maps, 336, 348
Aerial perspective, 121, 138
Aesthetic signs, 227, 228, 247
Age, and map schemata, 196–198, 267
AIDS maps, 187, 188, 369, 370, 430
Air navigation maps, dynamic, 372
Algorithm process tracking, 428
Alternating syntagm, 238, 417, 418, 447
Amacrine cells, 56
Ambiguous figures, 76, 108, 113, 114, 118–120
Ambiguous signs, 229, 323, 324
Analglyph plots, 139
Analogical motivation, 231, 243
Analogical pictures, 268
Analogical representations, 150, 171–173, 175
Animation, 71–74, 93, 126, 134, 236, 369, 371, 383
  dynamic variables in, 243, 281
  sound in, 431, 432
  space–time in, 429–431
  of weather maps, 192
Appraisive signs, 226, 227, 247–249
Apprisive signs, 246–249
Arbitrariness, 258, 259, 262–264, 323, 326
Argument sign, 222
Arrangement, as graphic variable, 267, 273, 275, 279, 387, 389

Art, 2, 9–11, 18, 19, 228, 336, 337
Articulation, 117, 240, 241
Artificial intelligence, 433
Artificial signs, 259
Assimilation, 133
Associative sign-vehicles, 257, 258, 262
Associativity, 36, 82, 83, 91, 92, 387
Astronomical maps, 378, 379, 449
Atlas, 275, 292
  maps, 200
Atlas Tour, 193, 304
Atmospheric simulations, 381, 388, 389
Atomists, 80, 203
Attack/decay, 288–290
Attention, 80–90, 101–104, 118
  divided, 87–90, 92, 96, 148, 149, 402
  global–local precedence in, 95–101, 380
  selective, 36, 81–87, 91, 387, 388, 402
  as sign function, 227, 247
  visual, 80, 81
  zoom lens analogy of, 95, 97, 99
Attitudes, 350, 351
Attributes, 35, 36, 163, 261, 263, 384, 385
  order of, 431
  of space–time position, 317–321
Audio variables. *See* Sound
Auralization of data, 379
Axiomatic semiotics, 238, 239
Axons, in eye-brain system, 63

Banded color palette, 391
Bar graphs, 180, 181
Behavioral theories, 7, 8, 23, 70
Bias, 337, 340–342, 447, 449, 450
Bidirectional information, 390

497

Binocular parallax, 139
Binocular vision, 105, 106, 122, 139
Biplots, 407–409, 430
Bipolar cells, 56
Bivariate data, 415, 416
Bivariate glyphs, 440
Bivariate maps, 90–92, 267
  color in, 415, 416, 418, 440, 441
Blinking, 73, 126, 127, 282, 286, 377, 378, 399, 417, 428, 434
Blurring, 117
Board of Geographic Names, 346
Bottom-up processes, 33, 39, 40, 43, 53, 117–120, 368
Boundaries, 251
  connotations of, 345–347
Brain, 63–68
  and eye system, 17, 24, 25, 28, 53–67
  and map-in-the-head, 171
  and picture-in-the-head, 171
Brightness, 111–116, 133, 145
Brushing, 418, 441
  geographic, 385, 386, 418, 419
  scatterplot, 417–419
  sonic, 421
  temporal, 385, 386, 419

Calico system, 391, 392
Cartographic rules, 342–343
(CARTOGRAPHY)^3, 362
Categories, 17, 50, 151–170, 210
  accuracy of, 436, 437
  basic-level, 162–167, 185, 190, 291
  classical approach to, 151–155, 167, 260
  of colors, 159, 160, 167
  and container schemata, 186–189
  context of, 157, 158
  exemplar members of, 153, 154
  family resemblance in, 155, 156, 160
  folk, 162, 166, 167
  fuzzy, 24, 156–161, 169, 170, 211, 357
  in geographic visualization, 392–398
  hierarchy of, 322
  interactional nature of, 168
  map, 151, 152, 154–157, 159–162, 169, 211, 308, 453
  measurement of, 253
  mental, 13, 24, 28, 151–170, 174, 210, 365
  and multiple representations, 168–170
  natural, 149, 151, 167, 168
  number of, 163, 164
  optimal, 151, 152, 156
  perceptual, 123–136, 147
  probabilistic, 154, 260
  prototype, 153–162, 260
  radial, 160–162, 263, 264, 308, 453

  of referents, 250–256
  scientific, 152, 168, 169, 218
  of signs, 243, 244, 259–261
  of space–time phenomena, 423, 424
  typical members of, 158–160
Cenological level, 239, 241
Center–surround cells, 59
Central Intelligence Agency, 9, 335
Chess expert, 42, 201, 365, 366, 395
Childhood schemata, 196–198, 267
Choropleth maps, 73, 153, 202, 210, 211, 265, 266, 369, 377, 446
  classed and unclassed, 163, 164
  color in, 62, 134, 416
  comprehension of, 233
  data in, 163, 166, 384
  gray tones in, 61, 62
  highlighting in, 282, 283
  interpretation of, 156
  legends of, 188
  of population density, 79, 87
  spatial texture of, 303
Chromostereopsis, 138, 145
Chunking, 40, 42, 79, 174, 209, 426
Clarity, 117, 275, 276, 279, 437
Classification
  biological, 436
  contiguity-biased, 166, 167
  interactive, 384, 385, 418
  optimal, 163, 164, 210, 384
Climate, 153
  global climate model, 220, 255, 284, 404, 437–439, 442
Closure, grouping by, 74–76
Clusters, 78, 384, 434
Codes, 88, 218, 229, 325, 331
  extrasignificant, 298, 332–336, 352
  iconic, 299, 300, 301, 312
  implicit, 338–351
  intrasignificant, 298, 299–302, 313, 332, 333, 352
  linguistic, 299–301, 312, 313, 316, 325, 345, 346, 349
  presentational, 299, 301, 302, 304, 312, 313, 320, 321, 325
  rhetorical, 333–336
  tectonic, 299–301, 312–314
  temporal, 299, 301, 312, 315, 316
  utilitarian, 333–335
Codification of signs, 229
Cognition, 8, 460
  and representations, 12, 14, 15, 47, 173
  and schemata, 40, 45, 193, 203, 448
  spatial, 25
  visual, 16–18, 23–50
Cognitive economy, 231
Cognitive gravity, 448–452

Subject Index   499

Cognitive model, idealized, 174–176, 211, 339
Coherent motion, 133
Color, 121, 270, 272, 279, 389–392, 411–416
  advance-and-retreat theory on, 21, 134, 135, 138, 145, 194
  assignment tool, 389
  banded palette, 391
  in bivariate maps, 415, 416, 418, 440, 441
  categories of, 159, 160, 167
  in choropleth maps, 62, 134, 416
  connotations of, 334, 353, 390
  constancy of, 133
  of contours, 126, 145
  cycling technique, 283–286
  and depth perception, 126, 138, 141–146
  discrimination of, 132, 133, 149, 440, 441
  of elevation maps, 21, 194, 195, 319
  in feature comparison, 411–416
  of fluorescent ink maps, 145, 146
  highlighting method, 428
  hue of. See Hue of colors
  lightness of, 295, 296
  in lithographic printing, 145
  matching of, 143, 144, 414–416
  meaning of, 329
  opponent-process theory on, 62, 143, 144, 148, 160, 194, 195, 293–295
  order of, 134, 135, 145, 188, 195, 292, 293, 318, 319, 390
  perception of, 60–63
  reflectance of, 133, 142
  and reliability, 437, 440, 441
  retinal sensitivity to, 55, 56
  saturation of, 62, 87, 121, 134, 135, 272, 279, 437, 440, 441
  scheme of, 294, 405, 411
  in scientific visualization, 413
  selection tools, 391, 392
  selectivity of, 84
  of symbols, 57, 125, 126, 132, 133
  syntactics of, 292–298, 391
  in temperature maps, 135, 318, 319, 325, 390
  in thematic maps, 295, 297, 298
  in trichromatic maps, 414
  value of. See Value of colors
  vision of, 63, 105, 106, 126, 133, 149
  wavelengths of, 21, 55, 84, 138, 194
Color path, 391–392, 416
Color sheet, 416
Color space, 62, 391, 392, 416
Communication function, 246
  of basic-level categories, 163
  of maps, 1–9, 11, 12, 14, 23, 40, 69, 124, 269, 459
    and visualization, 357, 358
  of signs, 227, 243
Communication paradigm, 2, 4, 6, 11, 459
Communicative function of art, 9, 336, 337
Competence, cartographic, 305
Compliance class, 199
Componential stand-for relationship, 196, 199
Comprehension, 314
  of signs, 232–234, 305, 306
Computational theories, 26, 27, 29, 33, 51, 54
Computer-assisted visualization, 50, 356, 416
Comsign, 228, 314, 321, 324
Concept representation (C-rep), 255, 256, 263, 308, 327, 328, 427
Conceptual message, 39, 179, 182, 206
Conceptual-proposition theory, 171
Condensation efficiency, 88
Cone cells, 54–56, 62, 63
Configural dimensions, 88–90
Configurational knowledge, 172, 173
Confirmatory bias, 449, 450
Conjunctions, 87–90, 106, 107, 129, 132
Connecting syntagm, 238
Connotations, 17, 229–232, 243, 331–351, 353, 448
  of color, 334, 353, 390
  degrees of, 350
  implicit, 452–456
  incitive, 337, 348, 349
  of integrity, 337, 340–342
  map-as-truth, 338
  measurement of, 349–351
  of power, 337, 338, 345–348
  of signs, 227, 229–231, 247, 249, 331–338
  valuative, 337, 342–345
  of veracity, 337–340
Conscious signs, 229
Container schemata, 186–189, 203
Contiguity-biased data classification, 166, 167
Continental drift, 450, 451
Continuous phenomena, 40, 252–254, 302–304
Contours, 54, 102, 108–110, 116, 117, 121, 139
  color of, 126, 145
  interpretation of, 200, 201
  lines of, 145, 372
  in Tanaka method, 146, 147
Contrast, 51, 52, 78, 111–116, 121, 122, 125, 133, 136
  simultaneous, 60–62, 133
Conventional signs, 256, 257, 262, 308
Convergence of eyes, 137
Convexity, 110, 111
Coordinate systems, 39, 48, 198

Crispness, 275, 276, 279, 309, 381, 382, 387, 399, 437, 458
Critical examination, 9
Critical theory, 350, 351
Cross-maps, 418, 420
Cyberspace, 396, 434

Data
 auralization of, 379
 bivariate, 415, 416
 classification of, 166, 167, 384, 385
  optimal, 163, 164, 210, 384
 exploration of, 18, 368, 384
 multidimensional, 401–422
 quality of, 186, 440, 458
 quantitative spatial, 302
 truth of, 440
Data models, 265–267, 302, 303
Database, 348, 352
 geographic, 251, 322, 323
 navigation of, 400
 space–time, 423
 spatial, 191, 260
Date, display, 281, 282, 287
Decision making, 68, 362, 363, 459, 460
Declarative knowledge, 171, 172
Deconstruction, 10, 11, 13, 190
Decoration, map, 347, 348
Defocusing, 382, 383
Denotations, 17, 230–232, 312–321, 331, 351, 352, 448
 of illustrations, 452
 of signs, 229–231, 331
 of space–time position, 317–321
 of spatial position, 313–315
 of temporal position, 315–317
Depth perception, 121, 136–247, 370
 accuracy of, 373, 374
 binocular, 122, 139
 color in, 126, 138, 141–146
 contour lines in, 126, 145, 372
 in fishnet maps, 139–141
 in fluorescent ink maps, 145, 146
 frame of reference in, 138, 370, 373
 graphic variables in, 374–376
 motion in, 138, 139, 371–373
 perspective in, 138–141, 370, 372, 373
 physiological cues in, 137, 139
 pictorial cues in, 137–139
 shading in, 138, 140–146
 shadows in, 138, 140, 146, 147, 373
 texture in, 138, 139, 141, 373
 in three-dimensional simulation, 370–376
 in two-dimensional scenes, 17, 136–147
 and visual levels, 122, 123
Descriptions, visual, 35–39, 46, 49, 150, 178, 179, 363

Descriptive syntagm, 238
Design, 1–16
 artistic, 2, 9, 11
 functional, 2, 3, 12
 historical, 1–6
 scientific, 9–12
Designative signs, 226, 247, 248
Detail, 138, 275, 381, 382, 437
 gratuitous, 453–456
Detection, 22, 123–127, 133, 148
Developmental issues, 195–198
Diagram hypoicon, 243, 257, 266
Diagrams, 209, 212, 223, 399
 reliability, 341, 342
Dicent signs, 222
Digital database, 348, 352
Digital relief map, 142
Disassociative variables, 91
Discourse, 225–228, 232
Discrete phenomena, 40, 252–254, 303, 304
Discrimination, 22, 43, 95, 123, 124, 127–134, 147
 of color, 132, 133, 149, 440, 441
Display
 date in, 281, 282, 287
 dynamic, 278
 as sign, truth of, 444–447
 signs in, truth of, 436–444
 trivariate, 403–405
 visual, 262, 263
Distributional form, 253
Divided attention, 87–90, 92, 96, 148, 149, 402
Domain-specific schemata, 446–448, 450, 451
Dot maps, 80, 93, 102, 103, 203, 413, 446
Dual coding theory, 47, 171
Duration, 134, 281–284, 287
 of sound, 288–290
 and temporal codes, 301
Dyadic signs, 219, 230, 242, 245
Dynamic maps, 18, 71, 186, 241, 278–290, 352, 372
 audio, 287–290
 visual, 278–287
Dynamic variables, 149, 243, 278–290
 display date, 281, 282
 duration, 281–284
 frequency, 281, 285, 286, 301
 order, 281, 284, 301
 rate of change, 281, 284, 285, 301
 synchronization, 281, 285, 287, 301

Earthquake epicenters, 73, 283
Ecological vision, 23, 26
Economic maps, 32, 34, 290
Edge detection, 382

Elevation maps, 21, 194, 195, 319
Ellipse symbol, 90
Emblems, 259, 261
Emotional connotations, 337
Emotive art, 9, 336, 337
Emotive signs, 227, 247, 249
Engineering of maps, 8, 18, 41
Environmental knowledge acquisition, 172, 173
Environmental psychology, 211
Episode, 238, 282, 424
Ethics, map, 340–342
Ethnocentricity, rule of, 343
Etymology, 329, 330
Euclidean spatial concepts, 197
Event, 309, 424
Event schemata, 151, 175, 190–193, 365
Evidence, visual, 392, 399, 450
Expert knowledge, 68
Expert–novice differences, 200, 393–398
 in categorization, 169
 in chess, 42, 201, 365, 366, 395
 in database navigation, 400
 in geographical information systems, 397, 398
Expert systems, 433
Explicit meaning, 13, 16, 214, 242, 312, 316, 351
 of signs, 225, 229
Exploration, 362, 368
Exploratory data analysis, 18, 384
Exploratory Visualization System, 387, 388, 401, 403, 409–411
 color in, 413, 415
 orientation in, 409–411
 sound in, 421
Expressionistic art, 9, 336, 337
Extrasignificant codes, 298, 332–336, 352
Eyes, 54–63
 and brain system, 17, 24, 25, 28, 53–67
 camera analogy of, 54
 movements of, 42, 43, 101–104
 structure of, 54–57

Family resemblance, 155, 156, 160
Features, 362
 comparison of, 359, 401–422
  color in, 411–416
  focusing in, 418–419
  orientation in, 409–411
  sound in, 419–422
  space in, 402–409
  time in, 416–418
 conjunctions of, 106–107
 discrimination of, 128–130
 global, 95–101, 362, 380, 382, 383

 identification of, 359, 362–369, 376, 377, 386–392
 integration theory, 106, 107
 local, 95–101, 362, 380, 382, 383
 maps of, 106
 matching of, 362–367, 457
 noticing of, 365, 383, 388
 space–time feature analysis, 359, 401, 422
 spatial, 364, 365, 386–392
 visual search for, 106, 107
Fechner's Law, 3
Figurae, 239–241
Figure–ground relationship, 21, 70, 77, 78, 104, 105, 107–120
 brightness in, 111–116
 contours in, 108–110, 116, 117
 convexity in, 110, 111
 heterogeneity in, 108–117
 orientation in, 109, 110, 115, 118, 119
 relative size in, 109, 111, 112
 reversal of, 118, 119
 surroundedness in, 109
Filmic syntax, 237, 423
Filtering, 54, 88, 381, 399, 426
Fishnet maps, 139–142
Fixation of focus, 57, 104, 148
Flickering technique, 417, 447
Fluorescent ink maps, 145, 146
Flybys, 32, 371, 372, 399
Focus, 117, 149, 308, 309
 fixation of, 57, 104, 148
Focusing, 309, 384, 418, 419, 441
 and defocusing, 382, 383
Fog, 275, 276, 278
Folk categorization, 162, 166, 167
Formalist art, 336, 337
Formative signs, 226, 227, 247
Fourth dimension, time as, 301
Fovea, 54–57, 59, 103
Frame, 174, 176, 282, 284
Frame rectangles, 438–440
Frequency, 281, 285, 286, 301
 spatial, 96, 116, 117, 377, 378
 temporal, 285, 377, 378, 399
Functional approach, 2, 3, 6, 7, 12–15, 17, 22, 25, 244–309
Fusion, 336
Fuzzy categories, 24, 156–161, 169–170, 211, 357

Ganglion, 56–65
Generalists, 80, 203
Generalization, map, 104, 322, 323
Geniculate nucleus, lateral, 65
Geodemographics, 348, 353
Geographic biplots, 407–409, 430
Geographic brushing, 385, 386, 418, 419

Geographic databases, 322, 323
Geographic features, 364, 365
Geographic hierarchy, 197, 198
Geographic regionalization, 164–166
Geographic visualization, 16, 18, 44, 139, 263, 308, 355–458
  color variables in, 389–392
  and feature comparison, 401–422
  and feature matching, 362–367
  as fuzzy category, 357
  importance of, 460
  and map syntactics, 367–392
  schemata in, 392–398
  and space–time processes, 422, 433
  truth in, 433, 435–458
  and visual thinking, 361–400
Geographical information systems, 1, 47, 152, 352, 397, 398, 451, 452
Geologic maps, 7, 200
Geological Survey maps, 324, 334, 335, 343–345
Geometric point sign-vehicles, 257, 258, 262
Gestalt theories, 69, 70, 77, 148
  on figure–ground relationship, 108–110, 117–119
  on grouping, 37, 38, 50, 52, 69, 70, 77, 81, 82, 108
  on visual cognition, 24
Global climate model, 220, 255, 284, 404, 437–439, 442
Global image map, 348, 349, 456
Global–local differences, 42, 362, 382, 383
  in attention, 95–101, 130, 131, 380
  in exploratory visualization, 380
  in syntactic analysis, 296, 298
Globes, connotations of, 347, 348
Glyphs, 91, 440
Goals, map, 269, 355
Graduated symbol maps, 77, 78, 88, 89, 93, 136, 188, 303, 318
Graph schema, 35, 39, 40, 176, 178–185, 205–209
Graphic icons, 267, 268, 324, 326
Graphic scripts, 304, 400, 417, 433
Graphic variables, 35–37, 84, 85, 147, 209, 240, 245, 269–276, 279, 308, 309
  arrangement, 267, 273, 275, 279, 387, 389
  clarity, 275, 276, 279, 387
  color hue, 121, 270, 272, 279
  color saturation, 121, 272, 279
  color value, 121, 270, 279
  and depth perception, 374–376
  differential, 274
  manipulation of, 386–392
  order of, 134, 135, 267, 274
  orientation, 270, 279, 386, 387
  shape, 270, 279
  size, 270, 279, 386
  texture, 121, 270, 272, 273, 279, 386, 387
Graphical user interface, 324, 326
Graphs, 452, 455
  bar, 40, 180, 181, 206, 207
  comprehension of, 33–41, 43–46, 178–185, 205–209, 363
  difficulty principle on, 205, 206
  line, 40, 206, 207
  schema on, 35, 39, 40, 176, 178–185, 205–209
Gratuitous detail, 453–456
Gray tones, 36, 37, 60–62, 136
Grouping, 40, 69–84, 87, 91, 147, 148, 151
  atomist–generalist, 80, 203
  by closure, 74–76
  by common fate, 72, 73
  by experience or habit, 76
  Gestalt theories on, 37, 38, 50, 52, 69, 70, 77, 81, 82, 108
  by good continuation, 74–78
  by motion, 74, 133
  of objective sets, 74, 100
  perceptual, 52, 71, 73, 77, 78, 108
  Prafnanzstufen factor in, 73, 74
  by proximity, 71, 72, 77, 81, 93
  by similarity, 72, 73, 77, 81, 93, 290
  by simplicity, 75, 76
  of sounds, 72
  by time, 377
Guided search theory, 107

Habit, grouping by, 76
Hermann grid, 59, 60
Hermeneutics, 331
Heterogeneity, 90, 108–117
Highlighting with color, 428
Highway maps, 7, 10, 17, 268, 325
Highway numbering system, 179, 180
Hill signs, 329, 330
Historical codes, 333
Holistic stand-for relationship, 196, 199
Homogeneity, 90, 153
Homological motivation, 231, 243
Horizontal cells, 56
Hue of colors, 21, 52, 62, 121, 143–145, 270, 272, 279, 441
  connotations of, 334
  discrimination of, 132, 440
  full-spectral, 319
  order of, 134, 135, 188, 292, 293, 319, 390
  selectivity of, 84
  visual acuity for, 126
Human factors, 41, 426
Human–map interactions, 8, 12, 18, 22, 150, 193, 363–365, 398, 434
  space of, 357

Hyperbole, 336
Hypergraphic environment, 193
Hypermedia, 389
Hypertext, 193
Hypoicons, 223, 224, 243, 257, 258, 262, 266, 447
Hypsometric map schemata, 201

Icons, 222, 256–268, 328, 330
  and codes, 299, 300, 301, 312
  graphic, 267, 268, 324, 326
  and memory, 35
  and signs, 224, 232, 256, 258, 259, 317, 323, 329, 330
  and sign-vehicles, 222, 223, 256–258, 291, 317, 323, 324, 328
  and symbols, 130, 224, 258
Idealized cognitive model, 174–176, 211, 339
Identification
  of features, 359, 362–369, 376, 377, 386–392
  of patterns, 176, 361, 362, 419
  as sign function, 226, 247
  of signs, 306, 309
  of symbols, 129, 130
Illusion, 59, 60, 68, 69, 417, 444, 449
Image, 48, 223, 230
  analysis systems, 54
  graphic continuum, 264, 340
  kinesthetic, 149, 185, 187
  master, 455
  mental, 187, 190
  radical image theory, 171
  regional, 165, 166
  retinal, 29, 137–139
  transformation of, 187
Image schemata, 151, 175, 176, 183, 185–191, 267, 365
  embodied, 185, 186, 189, 201, 202
  kinesthetic, 149, 185, 187, 201
  preconceptual structures in, 185, 201
Imagery, 46–49, 187, 211
  mental, 46, 48, 356
Imitative art, 9, 336, 337
Implicit meaning, 13, 16, 214, 242, 316, 452–456
  of maps, 338–351
  of signs, 225, 229
Incitive connotations, 337, 348, 349
Incitive signs, 227
Index, 222–224, 247, 263
Indicated entity, 243
Indicating signs, 247–249
Indispensable variables, 35, 36, 92–94, 98, 368–379
Inferential processes, 182, 366
Information graphics, 431

Information processing, 3–6, 8, 9, 12, 16, 17
  in graph comprehension, 43–45, 183
  and pattern matching, 44, 45
  vision in, 23–50, 68, 70, 118, 147, 430
Information theory, 69
Information transmission, 49
Informative signs, 227, 247
Injunctive signs, 247
Instantiated schema, 182, 365
Integral dimensions, 85–90
Integrity, connotation of, 337, 340–342
Interaction space, 357
Interactive classification, 384, 385, 418
Interactive maps, 18, 26
Interactive visualization, 132, 297, 358, 403
Interpersonal signs, 228, 313
Interposition of objects, 138
Interpretants, 219–222, 245, 290, 306
  as mediator, 256–269
  mediator of, 246–250
Interpretation, 12, 14, 219, 256
  of maps, 200, 201, 394, 395, 398
Interrogation process, 182
Interval information, 253, 255, 265, 270, 273, 316
Intrasignificant codes, 298–302, 313, 332, 333, 352
Intuition, 48
Isarithmic maps, 60, 403, 446
Isochrone maps, 200
Isolines, 60, 266, 267, 340, 405
Isometric projection, 370, 399, 445
Isomorphism, 263
Isopleth maps, 164, 201, 202, 265, 266, 303
Isotherm maps, 135

Jitter, 422
Judgment, 147, 337, 342–345
  of magnitude, 124, 135, 136
  of order, 134, 135

Kinesthetic image schemata, 149, 185, 187, 201
Knowledge, 4, 68, 171–173
  organization of, 163, 170
  of signs, 306
  structures of, 24
Knowledge-based systems, 245
Knowledge representation, 39, 150, 170–193
  analogical, 171–173
  procedural, 171–173
  propositional, 171–173
Knowledge schemata, 16, 49, 150, 151, 174–193, 215, 451
  development of, 18
  event, 190–193

Knowledge schemata (*continued*)
  and grouping by experience, 76
  image, 185–190
  and mental categories, 13, 24
  propositional, 176–185
  and sign systems, 246

Label, 115, 127, 128, 163, 247–249
Land use classification, 152, 159, 322, 444
Landform patterns, 395
Language, 217, 229, 230, 239, 276, 329, 330
  cartographic, 305
  color categories in, 159, 160, 167
  mapping as, 236
  natural, 152, 213, 214, 325
Lateral inhibition, 60–62
Layer tints, 60, 145, 149, 195, 201, 204, 209, 319
Learning, 171–173, 393, 395, 405
Least practical differences, 3, 22, 124
Legends, 49, 204, 205, 320
Legisign, 222
Lettering, 3, 104, 128, 258
  attention to, 97–100
Leukemia data, 429
Lexical approach, 12–15, 17, 310–353
Linear order schemata, 187–189, 195, 201
Linear perspective, 138–140, 372, 373
Linguistic codes, 299–301, 312, 313, 316, 325
  connotations of, 345, 346, 349
Link schemata, 187–189
Linked sign systems, 295–302
Literature, maps in, 345, 347
Lithographic printing, 145
Local–global differences. *See* Global–local differences
Location, 35–38, 93–96, 270, 279
  as audio variable, 288, 289
  temporal, 281, 318
Locational sign, 251
Logical pictures, 268
London Underground Map, 311
Loudness, 288, 289, 442
Luminance, 105, 106, 133

Mach bands, 60
Macula, 54, 55
Magnitude, 38, 124, 135, 136
Map-in-the-head, 171
Mapness, 155, 157
Mappaemundi, 343
Maps to be read, 206, 209
Maps to be seen, 206, 209
Marginalia, 347
Marks, map, 199, 246, 250, 258, 259, 305
Master image, 455
Match process, 182
Matrix manipulation technique, 403
Meaning, 23, 25, 27, 49, 68, 69, 186, 329, 330
  connotative. *See* Connotations
  denotative. *See* Denotations
  explicit, 13, 16, 214, 225, 229, 242, 312, 316, 351
  implicit. *See* Implicit meaning
  legitimate, 311
  in maps, 311–330
  of maps, 311, 312, 330–351
  multiple, 311, 433
  political, 312
  of signs, 245, 260, 261, 311, 325, 326
    connotative, 227, 229–231, 247, 249, 331–338
    denotative, 229–231, 331
    explicit, 225, 229
    implicit, 225, 229
  of space, 312–315
  of time, 312
Measurements, 253–255, 265, 270, 273, 274
  of connotations, 349–351
  of discrimination, 127
Memory, 29, 42–44, 46, 47, 171
  of categories, 164
  iconic, 35
  long-term, 43, 46, 49, 174
  representation, 39, 40, 47, 52, 118, 119
  short-term, 43, 211
Mental categories, 13, 24, 28, 151–170, 174, 210, 365
Mental models, 174, 395
Mental representation, 12, 28, 35, 38, 46, 210
Mental rotation of map, 173
Mental visualization, 255, 356, 422, 423
Message assembly, 182, 366
METAGIS, 441
Metalanguage, 230
Metalinguistic signs, 227, 247
Metaphors, 175, 176, 187, 189, 192, 223
  and hypoicons, 243, 258, 266
Meta-wayfinding, 396
Metonymy, 175, 176, 257, 336
Midwest region, 165, 166
Minimum instant of vision, 368
Mixtec–Aztec signs, 330
MKS-ASPECT™ system, 143, 144
Modeling, 255
Moire patterns, 389
Monochrome variables, 386–389
Monosemic signs, 229, 324, 325
Morse code, 239–241
Motion, 74, 105, 106, 126, 133, 134, 417
  depth perception in, 138, 139, 371–373

Motion parallax, 137, 371, 372
Motivation, 224, 231, 232, 243, 308
Mount Denali, 346
Multiresolution analysis, 380, 381
Multivariate analysis, 387, 388, 401–422
Multivariate symbols, 81, 85, 86, 88

Name
  connotations of, 344–346, 349
  as sign function, 225, 247
  visual search for, 104, 107
Narrations, 193
National Geographic Society, 9, 240, 349
National Park Service maps, 31, 56
  sign-vehicles of, 317, 318
  symbols of, 56, 57, 129, 177, 178, 232, 240
Natural categories, 151, 167, 168
Natural language, 152, 213, 214, 325
Natural legends, 204, 205, 320
Navigation, 126, 133, 173, 225, 339, 372
  of database, 400
  and wayfinding, 191, 225, 395–397
N-dimensional space, 430
Neurophysiology, 24, 26, 27, 116, 185, 211, 419
  of vision, 53, 63–65, 96, 147, 160
Neuropsychology, 26, 64
Neutrality of maps, 340, 341
Noetic field, 232, 233, 291, 305
Nominal information, 253, 255, 270, 273, 274
Notational schemata, 199, 200
Noticing, 363, 365, 383, 388, 450, 451
Novice–expert differences. *See* Expert–novice differences
Nuclear magnetic resonance spectrometry, 190, 202
Numerical measurements, 253, 270

Object integration, 90
Object recognition, 96, 100, 118, 119
Objective sets, 74, 100
Oblique projection, 138, 139, 370, 371
OFF-center cells, 59
ON-center cells, 59
Opponent-process theory, 62, 143, 144, 148, 160, 194, 195, 293–295
Optic chiasma, 63
Optic nerve, 63
Optimal categories, 151, 152, 156
Optimal data classification, 163, 164, 210, 384
Optimal maps, 7, 355
Order, 123, 147, 148, 287, 446
  apparent, 124
  audio, 288–290
  of colors, 134, 135, 145, 188, 195, 292, 293, 318, 319, 390
  dynamic, 281, 284, 301
  intuitive, 134
  judgment of, 134, 135
  linear, 187–189, 195
  temporal, 134, 284, 431
Ordinal measurements, 253, 255, 273, 274
Orientation, 37, 109, 110, 115, 118, 119, 270, 279, 386
  cells sensitive to, 66–68
  in EXVIS, 409–411
  of features, 387, 388, 409–411
  of symbols, 83, 84, 86
Orienteering, 395, 396
Oscillations, 429

Parallel processes, 53, 86, 106, 107, 120
Parallel syntagm, 238
Partonomy, 165
Part–part relationship, 327
Part–whole relationship, 327
Patterns, 18, 36, 85, 132, 267, 273, 275, 279, 364, 398
  discrimination of, 130–132, 147
  false, 388
  identification of, 176, 361, 362, 419
  and illusions, 59, 60
  landform, 395
  matching of, 44, 45, 362, 363, 457
  noticing of, 363, 365, 450, 451
  orientation of symbols in, 83, 84
  recognition of, 39, 96, 116, 125, 170, 201, 363, 387
  spatial, 369, 384, 386–392, 403
  spatial–geographic, 369
  task-specific viewing of, 103, 104
  temporal, 420, 431
  texture of, 125, 130, 149
  in visual evidence, 450
Perception, 2, 3, 22, 23, 25, 26, 53, 124, 363
  basic-level, 185, 190
  of categories, 163
  of color, 60–63
  of depth. *See* Depth perception
  opponent-process theory on, 62, 143, 144, 148, 160, 194, 195, 293–295
  and representations, 14, 52, 69
  thresholds in, 125
Perceptual categorization, 123–136, 147
Perceptual organization, 35, 44, 68–123, 147
  associativity in, 91, 92
  and attention, 80–90, 94–101
  figure–ground concept in, 107–120
  and grouping, 69–80
  indispensable variables in, 92–94, 368–379
  and map syntactics, 367–392
  of maps, 78, 79

Perceptual organization (*continued*)
  visual levels in, 120–123
Perceptual units, 35–37
Performance, cartographic, 305
Period, duration of, 282–284
Peripheral vision, 57, 58, 99, 104, 126
Personal navigation assistants, 225
Perspective, 138–141, 370, 372, 373, 399
  aerial, 121, 138
  linear, 138–140, 372, 373
Perspective projection, 370, 399, 445, 446
Persuasion, 337, 348, 349
Peters projection, 300
Phatic signs, 227
Phenomenon, 302–304, 308, 327, 328
  discrete–continuous, 40, 252–254, 302–304
  representation (P-rep), 255, 256, 263, 308, 327, 328, 427
  space–time, 423, 424
Photographs, 324, 453–456
Physiological–conceptual representation, 52
Physiology, 137, 139, 194, 195
Piagetian theory, 197, 198
Pictorial depth cues, 137–139
Pictorial point sign-vehicles, 56, 104, 105, 257, 258, 262
Pictures, 267, 268
  in-the-head, 171
  realistic, 268, 453
Pitch, 288–290, 421, 431, 432, 442
Placename. *See* Name
Plan-free map sign, 251, 253, 259
Plan map sign, 251, 252, 259
Plan views, 146, 196, 197
Plastic relief, 141, 142, 149
Plerological level, 239, 241
Poetic signs, 227, 228, 247, 249
Point features, 128–130, 294
Point sign-vehicles, 91, 92, 257, 258, 262, 292, 293
  pictorial, 56, 104, 105, 257, 258, 262
Point-cloud technique, 389
Polysemic signs, 229, 231, 324, 325
Population density, 79, 80, 87, 252
Position
  in space, 313–315
  in space–time, 93, 317–321
  in time, 315–317
Positivism, 3, 6, 19
Postmodernism, 6, 10
Postprocessing, 427, 429–432
Power, 13
  connotation of, 337, 338, 345–348
Pragmatic reasoning schemata, 394
Pragmatics, 234, 238, 242
Prägnanzstufen factor, 73, 74

Preattention, 70, 80, 118, 120
Precipitation maps, 86, 87, 92
Precognition, 30, 33, 35, 40–43, 363
Preconceptual structures, 185, 201
Prescriptive signs, 226, 247, 249
Presentational codes, 299, 301, 302, 304, 312, 313, 320, 321, 325
Primal sketch, 29, 31, 51, 65
Printing, lithographic, 145
Probabilistic model, 154, 260
Problem solving, 173, 366
  spatial, 396, 397
Procedural representations, 150, 171–173, 175, 211
Process steering, 427, 432, 433
Process tracking, 427, 428
Projections
  isometric, 370, 399, 445
  oblique, 138, 139, 370, 371
  perspective, 370, 399, 445, 446
  Peters, 300
Projective spatial concepts, 197, 198
Propaganda in maps, 10, 341, 348
Propositional representations, 150, 171–175
Propositional schemata, 151, 175–185, 191, 365
Props, 191, 192
Prototypes, 42, 153–162, 260
  maps as, 154, 155, 161, 196, 197
Proximity, 93
  grouping by, 71, 72, 77, 81, 93
Psychophysical research, 26, 53
Public aspects, 15–17
  of implicit connotations, 452–456

Quad-dimensional information, 421
Qualisign, 222
Quality
  of data, 186, 440, 458
  of visualization, 74
Quasi-symbols, 261
Quivering, 378, 379

Radial categories, 160–162, 263, 264, 308, 453
Radical-image theory, 171
Rank-ordering task, 446
Rate of change, 134, 281, 284, 285, 287, 301, 316
  audio, 288–290, 442
Ratio information, 253, 255, 265, 270, 273
Real time, 26, 173, 423, 427, 432
Realism, 268, 328, 451–457
Reasoning, 364, 366, 394
Recall. *See* Memory
Receptive field, 56–60
Receptor cells, 54–56, 62, 63

Recognition, 363
  and feature matching, 364, 365
  of patterns, 39, 96, 116, 125, 170, 201, 363, 387
Recreation maps, 317
Reductionistic approaches, 26, 33
Redundancy gains, 88, 89
Reexpression procedure, 431, 439
Reference
  direct, 262, 263, 268
  frame of, 138, 370, 373
  indirect, 262, 263, 267, 268, 326, 327, 331, 332
Reference maps, 78, 79, 95, 209, 242, 291, 292, 297, 340
Referential signs, 227, 247
Referents, 199, 213, 214, 219–222, 245, 290, 291, 306
  denotative, 324, 326, 327
  as mediator, 250–256
  mediator of, 246–250
Reflectance of color, 133, 142
Regionalization, geographic, 164–166
Regions, 71, 72, 107, 164–166
  images of, 165, 166
  segregation of, 85
Register of pitch, 288–290
Relating signs, 247–250
Reliability, 437, 445–447, 458
  diagrams of, 341, 342
  thresholds for, 441–443
Relief, 21, 141–146, 149, 197
Representations, 12–18, 25, 39, 40, 49, 150, 168–193, 460
  analog, 150, 171–173, 175
  cartographic, 214, 359
  cognitive, 12, 14, 15, 47, 173
  concept, 255, 256, 263, 308, 327, 328, 427
  functional, 12–15, 17, 244–309
  fuzzy, 169, 170
  lexical, 12–15, 17, 310–353
  map, 214, 217–353
  map-derived, 173
  memory, 39, 40, 47, 52, 118, 119
  mental, 12, 28, 35, 38, 46, 210
  multiple, 168–170, 398
  and perception, 14, 52, 69
  phenomenon, 255, 256, 263, 308, 327, 328, 427
  physiological–conceptual, 52
  procedural, 150, 171–173, 175, 211
  propositional, 150, 171–175
  retinal, 28
  and semiotics, 217–243
  spatial, 460
  stability of, 445
  terrain, 193, 201, 204, 205, 208, 209, 445

  theory of, 13, 28
  three-dimensional, 28, 30–32, 370
  two-dimensional, 28
  and vision, 14–16, 27, 28, 33, 51, 147
Resolution, 277, 302, 380–387, 399, 437, 458
  time, 424, 426
  as visual variable, 275, 276, 279
Restaurant script, 191, 192
Retina, 29, 54
  angle subtended on, 56, 58
  cells of, 54–57, 62, 63
  color sensitivity of, 55, 56
  image disparity, 137
  image size, 138, 139
  and optic nerve, 63
Retinal representations, 28
Retinal variables, 270
Rheme, 222
Rhetoric, 10, 11, 333–336
Rhythm, 72
River sign, 321
Road maps, 200
Rods and cones, 54–56, 62, 63
Roles, 191
Route learning, 173
Rules, 448, 449
  cartographic, 342, 343

Saturation of color, 62, 87, 121, 272, 279, 437, 440, 441
  order of, 134, 135
Scalar codes, 300
Scale, 96–101, 157, 203, 204, 252, 399, 426
  and resolution, 380–386
  of time, 424–426
Scaling, multidimensional, 369, 370
Scanning, 95, 101–107
Scatterplots, 417–419, 434
Scene, 237, 282–284, 309, 424
  segmentation of, 394
  visual. See Visual scenes
Schemata, 17, 42, 43, 174–210, 214, 215, 365, 366
  center–periphery, 187, 189, 201
  cognitive, 40, 45, 193
  container, 186–189, 203
  dark–light, 195, 201
  dark = more, 266
  down = less, 267
  down = south, 267
  event, 151, 175, 190–193, 365
  front–back, 189
  general, 183, 184, 196, 198–202, 448, 451
  in geographic visualization, 392–398
  graph, 35, 39, 40, 176, 178–185, 205–209
  hierarchical, 174, 191
  and highway numbering, 179, 180

Schemata (*continued*)
  image. *See* Image schemata
  knowledge. *See* Knowledge schemata
  linear order, 187–189, 195, 201
  link, 187–189
  map, 49, 193–209, 246, 456
  and National Park Service symbols, 177, 178
  origin–path–destination, 189
  part–whole, 187, 189
  pattern-indexed, 450
  photo, 456
  physiology of, 194, 195
  propositional, 151, 175–185, 191, 365
  and scripts, 191–193
  source–path–goal, 187–189, 201
  spatial, 451
  specific, 196, 199, 202
  standard, 192, 210
  in terrain representation, 193, 201, 204, 205
  up–down, 189, 201
  up = more, 195, 201, 266, 267, 320
  up = north, 267
  visual, 392, 393
Science, 190, 449, 450, 457
  of signs. *See* Semiotics
  society interface, 448, 455
Scientific approach, 2, 6, 9, 11, 12, 19, 366, 367
Scientific categories, 152, 168, 169, 218
Scientific code, 229
Scientific models, 308, 447
Scientific visualization, 45, 48, 190, 212, 278, 355, 356, 362
  color in, 390, 413
  realism in, 452, 457
Scripts, 174, 191–193, 304, 400, 417, 433
Sea surface temperature, 383
Search, 68, 86, 93, 101–107
  color in, 132, 133
  conjunction target in, 129, 132
  discrimination in, 127–130, 132, 133
  features in, 106, 107, 129
  figure–ground in, 104, 105
  guided, 107
  parallel, 106, 107
  regional, 107
  schemata in, 175
  serial, 86, 107
  symbols in, 126, 129, 130
Seeing
  errors in, 436
  and maps to be seen, 206, 209
  seeing wrong and not seeing, 436, 444–447, 451

Seismic event classification, 379
Seismic sounding map schema, 200
Selection problem, 393, 394
Selective attention, 36, 81–87, 91, 387, 388, 402
Selectivity concept, 82–85, 387, 388
Semantic differential, 350, 351
Semantics, 234, 238, 242, 245–269
Sematic field, 232, 291, 305
Semiology, 16, 217, 219, 245, 271
Semiotic systems, 238–242
Semiotics, 11, 13, 17, 41, 43, 213–215, 291, 459, 460
  axiomatic, 238, 239
  definition of, 213
  functional, 224, 239
  and map representation, 217–243
  terminology in, 218, 219
  texture, 272, 273
  theory of codes in, 218
  triangle of elements in, 221, 222, 244, 259, 260, 300
Separable dimensions, 81–90
Sequence, 237, 238, 316, 416, 417
Serial processes, 86, 107
Shading, 9, 114, 138, 140–146, 328
Shadows, 138, 140, 146, 147, 373
Shape, 29, 31, 32, 36, 38, 270, 279
  attention to, 91, 99, 100
  recognition of, 51, 119, 120
  selectivity of, 84, 85
Short-term memory, 43, 211
Short-term visual store, 35
Side-by-side maps, 402
Signa, 239, 240, 241
Signals, 225, 247
Signatures, spatial, 365
Signification, 230, 239
Signifier and signified, 219
Signifying, modes of, 226, 227, 246
Signs, 214, 217–243. *See also specific types of signs*
  arbitrariness of, 258, 259, 262, 264, 323
  aspects of, 224, 246–250
  bundled, 251, 260, 261
  categories of, 243, 244, 259–261
  comprehension of, 232–234, 305, 306
  concreteness of, 327, 328
  connotative, 227, 229–231, 247, 249, 331–338
  definition of, 218
  denotative, 229–231, 331
  direct, 262–265
  and discourse, 225–228
  functions of, 227, 246
  general, 228, 321–323

and icons, 224, 232, 256–268, 317, 323, 329, 330
indirect, 262, 263, 331
knowledge of, 306
levels of meaning, 228–232
maps as, 302–304
mediators of, 246–269
models of, 219–222
multiple linked, 295–302
nature of, 218–234
for phenomena, 327, 328
relating, 247–250
reliability of, 437
science of. *See* Semiotics
semantics of, 234
and signifying, 226, 227, 246
specificity of, 228–232, 321–325
and symbols, 224, 256
syntactics of, 234–236, 244, 245
systems of, 218, 234–242, 245
theory of, 219–221
truth of, 435–444
truth of display as, 444–447
typology of, 217, 222–225, 246, 260, 311
Sign-vehicles, 213, 214, 218, 219, 242–245, 306
appropriateness of, 318, 320
cultural differences in, 329, 330
and icons, 222, 223, 256–258, 291, 317, 323, 324, 328
as mediator, 246–250
of National Park Service maps, 317, 318
pictorial, 56, 104, 105, 257, 258, 262
semiotics of, 221, 222, 290, 291
sets of, 290, 295
syntactics of, 290, 291
SimCity and *SimEarth*, 155, 433
Similarity, 211, 403, 447, 448
grouping by, 72, 73, 77, 81, 93, 290
Simplicity, grouping by, 75, 76
Simulation, 191, 255, 427, 432
atmospheric, 381, 388, 389
forest-succession, 191
three-dimensional, 149, 370–376
Simultaneity, 240, 241, 296
Simultaneous bundles, 240
Simultaneous contrast, 60–62, 133
Singular signs, 228, 311, 321–323
Sinsign, 222
Size, 36, 38, 109, 111, 112, 270, 279, 386, 388
of symbols, 56, 57, 59, 86, 88, 89
Sketch
primal, 29, 31, 51, 65
2½-dimensional, 29, 30, 32, 51
SLICViewer, 404

Slide bar, temporal, 385, 386
Slope, 84, 142, 144
Small multiples, 36, 38, 74, 402, 403, 416
Social psychology, 217
Society, 15
science interface, 448, 455
Sociobiology, 455
Sociocultural issues, 10, 11, 14–16, 160, 211, 329–331
in categories, 166–168, 202
in connotations, 342–345
in mapping behavior, 198
in schemata selection, 202, 203
in sign-vehicle sets, 291
in symbolism, 261
Soil maps, 7, 38, 153, 186, 415, 416, 436, 442
Sound, 287–290, 309, 419–422
in animated maps, 431, 432
in data analysis, 379
in dynamic maps, 287–290
in feature comparison, 419–422
loudness of, 288, 289, 442
in map sonification, 379
pitch of, 288–290, 421, 431, 432, 442
in probe technique, 379, 421, 442
rate of change, 288–290, 442
rhythm of, 72
in soil maps, 442
Space, 35, 36, 48, 49, 380–383
autocorrelation, 381, 442
childhood concepts of, 196–198
clustering in, 384
of color, 62, 391, 392, 416
dimensions of, 252, 253, 302, 303, 405
display, 430
in feature comparison, 402–409
frequency in, 96, 116, 117, 377, 378
geographic, 369, 370, 431
in geographic biplots, 407–409
of human–map interaction, 357
and indexical signs, 223
as indispensable variable, 98, 368, 369, 430
map sign, 251, 252
meaning of, 312–315
patterns in, 364, 365, 369, 384, 386–392, 403
position in, 313–315
problem solving on, 396, 397
representations of, 460
scale of, 96, 101, 426
schemata on, 451
signature of, 365
and time. *See* Space–time
in trivariate mapping method, 405
two-dimensional, 369, 370
Space–time, 422, 433

Space–time (*continued*)
  in animated maps, 429–431
  categories of phenomena, 423, 424
  dimensionality, 252
  feature analysis, 359, 401, 422
  filtering, 426
  microscope, 425, 426
  position in, 93, 317–321
  postprocessing in, 427, 429–432
  process steering in, 427, 432, 433
  process tracking in, 427, 428
  slices of, 252
Spatial features. *See* Space
Spatialization of form, 189, 190
Spectrometry, 190, 202
Spontaneous looking, 102–104
Spot heights, 205, 208
Spotlight of attention, 94, 95
Stability, map, 445–447
Standardization, 290
Stand-for relationship, 195, 196, 213, 244
State features, 365
State of the World Atlas, 335
Static maps, 146, 236, 270–278
Statistical maps, 156, 265
Stereo balance, 421
Stereo pair maps, 139
Stereokinetic effect, 372, 417, 432
Stereoptic effect, color, 138
Steroscopic analysis, 399
Stimulative signs, 246–250
Stimuli, 3, 7, 8, 23, 26, 33–45
  compound, 81–87
  pie-wedge, 112, 113
Stimulus-processing model, 22
Strip maps, 188, 189
Structuring principles, 175, 176
Substantive sign, 251
Surroundedness, 109
Symbolization, 2–6, 8, 10, 11, 21, 41, 42, 44, 306
  stability of, 445
Symbols, 12, 13, 15–17, 40, 214, 218, 222, 224, 243, 263
  associative, 257, 258, 262
  attention to, 85, 86, 99, 100
  in cognitive approach, 14
  color of, 57, 125, 126, 132, 133
  continuum of, 261
  contrast of, 125
  detection of, 125, 126
  discrimination of, 124, 129, 132, 133, 147
  ellipse as, 90
  failure of, 27
  fixation on, 57, 148
  geometric, 257, 258, 262
  graduated, 77, 78, 88, 89, 93, 136, 188, 303, 318
  grouping of, 77, 78, 82, 83
  and icons, 130, 224, 258
  identification of, 129, 130
  indirect, 326
  integral dimensions of, 87
  just noticeable differences in, 3
  modular processing of, 31, 32
  multivariate, 81, 85, 86, 88
  of National Park Service maps, 56, 57, 129, 177, 178, 232, 240
  orientation of, 83, 84, 86
  pictorial, 56, 257, 258, 262
  precognitive processing of, 41
  quasi-symbols, 261
  recognition of, 57
  relative magnitude of, 135, 136
  separable dimensions of, 87
  and signs, 224, 256
  size of, 56, 57, 59, 77, 78, 86, 88, 89
  slope of, 8
  stability of, 434
  theory of, 199
  true, 261
  valuative connotations of, 345
  and visual acuity, 57, 58, 59
Symptoms, 225, 247
Synchronization, 281, 285–287, 301, 316
Synecdoche, 336
Syntactics, 149, 199, 234–238, 242–245, 269–306
  color, 292–298, 391
  and comprehension, 305, 306
  of map forms, 304
  and maps as signs, 302–304
  and multiple linked sign systems, 295–302
  and perceptual organization, 367–392
  sign-vehicle, 290–295
  and visual variables, 270–290
Syntagmatic types, 237, 238
  alternating, 238, 417, 418, 447
Syntax, 234, 236, 423
Systemic signs, 227
Systemology, 238–242

Tactile maps, static, 276–278
Tactual language for mapping, 276
Tactual variables, 277, 278
Tanaka method, 146, 147
Taxonomy, 165
  of depth cues, 137–139
  of pictures, 267, 268

Tectonic codes, 299–301, 312–314
Temperature
  global climate model on, 220, 255, 284, 404, 437–439, 442
  global sea surface, 383
  maps of, 86, 87, 92, 135, 314, 315, 318, 319, 325, 390
Temporal features. *See* Time
Tense, and temporal codes, 301
Terrain, 17, 138–146, 445, 456
  fishnet maps of, 139–142
  layer-cake model of, 146, 147
  representation of, 193, 201, 204, 205, 208–209, 445
  shading of, 9, 141–146, 328
Territorial control, 337, 338, 345–348
Text discrimination, 127, 128
Texture, 36, 105, 106, 114, 121, 270, 279, 308, 386, 387
  density of, 272, 274
  detection of, 124, 125
  directionality of, 272–274
  discrimination of, 130–132
  gradient of, 138, 139, 141, 373
  patterns of, 125, 130, 149
  semiotics, 272, 273
  of shading, 142
  size of, 272, 274
  of sound, 421
  of space, 285
  of time, 316
Theater metaphor, 192
Thematic codes, 333
Thematic maps, 2, 7, 12, 295, 297, 298, 329, 340
Theme sign, 251
Theoretical models, 447
Thinking, visual, 361–400, 435
  and cognitive gravity, 448–452
Three-dimensional maps, 26
Three-dimensional representations, 28, 30–32, 370
Three-dimensional simulation, 149, 370–376
Thresholds, 125
  reliability, 441–443
  uncertainty, 443
Timbre, 288, 289
Time, 35, 36, 368, 369, 376–379, 385, 386, 416–418
  in brushing technique, 385, 386, 419
  chronological, 430, 431
  code of, 299, 301, 312, 315, 316
  cycle of, 315, 424, 426, 427
  entities of, 316, 424–427
  and event schemata, 191
  existence in, 316
  as fourth dimension, 301
  frame of, 424
  frequency in, 285, 377, 378, 399
  grouping by, 377
  instant of, 315
  interval of, 316
  layers of, 253
  location in, 281, 315–318
  map sign, 251, 252
  meaning of, 312
  order in, 134, 284, 431
  patterns in, 420, 431
  phase of, 424
  real time, 26, 173, 423, 427, 432
  resolution, 424, 426
  scale of, 424–426
  sequence of, 237, 238, 316
  series of, 315, 424, 426, 427
  slice of, 252, 315
  slide bar representing, 385, 386
  and space. *See* Space–time
  and syntax of animated maps, 236
  texture of, 316
  thickness of, 301
Top-down processes, 30, 39, 40, 42, 43, 51, 68, 117–120, 363, 366
Topic codes, 333
Topographic maps, 6, 7, 200, 201, 220, 242, 256, 324, 329, 395
Topographic relief, 197
Topological codes, 300
Topological spatial concepts, 197
Tourist maps, 130, 160, 196, 317
Track-up display, 372
Training in schemata, 210
Transparency, 275, 276, 279, 387, 437
Travel maps, 7, 95, 133, 173, 188
Triadic signs, 219–221, 242, 245, 435
Trichromatic maps, 414
Trivariate display, 403–405
Trivariate maps, 93, 415
Truth, 337, 338–340
  of data, 440
  of maps, 436
  maps as, 338
  of signs, 435–444
  in visualization, 433, 435–458
Two-dimensional representation, 28
Two-dimensional space, 369, 370
2½-dimensional sketch, 29, 30, 32, 51
Type I error, 436, 445
Type II error, 436, 445
Type placement, 104, 258
Type size, 3, 128
Typicality effects, 158–160

Unambiguous signs, 212, 228, 323, 324
Uncertainty, 441, 443, 458
  visualization of, 200, 275, 276, 340
Unclassed maps, 163, 164
Unconscious signs, 229
United Nations, 345, 349
User interface, graphical, 324, 326
Utilitarian codes, 333–335

Vague signs, 212, 228, 229
Valuative connotations, 337, 342–345
Valuative signs, 227
Value of colors, 36, 37, 52, 84, 87, 121, 270, 279, 440, 441
  and brightness, 111–116
  order of, 134, 135, 318, 319, 390
  and symbol size, 88, 89
Vase–face figure, 113, 114, 118
Vegetation maps, 17, 319, 322, 334
Veracity, 435
  connotation of, 337–340
Virtual maps, 193, 348, 352, 360
Virtual reality, 26, 256, 433
Vision, 16, 17, 21–24
  acuity of, 57, 58, 59, 126, 133
  central, 54, 57, 58, 99
  and cognition. See Visual cognition
  color deficiency in, 63, 126, 149
  computational theory of, 51, 54
  ecological, 23, 26
  and eye–brain system, 17, 24, 25, 28, 53–67
  and eye movements, 42, 43, 101–104
  and focus, 57, 104, 148
  fovea in, 54–57, 59, 103
  and graph comprehension, 33–41
  as information-processing system, 23–50, 68, 70, 118, 147, 430
  and maps to be seen, 206, 209
  Marr approach to, 27–33, 51, 54, 64, 65, 70, 101, 118
  minimum instant of, 368
  neurophysiology of, 53, 63–65, 96, 147, 160
  opponent-process theory on, 62, 143, 144, 148, 160, 194, 195
  pathways in, 96, 105, 106
  peripheral, 57, 58, 99, 104, 126
  precognitive processes in, 33, 35
  processing of stimuli in, 33–45
  reacting to information, 26
  and representations, 14–16, 27, 28, 33, 51, 147
  and seeing wrong, 436, 444–447, 451
  and shape descriptions, 29, 31, 32

Visual array, 35–39, 46, 49, 137, 150, 363
Visual attention, 80, 81
Visual cognition, 16–18, 23–50
  and imagery, 46–49
  and visual stimuli, 33–45
Visual cortex, 65–68
Visual descriptions, 35–39, 46, 49, 150, 178, 179, 363
Visual dimensions, 81–87
Visual display, 362, 363
Visual evidence, 392, 399, 450
Visual grouping, 77, 78, 81–84
Visual hierarchies, 121, 122
Visual imagery, 48
Visual levels, 120–123
Visual maps
  dynamic, 278–287
  static, 270–276
Visual scenes, 25, 49, 51, 147
  edge of, 101, 102
  figure and ground of, 107–120
  grouping of stimuli in, 151
  interpretation of, 42, 150
  location of items in, 93–96
  meaning of, 23, 25, 27, 101
  perceptual organization of, 69, 70
  perceptual units in, 35, 36
  and primal sketch, 29
  scanning of, 101–107
  segregation of, 93, 94
Visual search. See Search
Visual stimuli, 33–45
Visual thinking, 361–400, 435
  and cognitive gravity, 448–452
Visual variables, 35–37, 53, 82, 87, 88, 182
  associativity of, 91, 92, 279
  attention to, 84–86
  dynamic, 278–287
  graphic. See Graphic variables
  location as, 270, 279
  retinal, 270
  selectivity of, 91, 279
  and syntactic rules, 270–290
Visualization
  cartographic, 170, 176, 255, 356
  and communication function of maps, 357, 358
  computer-assisted, 50, 356, 416
  of data quality, 186
  dynamic, 93
  Exploratory Visualization System. See Exploratory Visualization System
  geographic. See Geographic visualization
  interactive, 132, 297, 358, 403
  mental, 255, 356, 422, 423
  quality of, 74

scientific. *See* Scientific visualization
truth in, 433, 435–458
of uncertainty, 200, 275, 276, 340
Visualization task, 360, 363–365
Visualization tools, 149, 170, 203, 356, 357, 359, 377, 379
  realism in, 457
  scale and resolution in, 381
  schemata in, 393
Visuospatial scratch pad, 35, 50

Wavelength of colors, 21, 55, 84, 138, 194
Wayfinding, 191, 225, 395–397
Weather maps, 86, 87, 92, 200
  animated, 192
Weighted isolines, 405
World political maps, 154, 196, 197

Zero-cost information, 48, 209
Zero-crossings, 29, 30
Zeugma, 336

Printed in the United States
103061LV00002B/28/A